The Deconstruction of Time

CONTEMPORARY STUDIES IN PHILOSOPHY AND THE HUMAN SCIENCES

Series Editors:
Hugh J. Silverman and Graeme Nicholson

THE DECONSTRUCTION OF TIME

David Wood

Humanities Press International, Inc.
Atlantic Highlands, NJ

First Published in 1989 by Humanities Press International, Inc.,
Atlantic Highlands, NJ 07716

Reprinted in paperback 1991

Library of Congress Cataloging-in-Publication Data

Wood, David (David C.)
    The deconstruction of time.
    (Contemporary studies in philosophy and the human sciences)
    Bibliography: p.
    Includes index
1.   Time—History.   2.   Derrida, Jacques—Contributions in
concept of time.   3.   Deconstruction.   I.   Title.
  II.   Series.
BD638.W654   1989   115   87-26164
ISBN 0-391-02743-3
ISBN 0-391-03704-8 (pbk.)

The following material has already appeared in print:

Part One has appeared, somewhat changed, as "Nietzsche's Transvaluation of Time" in *Exceedingly Nietzsche*, eds. David Farrell Krell and David
Wood (London: Routledge and Kegan Paul, 1988).

Part Four, chapter 1 is a slightly revised version of "Time and the Sign,"
Journal of the British Society for Phenomenology 15, no. 2 (May 1982).

Part Four, chapter 2 is a slightly revised version of "Derrida and the
Paradoxes of Reflection," *Journal of the British Society for Phenomenology*
11, no. 3 (October 1980).

Part Four, chapter 3 was previously published as "Style and Strategy at
the Limits of Philosophy" in *The Monist* 6, no. 3 (October 1980), and in an
extended form as "Metaphysical Textuality," in *Heidegger and Language*,
ed. David Wood (Warwick: Parousia, 1978).

The *Postscript* to Part Four, chapter 3 is an extract from "*Différance* and
the Problem of Strategy," in *Derrida and Différance*, eds. David Wood and
Robert Bernasconi (Warwick: Parousia, 1985), and later republished
(Evanston: Northwestern University Press, 1988).

Part Four, chapter 4 was published as "Time and Interpretation," in
Time and Metaphysics, eds. David Wood and Robert Bernasconi (Warwick:
Parousia, 1982).

Part Four, chapter 5 in part appeared as "Prolegomena to a New Theory
of Time," *Research in Phenomenology* 10 (1980).

Printed in the United States of America

Contents

Dedicated to
JACQUES DERRIDA
in friendship and admiration
Warwick, England
15 February 1987

Acknowledgments

I would like to thank all those who even in the darkest hours still offered hope and encouragement, and those whose skepticism about this project's completion spurred me to greater efforts.

I have profited enormously from discussions with friends and colleagues—in particular Robert Bernasconi, Tina Chanter, David Boothroyd, David Krell, Hugh Silverman, Andrew Benjamin, David Allison, Ed Casey, David Holdcroft, Susan Haack, and many others. I am indebted also to various institutions and their organizers for offering me the occasion to address them on these matters—the Society for Phenomenology and Existential Philosophy, the British Society for Phenomenology, the International Association for Philosophy and Literature, the Collegium Phenomenologicum at Perugia, and numerous societies and philosophy departments in Britain and the United States. I have also been greatly stimulated by the other participants in our series of summer Workshops in Continental Philosophy, especially that on Time and Metaphysics.

Special thanks are due to Wolfe Mays for showing me the way into phenomenology, to Alan Montefiore for introducing me to the work and person of Jacques Derrida and for nourishing my Continental proclivities, to John Sallis for his philosophical example, his friendship, and his persistent editorial encouragement, to my students who have sometimes had to take parts of chapters in lieu of lectures, to my wife, Mary, for her constant encouragement and infinite patience, and to my children for putting up with an absentee father during the later stages of my pregnancy.

Finally, I am grateful to the generosity of the Sabbatical Leave Committee of the University of Warwick for giving me a year in which to write.

Summary

There are philosophical topics other than time. Some of them engage our attention for extended periods. But sooner or later each is sucked down into the vortex of that one problem that is truly permanent: time. Prediction is an uncertain art, but I would venture the suggestion that our century-long "linguistic turn" will be followed by a spiraling return to time as the focus and horizon of all our thought and experience. For this to happen time has to be freed from the shackles of its traditional moral and metaphysical understanding. The signs of this transformation have already become visible in numerous, disparate fields, but nowhere more so than in philosophy.

Indeed, there is something approaching a tradition of philosophers who were not only centrally concerned with time, but who came to see the nature and possibility of philosophy itself as linked to questions about time. And with Nietzsche, Husserl, and Heidegger, that tradition joins the twentieth century. Each of the philosophers made the most strenuous efforts to think time outside of metaphysics, to *rethink* time. Derrida, however, takes the whole question a step further by arguing that time is an essentially metaphysical concept, and that no alternative concept of time can circumvent this necessity. Why does he make this claim? And how then are the efforts of Nietzsche, Husserl, and Heidegger to be judged?

After presenting Nietzsche's transvaluation of time as a proto-deconstructive strategy, this book undertakes a critical reading of the two major works on time by Husserl and Heidegger, arguing that for all their repetition of metaphysical themes and values, they each contribute at the same time to our release. While the necessity of this repetition is both evidenced and explicitly confirmed in Derrida's own writing, his denial of the possibility of an alternative concept of time is compromised by his own endorsement of a pluri-dimensional temporality, and his proper insistence that it is not concepts per se that are metaphysical, but their mode of textual articulation.

These more subtle claims allow the development of a double strategy. First, the "moment" can be redescribed and rethought in ways that break utterly with representation. This move is located first in Nietzsche and Husserl, and then in the time-dissolving moves of the later Heidegger and

Derrida. Second, deconstruction's refusal of an original primitive time, coupled with its reading of metaphysics as a form of textual articulation, licenses a program for the analysis of temporal structures and representations of time, one that abjures foundationalist pretensions and resists reductively spatializing interpretations.

No theoretical synthesis of this forked strategy is attempted. "Time as absolute openness to the other" and "time as complex textuality" displace the value of presence at two distinct levels of articulation: that of primitive event and that of structure. If a desire remains for an account of their inner unity, that very possibility remains a matter of speculation.

Introduction

It is perhaps a common and recurrent claim among philosophers and others who think about time that there is something wrong with "our ordinary concept of time." The fault may be diagnosed in different, even incompatible ways. There are those who think that the linear representation of time is responsible for its mistaken assimilation to space. There are those who argue that the metaphorical comparison of time with a river is responsible for the illusion that time "flows." There are those who think that our very possession of namelike words for "past," "present," and "future" erroneously leads us to suppose that these each in some sense still or already exist, like distant parts of space. More recently, there has been an attempt to rescue our understanding of time from "metaphysics," a reading that treats most of the history of philosophy as the history of metaphysics. It is a reading announced in this radical form by Nietzsche, developed by Heidegger, and further radicalized by Derrida. Quite what is meant by "metaphysics" and why it might be thought valuable or necessary (or possible, or impossible) to overcome it, is a matter I discuss in detail later. For the moment what is especially noteworthy is the suggestion made by Derrida that if we do succeed in overcoming our ordinary (that is, metaphysical) conception of time, it will involve the elimination of the concept of time altogether, that it is not possible to purify the concept of time of the metaphysical, for there is an intimate connection between time, however we conceive of it, and metaphysics, such that time and metaphysics would stand or fall together. For someone for whom the history of philosophy is a series of footnotes to Plato, his linking of truth to the eternal might seem to lend support to this view. But my argument here has a more general form.

The argument in its most general form is simple. It is that qua metaphysics, the history of philosophy is the history of the privileging of a certain temporal/evidential value, that of "presence." This value both supplies and is confirmed by an interpretation of time, one that privileges the present. It can already be found at work in Aristotle's treatment in the *Physics*, and thereafter in various different forms in all subsequent treatments of time, until perhaps Nietzsche.

For reasons of influence as well as objective historical importance, an assessment of this claim could not be attempted without offering an account of the theories of time supplied by Husserl and, later, Heidegger. Each of them made a sustained and brilliant attempt to correct our ordinary understanding of time, each saw this understanding as historically sedimented in the history of philosophy, and each saw its correction as requiring a radical change in general philosophical method and practice. Yet despite these self-conscious attempts at radical new departures, it has been claimed that they do nothing but reproduce the very values that they are trying so hard to question. In particular it has been claimed that their revision of our understanding of time moves *within* the limits they take themselves to be transcending. For Derrida there is a very general reason for this failure: "time in general belongs to metaphysical conceptuality."

This is a powerful challenge. There can be no adequate philosophical reflection on time that could successfully retain the concept of time. The precise relation of metaphysics that we are being recommended to take up, by Nietzsche, by Heidegger, or by Derrida, is nothing as simple as "rejection," or "transcendence." Indeed, most if not all the critical terms traditionally used to describe the relation of one account to another account (even "more or less adequate," even supposing them to be "of the same thing") could be claimed to depend on assumptions that are themselves no less "metaphysical." The strategies available here will have to be discussed in some depth. But there is no doubting the significance of an analysis that succeeds in showing that a supposedly radical account actually reproduces at one level or other the lineaments of the position from which it was attempting to distance itself.

I have just referred to Derrida's insistence on the metaphysical status of the concept of time. Let me quote it in full [my translation]:

> The concept of time belongs entirely to metaphysics and it designates the domination of presence.
> . . . if something connected with time but which is not itself time needs to be thought of outside the determination of Being as presence, we are no longer dealing with anything which could be called *time*.
> . . . it is not possible to oppose to it [that is to the whole historically developed system of metaphysical concepts] a different conception of time, because time in general belongs to metaphysical conceptuality.[1]

These remarks first appeared in 1968 in an essay by Derrida entitled "*Ousia* et *Gramme*: note sur une note de *Sein und Zeit*" as one of a collection of essays published in honor of Jean Beaufret, a respected French Heideggerean. What Derrida tries to do in this essay is to show that Heidegger's attempt to develop a new existential concept of time, in opposition to the metaphysical one traceable to Aristotle, does not face up to the more radical possibility that there may be an essential connection between the concept of time (and *any* concept of time) and that consolidated collection of Greek concepts that has determined the history of philosophy as the history of metaphysics.[2] If such a connection can be established, Heidegger's revisionary project in *Being and Time* would have to be rethought. Derrida seeks to do just this.

The connection with Heidegger is only one of a number of features of Derrida's general philosophical position that makes his approach to the specific discussion of time for our purposes exemplary and critical. In a massive output that began with his book-length introduction to the French translation of Husserl's short late essay "The Origin of Geometry," Derrida has addressed himself directly or indirectly not only to most of the important traditional philosophical discussions of time, but also to the possibility of understanding the history of philosophy as a single tradition marked by allegiance to certain determining values that, historically speaking at least, define it and limit its scope. Derrida's name for the most dominant of these values is *presence*, understood as a fusing together of evidential, spatial, and above all temporal motifs. Derrida's position is for me exemplary because he poses in the most general and yet scholarly way the question as to the relationship between time and philosophical conceptualization. Instead of offering a (new) philosophy of time, he explores the dependence of the philosophical tradition itself on our interpretations of time. These interpretations are claimed to exhibit a superficial plurality but an underlying unity: the commitment to the value of *presence*. His position is critical in that it takes up and works with, even within, a most original, productive, and far-reaching tradition. It includes Kant, Hegel, Nietzsche, Husserl, Heidegger, and recent structuralist thought. The importance of this tradition is not so much directly defended by this book as presupposed by it. But that there is such a tradition is something that even those who have not felt the urgency of its problems or tasted its delights should be able to agree on. In discussing the problems connecting time and metaphysics I shall remain, (and for positive reasons), largely constrained by the way these problems have been developed within this tradition.

Within this tradition, for example, there is a keen awareness of the difficulty of doing justice both to our intuitive sense of the intellectual transparency of time as experienced—we seem to directly witness its shape, its action every moment of our lives—and to a growing awareness that the concept of time[3] can be understood only by reference to the role it plays in a range of different theories, and indeed that it only ever appears with some such tacit or explicit theoretical implication. This problem is at least a stage beyond the problem of producing a theory of time compatible with our intuitions.

Once one is convinced of the variety of the theoretical complicities of the concept of time, one is confronted with the further question as to whether there is some order of priority among these theoretical dependencies (Is physical time more basic than literary time?) such that one could arrive at the fundamental features of the concept of time by way of such a hierarchization. Derrida's answer, as will become apparent, works at three levels. First, he credits a certain physical conception of time (derived from Aristotle) with having enjoyed a historical privilege in relation to other concepts such that they were typically modeled on it. Second, he questions the legitimacy of that privilege. Third, he claims that there are fundamental features of any concept of time by which it has a certain complicity with metaphysics. This last level will raise the important question as to whether Derrida does not prematurely reunify our understanding of time, albeit at a formal level. ("Formal" here points to the general theoretical functioning of any *concept* of time.) This would leave undeveloped the insight into the plurality of our concepts of time. I will suggest that there is a conflict within Derrida's own account between the recognition of this multiplicity and the underlying reunification wrought by his critical method.[4]

This tradition is also peculiarly sensitive to what could be called "reflexive" problems (where the sense of reflexivity is neither purely logical nor psychological, but perhaps best understood as formal). And this is particularly important for the treatment of time. The history of the philosophical treatment of time (as of other key concepts—truth, language, the subject, and so forth) throws up from time to time problems in which the proposed account of such concepts seems in the very form or fact of its existence to presuppose, or at least to involve or suggest, answers to the question at issue. For instance, does not any *theory* of truth necessarily give a privileged status to the truth of theoretical statements (of which it itself consists) that would undermine any attempt to explain the truth of complex statements as a product of the truth of primitive or atomic propositions? For the theory would have no legislative value if it were merely a summary of a set of primitive truths. And any account of the *concept* of time would have to allow one to explain and endorse any temporal presuppositions built into the term "concept." That this is not empty possibility will be seen when Derrida's radically anticonceptual "theory of language" is discussed. Derrida and the tradition within which he can be located—particularly Nietzsche and Heidegger—are all acutely aware of such reflexive problems.

The key questions that open and guide this book are related directly to the initial quotations from Derrida: (1) Is there, as he suggests, a necessary link between time and metaphysics? (2) Is the account of metaphysics presupposed by this question acceptable? (3) Is Derrida's account of the consequences of such a complicity between time and metaphysics adequate? (4) Can Derrida's remarks about there being no other concept of time than a metaphysical one be reconciled with his suggestion[5] that time be thought of as multidimensional? (5) Does the level at which Derrida discusses the question of time (which is quite as abstract as any metaphysics could be) occlude the possibility of developing a more concrete account of the diverse modes and variety of temporal structure?

These questions have a clear analytical side to them. Is an alleged connection necessary? Is a certain account adequate or acceptable? Can two claims be held consistently? Does a theory inhibit the development of a certain line of thought? The historical development of the concepts and theories to which the analysis is to apply cannot be ignored, even if it *is* possible to pretend of certain concepts that they have no important history, and certain theories (for example, those with an immediate practical import) can be treated in abstraction from their past. But there is always the risk of intellectual naivety, and in no subject is this truer than in philosophy. Philosophy does not have a tradition, it *is* a tradition. It does not follow from this that there is any guaranteed unity of problems throughout the ages. Tradition need not imply the continuous development of concepts of theories. There may be jumps, gaps, leaps, discontinuities, losses, as well as continuity, development, enrichment, preservation. What all this suggests is that the method employed will have to be historical as well as analytical.

In this respect, Derrida's position can be used as something of an ordering device. On the assumption that his account of the relation between time and metaphysics in some way takes in, or is responsive to, all of his significant precursors, then the historical part of our thesis can be ordered by asking the question: What reading of the history of philosophy leads Derrida to make this claim about the time/metaphysics relationship? This makes it possible to reconstruct, indeed perhaps to construct, for the first time Derrida's own position on time, but also to traverse the territory that would have to be passed through were the question about time and metaphysics to be pursued independently. And in traversing that territory with these critical questions in mind, the answers arrived at will have some historical depth. It might turn out, for example, that the Nietzsche-Heidegger-Derrida account of metaphysics simply will not adequately cover the whole of the history of philosophy, which would radically affect the significance of any assessment of the relation between time and metaphysics.

This historical approach coth determines and is determined by the tradition chosen. First of all, it suggests that whatever the universality of the wellsprings of philosophical inquiry (wonder, astonishment, despair, joy, puzzlement), this inquiry is seriously developed only within a written tradition. And in such traditions, concepts and theories will often be elaborated in ways not immediately translatable into other traditions. The assumption that all philosophical traditions are dealing in their own way with the same problems is easier to cast doubt on than it is to support. And this serves as a kind of justification for the self-imposed limitations of this study. The particular tradition chosen also naturally suggests taking a historical approach to it, for all of its members were themselves keenly aware of and concerned with their own historicity, as an essential feature of their philosophizing. It would be either an enormous arrogance or naivety to try to understand those thinkers without taking seriously their own historical sense of their activity. Of course they did not share a single understanding of what might be meant by that historicity. For Husserl, the lesson of history was the need for a radically new (and originally unhistorical) approach, and it was only later that he attempted anything like a serious treatment of the history of philosophical problems and outlooks.[6] For Hegel and Heidegger, on the other hand, it is only the history of thought that makes thought possible, even if, for Heidegger, it also sets limits to thought, limits that seem to require to be surmounted. This general concern with the history of philosophy is not merely something to be mentioned once and then forgotten. For Derrida's method of reading philosophical texts to become clear, it will have to be related to precisely those ways of thinking about and working through philosophy's past that his predecessors developed. So even one's method of reading history has a history that needs to be explained. For example, Derrida calls his own method deconstruction, which has clear filiations with Heidegger's positive account of "destruction" in *Being and Time*.[7] And some attempt is made to deal with these questions in the last chapter of this book.

My general critical claim about Derrida's views on time is that unless *all concepts* are metaphysical (in which case the claim about the concept of time in particular is much less interesting), his conclusion about time does not follow from his premises without endorsing in his writing a level of analysis (quasi-transcendental) that he himself supplies us with the tools for dismantling. Derrida does indeed liberate us once and for all from the quest for an "original" time, from one that would contain within itself some fundamental power and evidential primacy. But far from ruling out "another concept of time," he actually opens the way for one.

The reconstruction of "the argument"—the series of analyses on which Derrida's general remark about time rests—opens with a discussion of what I have called Nietzsche's transvaluation of time. The aim here is strictly limited. Nietzsche's importance for the general project of deconstructing metaphysics is one that is taken for granted rather than explored. And Nietzsche's impact on the wider thought of Heidegger and Derrida seems relegated to the status of that of a Hegel or a Kant in terms of the weight subsequently given to it in the book. Furthermore, using Nietzsche to set the stage for twentieth-century treatments might suggest that we had somehow gone beyond him. Despite the risk of such misunderstandings, it seemed that a more sharply focused discussion of Nietzsche's theory of eternal recurrence was the most effective way of introducing the project of deconstructing time, and its wider implications.

What follows are the two most obvious and plausible candidates for an "alternative concept of time"—the phenomenological version offered by Husserl, and the existential account(s) offered by Heidegger. These present themselves as pivotal paradigms because they do so in the course of a general critical reading of the whole history of philosophy (as metaphysics), because there are clear and strong intellectual biographical links between Heidegger's account and Husserl's (Heidegger edited, perhaps somewhat perfunctorily, Husserl's lectures on time-consciousness from 1905 and 1905–10, published eventually in 1928 as *Vorlesungen zur Phänomenologie des inneren Zeitbewusstseins*), and because Derrida has devoted much of his writing to their general philosophical positions and to their specific claims about time. Indeed, it would not be entirely misleading to treat Derrida's refusal of a nonmetaphysical conception of time as a conclusion drawn from, and certainly well-exemplified by, the difficulties encountered in and by Husserl and Heidegger in trying to formulate one.

Part 4, *Time Beyond Deconstruction*, weaves discussions of time with discussions of textuality, style, and strategy. The rationale for this is to be found in the belief that (a) questions of time and signification cannot "finally" be separated, and that the readier access we have to textual structures of signification affords us powerful insight into the complexities of temporal organization, and (b) in both strands of the double movement traced by this work—time as structure, time as event—the question of style and strategy, indeed of writing itself, has repeatedly become problematic. Indeed, Derridean deconstruction *necessarily* problematizes writing, his own included.

The apparently untroubled innocence of my own style reflects in part a sense of the limits of alternative styles and strategies while fully appreciative of their value.

In chapter 1, I argue that Derrida's concern to expose the metaphysical determination of our ordinary concept of time as "presence" opens the possibility of a "positive" account of the temporality of language. But by lodging his residual sense of the temporal in such a quasi-transcendental term as "différance," this possibility is unfortunately closed off.

Chapter 2 opens a discussion of style and strategy. Husserl admits that his word *Fluß*—for the Absolute Flux of time — is a metaphor. Perhaps the understanding of what is primitively temporal puts a strain on our ordinary linguistic resources. In this light some of Derrida's innovative "philosophical" styles and strategies of writing are dealt with, and some alternatives suggested. Chapter 3 compares Heidegger's style with Derrida's "strategy" and argues that the success of Derrida's philosophical writing depends on our illicitly privileging his strategic (meta-) intentions in the guidance of our reading. In the postscript to chapter 3 it is further suggested in relation to Derrida, that "différance" and "presence" are not transcendentally but "dialectically" related and that deconstruction is open to the charge of a certain formalism. Again, serious issue is taken with Derrida's use of the "sous rature," claiming that he sails too close to the winds of metaphysics for his own good.

Chapter 4 illustrates the consequences of restricting the scope of a range of philosophical views of time to regional applications, serving more modest hermeneutical goals. Cosmic, dialectical, phenomenological, existential time, and the "time of the sign" are each considered.

It is suggested that something like the existential framework of Heidegger's *Being and Time*, with radical pluridimensionality replacing considerations of authenticity, might (if only for methodological reasons) have a privileged position here. Chapter 4 argues that an alternative model of time can be drawn from the structures of *temporal textuality*. I begin by looking at more or less structural features of discourse, defend the role of representation and diagrams of time against the charge of spatialization, and discuss the exemplary structure of narrative. At a certain distance from traditional "structural analysis," six levels of temporality in the narrative are distinguished (that of the reader, the narrator, the plot, the action, the characters, and of the narrative discourse itself). I argue that these sustain a strong analogy between text and experience.

In chapter 5 two traditional grounds are distinguished for giving the concept of "the future" a special place in philosophical thinking: hermeneutical and ethical. Poststructuralism, it has been claimed, "has no vision of the future." Derrida's writing is deployed to show why this is so, and to try to *interpret* it more positively. I suggest, reflexively, that philosophy be treated as a fundamental event.

This book offers both a critical assessment of Derrida's general argument insofar as it draws conclusions about time, and a constructive attempt to supply, if only in outline, and develop certain suggestions Derrida himself makes, an alternative, yet nonmetaphysical account of time and the temporal.

Part 1

Nietzsche's Transvaluation of Time

1   Nietzsche's Transvaluation of Time

The position of Nietzsche in this work is somewhat complex. He was undoubtedly influential in guiding Heidegger's formulation of the idea of the "end" of metaphysics, and in setting up the whole problematic of its "overcoming." His radical skepticism about unity, identity, integrity will play a critical role at a number of points—in particular in readings of both Husserl's and Heidegger's ways of reformulating the status of time and temporality. I will endorse Derrida's general attack on the idea of a more authentic "primordial" time that would underlie our everyday understanding of time as a succession of moments, and I will argue that this opens the way for a more complex investigation of the variety of temporal phenomena, in which concepts, theories, and representations "of" time are embraced as forming part of the temporal itself.

In this context, the interpretation of Nietzsche's thought of eternal recurrence offered here may seem somewhat anomalous. For it may read like the strongest possible reaffirmation of the "instant," in which it receives a new kind of purity, freedom from representation. Certainly Nietzsche might question the theoretical motivation behind the pluralization of time and its structures that does not readily mesh with what I have called "the intensity of the moment." Nonetheless, Nietzsche's account is instructive, and what is offered here is an interpretation of his self-avowedly "highest thought," the eternal recurrence.

This presentation relates to the general argument of the book in the following ways: Nietzsche offers us a brilliant example of how to displace a frame of reference (in this case, the ordinary "metaphysical" concept of time) from within—It retains the centrality of the instant, only to explode the traditional value of such a primitive concept. For Nietzsche, the instant opens out onto what is other. But at the same time as the instant is a nonrecuperative opening out, it is also an embodiment of all those identical moments it repeats. This "repetition" could be said to be a most primitive kind of "representation." If so, Nietzsche's "moment of intensity" builds in representation in its most primitive "element" (which is primitively *not* self-contained). Broadly speaking Nietzsche's account is treated as a reflective intensification of our understanding of the "present," one that successfully subverts our everyday "recuperative" understanding. As such, it belongs (in general) to the deconstructive phase of the rethinking of time, which, however interesting, cannot be the end of the story.

Despite doubts about Derrida's claim that there can be non-metaphysical concept of time, it is quite true that it is often more productive to pursue the goal of rethinking time by focusing on terms and concepts other than that of Time itself. One is reminded of the way in which Austin sought to avoid the quagmire of questions about freedom and determinism by looking instead at what at first seem more marginal concerns, such as the difference between excuses and justifications.[1] Nietzsche does not say all that much about Time per se, and yet his most fundamental concept is an essentially temporal one: that of eternal recurrence.[2] And as is true of both Husserl and Heidegger, the temporal here is linked with the ontological. However, I shall argue that Nietzsche does not, for all that, repeat the metaphysical motif of presence; he subverts it instead. His account of the *moment* is fundamentally at odds with any such value as presence.

I begin by showing how Nietzsche radically questions, subverts, and displaces "our ordinary concept of time," at number of different levels. In terms of the Derridean formula, eternal recurrence is an undecidable term, and I shall use this suggestion as a way of clarifying the various otherwise incompatible ways in which he explains it.

How, one might ask, is it in principle possible to avoid Derrida's strictures? I show later[3] that Derrida has to admit that it is not concepts themselves that are or are not metaphysical, but rather their textual exploitation and functioning. This opens the way, I argue, to Nietzsche's use of such concepts. It may be, if we stick to some rather narrow sense of the term, that Nietzsche has not produced a new *concept* of time, but that he has replaced its status as a concept with something else. The impossibility of conceptualizing eternal recurrence may turn out to be a positive feature.

It is however worth asking, to begin with, whether one can properly talk about "our ordinary concept of time." This phrase can itself be understood in a common-sense way as our taken-for-granted understanding of time, but, as St. Augustine showed long ago, the attempt to transform this tacit understanding into something more conceptually rigorous is somewhat problematic. The other way of understanding the phrase would then rest on some such view (shared by Heidegger and Derrida) as that the one ordinary concept of time we all share is that derived from Aristotle. But are there not difficulties attached to the idea? Who are "we"? Is it to be taken for granted that we all share a common concept of time? Might there not be a hidden complicity between the idea of a concept and some particular interpretation of time? Do we suppose there is but *one* concept of time? Let us develop this last question.

Far more "ordinary" than having a unitary concept of time is the common distinction between different kinds of time: subjective/objective, existential/cosmic, qualitative/quantitative, time as experienced/time as measured, and so on. But what if each of these oppositions merely *distributed* according to an unanalyzed schema (such as inside/outside) the same fundamental value (such as presence). What if such a distribution functioned to *preserve* that value? It could be argued, for example, that the distinction between subjective and objective time is a conceptual labor that ensures under each heading the preservation of the unity of the temporal series, and that this is achieved precisely by making this distinction. All events can then be *located* in one or other category by distinct rules of integration (such as narrativity for subjective time, seriality for objective time).

On such an argument, the fundamental unity of time would have been preserved precisely by multiplying the frames of reference within which it operates. Undoubtedly one then needs a further story to integrate these multiple frames. The usual one, with only two frames, is a story of derivation. Objective time is shown to be dependent on, derivative from, subjective temporality, or vice versa. The point of these remarks is to suggest that talk about "our ordinary concept of time" in the singular need not be undermined by the fact that everywhere a duality or plurality of such concepts is to be found. But it might perhaps make sense to allow that there *are* different ordinary concepts of time, and fall back on a more fundamental claim—that they each embody a common basic set of values, which could be called "unity," "integration," "identity," or, even more fundamentally, that of "presence" as the condition of such values being realized. On Nietzsche's account, these would be the values pertaining to Being. The traditional betrayal of time would consist of subjugating the values associated with Becoming to those of Being.

Nietzsche, Heidegger, and Derrida share (and they are not alone) the view that the ordinary concept(s) of time embodies values that are more ontological than temporal. But in each case the lesson this insight teaches is that questions about time and questions about Being collapse into one another and cannot be separated, as Plato long ago made so apparent.

Each of them is also engaged not merely in a *critique* of the ordinary concept of time, but in what might be called, borrowing from Nietzsche, a *transvaluation* of time. And for Nietzsche especially, this is presented as the key to a human transformation, indeed a transformation that would point beyond "man." The focus of this transformative projection is the thought of eternal recurrence, announced in the *Gay Science*, claimed (in *Ecce Homo*) to be "the fundamental conception" behind *Thus Spake Zarathustra*, and the title of an incomplete book project bequeathed to us in his Nachlass. One could avoid the analytical task of separating the various versions by saying that the thought of eternal recurrence is not a single thought at all, but a *constellation* or *family* of "thoughts." But this would only be a temporary, because largely uninformative, way of legitimating the diverse and seemingly contradictory accounts Nietzsche offers. Moreover, there are other considerations.

First, Nietzsche's mode of discourse and style of presentation varies from from account to another, which makes a comparison of the abstract "thesis" in each case problematic. How, for example, should one compare the "poetic" versions (for example, in later parts of *Thus Spake Zarathustra*)[4] and the scientific proofs in *The Will to Power*.[5] Second, the *rhetorical strategy* varies, and may indeed be the subject of critical disagreement. Again, that there was a "content" to be drawn out for comparison would be made questionable in principle. And third, the audience/level of exposition varies. Would one necessarily expect esoteric and exoteric versions to be consistent?[6]

One can make a virtue out of contradictions (vide Walt Whitman, for whom contradiction was a sign of his spiritual abundance). Jaspers' reading of Nietzsche does just this. Or one can offer general arguments drawing on the thesis of semantic indeterminacy or of the fictionality of truth and hence of consistency (both of which Nietzsche held). But these are mere slogans, signposts, the husks of thought; they are not in themselves arguments. And constructively to sieve through Nietzsche's various presentations of eternal recurrence, a discriminatory classification of the various accounts is essential. This will not be free from the theoretical interests of this treatise, but these frequently coincide with Nietzsche's. That theoretical interest, to make it explicit, is in understanding the role of eternal recurrence in Nietzsche's transvaluation of time. The value of such a guiding interest remains to be seen. It is also an open question whether it can be sustained as such, as a "theoretical interest," or whether the cool analytical security that such a phrase suggests is not itself fundamentally at risk. Placing what Nietzsche called the "greatest weight" on the scales is an act not free from danger. Nonetheless, I shall begin by arguing that the various ways in which eternal recurrence is formulated reflect the different modes (levels, dimensions) at which time needs to be transvalued or "deconstructed,"[7] that the question of the *status* of eternal recurrence (whether it is a concept, a theory, a thought, an experience, a test) will vary with the level in question above, and that one ought to take seriously the suggestion that eternal recurrence is "undecidable" in terms of the conceptual framework it subverts. (What that implies is that eternal recurrence not only would be a "moral" test [try to affirm that!] but an intellectual test [try to think that through!]).

To give some structure to this account, I propose, perhaps unoriginally, to distinguish three levels or ways in which eternal recurrence is presented: cosmological, psychological, and ontological. It might be objected that this framework is not complete (what about the historical, for instance?), but I would not accept that it is "naive." Of course, Nietzsche might put these very distinctions in question. But in my view the different presentations of eternal recurrence do work within such traditional categories, *if only to dissolve them.* And these categories also correspond to three different dimensions of time: universal (including historical) time; what I will call "motivational" time; and time as linked with identity (both of things and persons). If the ordinary schemas, concepts, of time can be understood to rest on the value of *unity*—guaranteeing it, reappropriating it, making it possible, and so forth— then the deconstruction or transvaluation of time will invert and displace this value as it is specifically embodied in each mode, level, or dimension.

Thus, corresponding to the category of universal time there is the idea that time stretches infinitely in both directions, that it is one-dimensional, and that every event has a uniquely determined place in it. Ordinary motivational time can be glossed as a structure of asymmetry between past and future, in which the past is complete and unaffectable, and the future the scene for the projection of one's freely chosen ends. Time understood ontologically, in relation to identity, is understood either as a neutral container in which things or persons endure as they are (substances), where their identity is independent of time, or as productive of identity by making possible the development of a being's identity to fulfillment or completion.

It is a sign of the polysemic depth of the idea of eternal recurrence that it can be transformed to transvalue each of these settings. I shall now look at some of Nietzsche's remarks that would bear out this reading. First, the cosmological version of eternal recurrence.

The Cosmological Version

This is most explicitly formulated in *The Will to Power*, sections 1062 and 1066, and in various notes dating from 1885–88. Let me quote, as countless others have done, Nietzsche's "most scientific" argument for eternal recurrence:

> If the world may be thought of as a certain definite quantity of force and as a certain definite number of centres of force—and every other representation remains indefinite and therefore useless—it follows that, in the great dice game of existence, it must pass through a calculable number of combinations. In infinite time, every possible combination would at some time or another be realised; more: it would be realised an infinite number of times. And since between every combination and its

next recurrence all other possible combinations would have to take place, and each of these combinations conditions the entire sequence of combinations in the same series, a circular movement of absolutely identical series is thus demonstrated: the world as a circular movement that has already repeated itself infinitely often and plays its game *in infinitum.*[8]

I will begin with two observations. First, it is not convincing, and second, it is not clear what difference it could possibly make if it were true, and consequently, it is not entirely clear what it would mean for it to be true.

There are a number of different claims to be distinguished in this passage, some more plausible than others, even allowing for his premises (a finite number of forces and infinite time). As it stands, I am not persuaded either that a finite number of elements could not generate an infinite number of qualitatively different states, thus making the necessary infinite repetition of each state a non sequitur. Nor, if everything were to be repeated infinitely, is it clear why it must take the form of exact cycles of complete sequences of permutations. I am not the first to have had doubts about this proof. Others have found it more acceptable; Danto, for example,[9] believes that with the addition of various other plausible premises it can be made to work. I shall argue, however, that Nietzsche does not need the precise repetitions that this argument, if successful, would prove. But first I shall look at another problem with the idea of eternal recurrence as so conceived: What would it mean for it to be true?

My worry could be put like this: What would it be to be in one cycle rather than another? The request for some distinguishing feature by which one cycle be distinguished from another must always be refused, for *ex hypothesi* there can be none without violating the principle that what returns returns identically. There is no point outside such a series to mark one's place. The hypothesis could not, then, be confirmed or refuted by ordinary experience. Yet unlike the hypothesis of a divine being, it does not posit a transcendence, a higher plane of existence, only a horizontal extension of this one. Eternal recurrence is minimally, it seems, an untestable hypothesis, consistent with ordinary experience, requiring no higher being, and one the mere truth or falsity of which has no effect on experience of the real world. This sounds like an empty, uninteresting idea that might at best occupy a marginal place in the dreams of an idle mind. And yet Nietzsche himself claims it to be his most powerful insight! Thus: First, he claims (*The Will to Power*, section 55) that it is "the most scientific hypothesis." I shall have something to say about this shortly. Second, in the plan of the book on eternal recurrence outlined in *The Will to Power*, section 1057, section 2 is entitled "Proof of the Doctrine." He clearly takes the cosmological argument seriously. And third, he clearly contends that belief in eternal recurrence does make an enormous difference. In a letter to Overbeck, he wrote:

> If it is true—or rather if it is believed to be true—then everything changes and turns around and *all* previous values are devalued.[10]

What is proposed here is a strategy of reading Nietzsche's cosmological presentation of this doctrine that will both explain the importance with which he credits it, and resolve some of its difficulties. First, Nietzsche is not offering this proof of eternal recurrence in an intellectual vacuum. He is opposing it both to the traditional conceptions of nineteenth-century mechanism and to the teleological conceptions of traditional theology. Take mechanism first. Nietzsche has a number of reservations about the idea of the world as a network of causes and effects. He is skeptical about the very concepts of "cause" and "effect" (as varieties of fiction). And if (as Kelvin argued) mechanism leads to a final state (entropy), it must, in infinite past time, have already reached it. As it clearly has not, mechanism must be false.

Coming from Nietzsche, and especially where they concern eternal recurrence, which was after all the subject of a vision (see below), these arguments raise certain questions of evaluation. Nietzsche is not parodying scientific discourse,[11] but neither is he committed to its concepts, its assumptions, its standards. Indeed, he elsewhere pours scorn on those conclusions that need proving. What he is doing, surely, is showing how eternal recurrence can be argued for even in terms that he did not himself endorse.

Some sort of confirmation of his willingness to adopt such a strategy can be found in the way he handles the other main view to which eternal recurrence is opposed: theological teleology. For example he offers the following argument:

> If the world had a goal, it must have been reached. If there were for it some unintended final state, this must also have been reached. If it were in any way capable of a pausing and becoming fixed, of "being," then all becoming would long since have come to an end, along with all thinking, all "spirit." The fact of "spirit" as a form of becoming proves that the world has no goal, no final state, and is incapable of being.[12]

The language here is surely that of theology. The argument can be constructed by substituting the opposition between Being and Becoming for that between Being and non-Being in Aquinas's Third Way. Where Aquinas argues that a necessary being must be posited to explain how, in infinite past time, we have not been swallowed up by a coincidence of non-Being, Nietzsche argues that the absence of a goal must be posited to explain how, in infinite past time, that universe did not cease all becoming. Aquinas did not think that such proofs were necessary, only convenient for some. No more did Nietzsche. That Nietzsche's attitude to the frame of reference in which such arguments are constructed is something short of total commitment can surely be seen in the next paragraph, in which he continues:

> The old habit . . . of associating a goal with every event and a guiding, creative God with the world is so powerful that it requires an effort for a thinker not to fall into thinking of the very aimlessness of the world as intended.[13]

However, there is clearly a sense in which the cosmological version of eternal recurrence moves within this frame of reference. In our view, the cosmological argument for eternal recurrence can be seen as subversive of both teleology and mechanism in that it can be shown to be no less plausible in the very same terms. And that is the point of saying that this proof is not offered in an intellectual vacuum. It may be quite as important that it challenge the existing contenders for our intellectual assent as that it finally convinces us in its own right.

The second consideration one could offer toward a more receptive reading of the proof of eternal recurrence is that it can be seen as an attempt to give a rigorous scientific justification for believing something that in a more general form might be found far more plausible. And Nietzsche himself, at the beginning of the very passage from which this proof comes, offers just such an account of his "new world conception" that depends not on any detailed argument, but on a model of a closed economy of forces. Does Nietzsche really need more than this? He writes:

> The world exists; it is not something that becomes, not something that passes away. Or rather: it becomes, it passes away, but it has never begun to become and never ceased from passing away—it maintains itself in both. It lives on itself: its excrements are its food.[14]

It might, however, seem that repetition is entirely missing from this picture, which would make it deficient in a vital respect. As will be seen later, the abandonment of exact cycles is a price some think Nietzsche has to pay to give eternal recurrence psychological force. But for this picture here to be sufficient, one would have to find in the fact that becoming "never ceases" a sufficient embodiment of becoming. I shall come back to this question.

I have cited Nietzsche's remark in a letter that if it is believed true, eternal recurrence turns everything round. It might be suggested that a comparison with heavenly salvation would be appropriate here. For that too, true or not, makes an enormous difference if one believes it possible. And it shares the feature of unverifiability.

Heavenly salvation is usually thought of as a reward, or at least as a consequence of having led a certain sort of life. The reward is to be led to another life distinct from the first, one in which, one presumes, suffering is absent (or, in the case of hell, in which suffering is eternal and hope is absent). In this case too, it is arguable that the truth or falsity of the existence of an afterlife has no direct bearing on the facts of this life, while belief in such a state can have an enormous "effect." Is eternal recurrence just a kind of inversion of salvation? A kind of upside-down theology?

Surely not on this first cosmological version, because on this account it can be shown that the connection between one cycle and the next is neither causal nor moral, so nothing I do will have the slightest effect on the next round. This is true whether I believe in eternal recurrence or not. And furthermore, if one did suppose the relationship between one cycle and the next to be one of cause/effect in whatever loose form, one could not call it salvation (or damnation), for what one does is merely to be *repeated* infinitely. Eternal recurrence does not offer a version of heaven or hell. It can offer in itself neither punishment nor reward. It is a horizontal move, not a vertical one. It is this feature—the horizontality of eternal recurrence—that, as I shall now explain, allows the cosmological version of eternal recurrence to play its special role in the transvaluation of our ordinary concept of time.

It seems to be possible to think of eternal recurrence as a loopy arrow, as a single temporal series of a strange loopy shape. And yet, without departing from seriality, Nietzsche has thereby generated a structure that threatens it. When one talks of a loopy shape, one thinks representationally of something like a coil, a spring shape. On such a model there is a way of distinguishing between the different cycles, even if there were no clear point at which one cycle begins and ends. But on the cosmological model, there is no external point of vision, and indeed one is not simply dealing with a model. There is no way of distinguishing between one cycle and the next, etc. I am, or am inside, the coil, and to me the entire history of the universe *could* just seem like a straight line, the arc of an immense circle.

Why labor this literal interpretation of the eternal recurrence as a cosmological hypothesis? Should one not simply feel its unfathomable power, and not pause to analyze its reflective implications? I claim that this cosmological version is the first stage (or can be so construed) of Nietzsche's deconstruction of ordinary serial time. The key to this operation is that Nietzsche has managed to *construct*, using ordinary serial time, a plausible account of the structure of time that provides the basis for the transvaluation of that "ordinary time." Eternal recurrence seems merely to be a very powerful *modification or extension of seriality*. But in fact it puts in question the assumed self-contained status of its units—"nows"—and their fundamental successivity. For as well as being located on a horizontal axis of succession, each "unit" also apears to be a member of a vertical series of repetitions.

PSYCHOLOGICAL TIME

It is hard to contain this version, as it appears in so many guises. Perhaps the key to it is the role of eternal recurrence in undermining and displacing the idea of *purpose*—be it personal or historical. As such, however, it is merely negative, and much more needs to be said about the idea of affirmation and the transformation of the will. I begin with an account of what might be called the "intensity of the moment," which provides something of a template for self-affirmation.

THE INTENSITY OF THE MOMENT

The cosmological version of eternal recurrence functions in such a way as to extend serial succession (in "loops") so that each individual moment acquires membership of an additional, vertical series.[15] Consequently—and this is taken up more fully in the next section—the identity of such units will also be divided along these two axes.

> Has time flown away? Have I not fallen . . . into the well of eternity?[16]

This second, vertical dimension to each moment opens up the possibility of depth and, experientially, intensity. I siall first look at some of the formulations Nietzsche has given of "the intensity of the moment." Writing of the aesthetic moment of "rapture," Nietzsche says:

> . . . in this condition one enriches everything out of one's own abundance: what one sees, what one desires, one sees swollen, pressing, strong, overladen with energy. . . .[17]

He goes on:

> . . . the entire emotional system is alerted and intensified so that it discharges all its powers of representation, imitation, transfiguration, transmutation, every kind of mimicry and play-acting conjointly . . .[18]

And the first account Nietzsche gives of eternal recurrence is the occasion for another account of a "privileged moment." He imagines a demon suggesting that the "eternal hourglass of existence is turned upside down again and again" and suggests two possible reactions: (a) utter despondency and misery, (b) exhilaration. The latter he puts like this:

. . . or have you once experienced a tremendous moment when you
would have answered him: "You are a god . . ."?[19]

And elsewhere[20] he tells of "attaining" the *Übermensch* "for one moment" and
of the immortal "moment . . . in which I produced return."

The question I would like to put is this: How far are we confronted here
with a new *general* account of the moment—and how far only of certain
privileged moments? And if the latter, in what way is Nietzsche offering a
more general transvaluation of time? One thing must be true: any "general
account of time" must allow the possibility of these self-expansive moments.
But are they special cases or the general rule? Clearly at one level they must
be special cases, or they could not exist distinctively, or in contrast, e.g., to
the depressive moments of anguish at the possibility of eternal recurrence.
And yet I think it is clear that for Nietzsche they are special in a special
sense; they realize the highest possibility of temporal experience—an
intensity[21] of vision and/or self-mobilization and affirmation. But what of all
the *other* moments that never achieve this status? Are they not too part of
time, so that Nietzsche's account of time *in general* cannot be said to be
ecstatic and affirmative? Should one *just* say that he offers an ideal way of
living (the moment) based on possibilities intermittently realized?

This would not be a dull view. By itself, it would require and justify
some account of existential temporality merely to accommodate these peak
experiences.

But one cannot rest with this picture, for a number of reasons: First, the
cosmological version of eternal recurrence has already compromised the idea
of time *simply* as a sequence of distinct moments. The addition of a depth to
each moment redirects attention away from relations of succession toward
the possibilities of intensification. When Nietzsche writes of the "eternity" of
the moment, this is not simply a reference to its infinite recurrence at other
times. The picture one might have of life as a sequence of moments, some
high, some low, still depends on giving seriality the last word. *Given that model,*
leaping from peak to peak would correspond more closely to Nietzsche's
picture. But Nietzsche does not retain that picture. The variable intensity of
the eternally recurring moment constitutes an alternative. I return to this
shortly.

Second, there is another crucial reason why these peak moments are not
just different, but privileged in relation to those of the plains or the valleys.
Part of what makes for the experience of exhilaration is that they project their
own ecstatic affirmative vision onto the rest of time. If time is thought of from
within the ecstasy of such moments, then those moments that do not
actualize this peak possibility simply do not figure. This offers a way of

understanding the eternity of the moment (" . . . joy wants the eternity of all things, wants deep, deep, deep eternity") (in a way quite different from, say, Goethe)[22] and in a way that seems to *conflict* with the supposition that in eternal recurrence *everything* returns, even the meanest. This approaches Klossowski's[23] reading of the eternal return as a *selective* operation.

Third, it is *possible* to argue that it is *not* the privilege of peak experiences that they project their own temporality onto the rest of time. Do not such experiences as boredom, depression, and self-destructive feelings do the same? But if they do so, there is a major difference. Only this affirmative rapture *wills itself enough* to will its own *infinite repetition*. Only this total self-affirmation wants itself more and again.

I now turn to the role of eternal recurrence in undermining and displacing the idea of purpose—the transformation of the will.

THE TRANSFORMATION OF THE WILL

When one considers the structure of action (and the temporality that underlies it) in which goals are pursued, ideals aimed at, and so on, it is hard to see it as avoidable, let alone as being flawed in some way. For Nietzsche, this simple structure—the pursuit of values—very easily takes a pathological form, which he calls nihilism. For the positing of ideals—especially those that could never quite be realized—is a tacit *devaluation* of this world. In this sense Plato is a nihilist. Similarly, the belief in the perfectibility of man, utopian thinking, the search for truth can be seen as valuations of the future that negatively devalue the present. But the mere recognition of the negative nature of ordinary human values simply leads to a second stage of depressive nihilism, in which the world has no value and there are no values for it to have. The thought of the eternal recurrence is meant to be able to take this crushing idea to its limit, and in so doing to set a challenge to our attitude to the past. For the will finds in the past an obstacle that it finds impossible to overcome: the past, it seems, is over, is not subject to my will, cannot be changed.

However, in the section "Redemption" in *Thus Spake Zarathustra* Nietzsche charts the course of the liberation of the will. In his *What Is Called Thinking?* Heidegger devotes some five pages to a discussion of this section, in which, broadly speaking, he tries to show that the will's liberation consists in a triumphant victory over all obstacles, by which victory it instates itself unchallenged in the seat of metaphysical subjectivity.

Zarathustra speaks to his disciples:

"Will—that is what the liberator and bringer of joy is called: thus I have taught you my friends." But now learn this as well: the will itself is still a prisoner.[24]

Why is it still a prisoner?

> Powerless against that which has been done, the will is an angry
> spectator of all things past . . . it cannot break time and time's desire.[25]

There follows an account of a sequence of moves the will makes in revenge, as
Nietzsche puts it, for this obstacle to its power, concluding with the position
we could attribute to Schopenhauer and to Buddhism: "willing become
not-willing." Zarathustra's comment on this is:

> I led you away from these fable-songs when I taught you: "The will is a
> creator."
> All "It was" is a fragment, a riddle, a dreadful chance—until the
> creative will says to it: "But I willed it thus."
> The will that is the will to power must will something higher than any
> reconciliation—but how shall that happen? Who has taught it to will
> backwards too?[26]

The question is left unanswered. Zarathustra "looked like a man seized by
extremest terror." It will turn out, of course, that the *Übermensch* is the answer
to the question.

Surely this sentence: "The will that is the will to power must will
something higher than any reconciliation"—which Heidegger does not to my
knowledge allude to (in *What Is Called Thinking?*)—poses a difficulty for his
reading of Nietzsche's account as a metaphysics of will. What Nietzsche is
insisting on is that the aim of this new *creative* willing is not merely reconciliation to the necessity of time's passing away, but an active affirmation of that
fact. The question is whether this affirmative willing can importantly exceed
the identity-seeking, reconciling, appropriating gestures characteristic of the
rationalizations that had preceded it.[27]

Clearly, this transformation of the will already subverts what I have
called "ordinary motivational time." One is to will—not just accept, but
will—what has been, and the fact that time will continue to pass over into the
past. This is an idea it is almost impossible to think. It might in theory satisfy
the condition of going beyond mere reconciliation, but what can it mean?
The idea *sticks* there, undecidable in terms of any ordinary understanding of
willing, and will not go away.

I mentioned before the importance for Nietzsche of a transformation of
nihilism—he saw himself as moving beyond nihilism only in his latest work
(*Ecce Homo*). And in this move lies the greatest challenge to our ordinary
understanding of motivation. The thought of eternal recurrence has two
faces: a face of terror and a face of exhilaration. The *test* that eternal
recurrence poses is whether one can say yes to it, and transcend terror.
"Courage," Nietzsche writes, "destroys even death, for it says 'Was *that* life?

Well then, once more.' "[28] But the passage I have just been alluding to is Nietzsche's first real announcement of the eternal recurrence.

When one first grasps eternal recurrence, all hope is extinguished. Everything incomplete, ill-formed, and unwanted will return. But there follows an *expansion*, "a tremendous moment" in which one finds oneself strong enough to affirm *all* this. Consider Nietzsche's precise words: "*If this thought gained possession of you, it would change you* as you are, or perhaps crush you," or "What does not kill me makes me stronger"—as he wrote in *Twilight of the Idols*.[29]

The idea of eternal recurrence is both a test of one's strength and a source, an inspiration. The demon, on the second reading, is recognized as "a god." The idea can "take possession." At this most vital point, Nietzsche is taking as seriously as he could the idea that the consequences of believing something are more important than its truth (if indeed they can be distinguished from it). The eternal recurrence may be a trial; it is also a ladder.

But can it really *make sense* to believe it? In this form? And could it, after all, really make any difference, except by a misunderstanding? These questions are important in their own right, but also for judging whether Nietzsche has really succeeded in displacing the ordinary motivational time usually taken for granted.

There are at least two ways of articulating a critical attitude to the value of significance of eternal recurrence: First (as was seen above), that eternal recurrence cannot make any difference because we will never have any memory of past cycles, nor will our actions ever *affect* future cycles. Indeed, it could be said that experientially it makes no difference whether I do this self-same thing once or an infinite number of times, if each time is independent of the other, and each occasion is identical; and second, that it makes an enormous difference, because if the doctrine is true, then I have no real possibility of transforming, or transvaluing, anything, unless, of course, I have already done so before. This view Nietzsche calls "Turkish fatalism."

When one reads commentaries on Nietzsche, and indeed Nietzsche himself, it might seem that my questions have sadly "missed the point." Two different general responses are offered, which are not, in my view, mutually consistent. The first, Nietzsche's own, which applies both to fatalism and also to the indifferentism of the first objection, is basically a form of compatibilism.

> The truth is that every man himself is a piece of fate; when he thinks he is stirring against fate in the way described, fate is being realized here, too; the struggle is imaginary, but so is resignation to fate—all these imaginary ideas are *included* in fate.

> You yourself, poor frightened man, are the invincible Moira reigning far
> above the gods . . . In you the whole future of the human world is
> predetermined—it will not help you if you are terrified of yourself. *The*
> *Wanderer and His Shadow*[30]

But the second explanation of why they are misplaced involves a further step
away from the idea of eternal recurrence as a scientific hypothesis about
repetitive cycles. The eternal recurrence is neither a mechanical, nor a
"logical," nor a mathematical "repetition." Rather, it can exercise a *selective*
power.

This selectivity can be understood in two different ways, one "moral"
and the other ontological. The first, "moral," way is the more obvious. On
this account, only what can affirm its infinite repetition *deserves* to return.
Clearly, such an idea can function as a principle by which one regulates one's
words and deeds, a kind of supervigilant conscience. The ability to affirm
eternal recurrence is a moral test. Nietzsche clearly offers us this version, but
it has its difficulties. It seems to leave open the prospect of a *gap* between
what ought to happen and what does. This is especially true if we take
seriously the idea that actually *everything*, deserving or not, returns. And the
problem is that for Nietzsche, such a gap *invites nihilism*.[31] The second, less
obvious, way I will call "ontological." On this account, willing eternal
recurrence *actually does* operate as a selective procedure, such that certain
events, objects, relations, moments will return, and some will not. Clearly
this is a significant modification of the original cosmological account of
eternal recurrence. And there are clearly difficulties with it. For a start, it
undercuts the heroism of the original affirmation that wills the return of even
the lowliest and meanest thing. For that would no longer return at all.

There are two rather different ways of thinking through this second
alternative. The first would concentrate on such Nietzschean remarks as that
the thought of eternal recurrence licenses most men to self-erasure (crushed
by "the greatest weight"). There will be some people who continue to affirm
themselves and the world, and some also who do not. (I take it this can be
understood either physically or existentially.)[32] But here "continue" is
relative to a particular life, and on the *standard* version of eternal recurrence,
those acts (or conditions) of self-erasure would themselves return eternally
just as much as the acts of self-overcoming. And that would surely spoil the
story. The second way of pursuing the *ontological* version of selectivity is not
wholly distinct, but conceptually far more subtle.

Gilles Deleuze[33] argues that if we take seriously the idea that eternal
recurrence is "the being of becoming" (as Nietzsche himself claims[34]), and
distinguish (as Nietzsche does) between "active"and "reactive" (forces) and
apply this distinction to kinds of "becoming," then we can say that only

"active becoming" "has being" in the sense of embodying the principle of eternal recurrence. So only *active* becoming returns. Why?

> The eternal return would become contradictory if it were the return of *re*active forces."[35]

The work here is obviously being done by the relation between being and becoming, and a proper assessment of this position must be deferred until the next section. What can be said is that in Deleuze can be found for the first time an account of how eternal recurrence would function as a selective principle that does not obviously presuppose the ordinary model of time as a succession of point instants. The question to be posed is this: What understanding can we have of the moment that will allow the return of only those moments that affirm themselves, or perhaps better, in which one affirms oneself? The clue must lie in a way of thinking of time that leaves seriality behind.

ONTOLOGICAL TIME

I have suggested above that the clue to how we can understand "the moment" in such a way as to allow the return only of those moments in which one affirms oneself lies in a way of thinking about time that leaves seriality behind. But how is such a thought to be realized? Would one not be trapped in a perpetual present?

But perhaps what is at issue, what is valuable, what is at *stake* in any experience is *always the same*: it is the intensity of self-affirmation that it contains. In this sense, other moments, moments that don't make the grade, just don't come into the reckoning.

Moreover, if one asks of "the moment" whether it is one or many, one finds oneself embroiled in the most difficult thoughts. It is not enough to distinguish qualitative and numerical senses; one is left with, and I think Nietzsche means to leave us with, a genuine "undecidable." I would like to spend a little time on this, focusing on a famous passage from *Thus Spake Zarathustra* ("Of the Vision and the Riddle") in which Nietzsche explains the eternal return in relation to the gateway called "Moment," where the paths of the infinite past and the infinite future meet. The account proceeds in increasing tones of horror, and then turns to the vision of the shepherd choking on a snake, who bites off its head and is a transformed being . . .

> Then something occurred which lightened me: for the dwarf jumped from my shoulder, the inquisitive dwarf! And he squatted down upon a stone in front of me. But a gateway stood just where we had halted.

"Behold this gateway, dwarf"! I went on: "It has two aspects. Two paths come together here: no one has ever reached their end.

"This long lane ahead of us—that is another eternity. They are in opposition to one another, these paths; they abut on one another: and it is here at this gateway that they come together. The name of the gateway is written above it: 'Moment.'

"But if one were to follow them further and ever further and further: do you think, dwarf, that these paths would be in eternal opposition?"

"Everything straight lies," murmured the dwarf disdainfully. "All truth is crooked, time itself is a circle."

"Spirit of gravity!" I said angrily, "do not treat this too lightly! . . .

"Behold this moment!" I went on.

"From this gateway Moment a long eternal lane runs *back*: an eternity lies behind us.

"Must not all things that *can* run have already run along this lane? Must not all things that *can* happen have already happened, been done, run past?

"And if all things have been here before: what do you think of this moment, dwarf? Must not this gateway, too, have been here—before?

"And are not all things bound fast together in such a way that this moment draws after it all future things? *Therefore* draws itself too?"[36]

Three sections of this passage deserve special comment:
"They are in opposition to one another, these paths . . . and it is here at this gateway that they come together";
" . . . what do you think of this moment, dwarf? Must not this gateway, too, have been here—before? And are not all things bound fast together in such a way that this moment draws after it all future thing? *Therefore* draws itself too?" and
"The name of the gateway is written above it: 'Moment.' "

I first note that the moment is the *coming together* of two paths (past and future) in opposition. And yet this *coming together* is not as such a reconciliation, but a tension. It is left to the possibility of eternal recurrence to alleviate their oppositional character. Second, the question of the gateway. When Zarathustra asks, "Must not this gateway, too, have been here—before?" is he talking of *this particular* moment, or is he talking of the moment as the basis of a structure of repetition? What would it be to talk of *this* moment? Nietzsche has already given it metaphorical substantiality by calling it a *gateway*. And one can suppose that the vertical parts of the gateway will symbolize the vertical dimension of time. But there is a third point.

It is a commonplace of a structuralist view of language to suppose that it
has to be understood as comprising two axes—syntagmatic/paradigmatic, or
metonymic/metaphoric.[37] The serial articulations of a word are supplemented by paradigmatic relationships of substitutability—relationships in
which substitutions of other words would retain some important feature,
same meaning, same grammatical category, still forming an intelligible
sentence, and so on. Now it might be thought gratuitous to suggest a parallel
between the double axis of language and the double axis of time that eternal
recurrence would generate. But consider the sentence "The name of the
gateway is *written* above it: 'Moment'." At this critical point, Nietzsche
introduces not just "language," but writing. What is the force of this
appearance of writing? It is tempting to compare it to that point in the first
chapter in Hegel's *Phenomenology of Spirit* at which he is demonstrating that
"the Universal is the truth of sense-certainty." He writes:

> It is as a universal . . . that we give utterance to sensuous fact. What we
> say is "this" i.e. the general this, . . . we do not present before our mind
> in saying so the universal this, or being in general, but we utter what is
> minimal . . . language . . . as we see is the more truthful.[38]

For Hegel, the moment of language is the moment of universality. Can the
same be said of Nietzsche? Is not Nietzsche suggesting that the structure of
"moment" is as independent of any particular experience as is writing of any
particular intentional context? Might it not then be that it is precisely the
moment that eternally returns the same? But what is the moment? When he
writes "*this* gateway," should we read it as one or as many? It is tempting to
read him as saying there is just one moment, but then we have to add "which
gets repeated." So there are many.

I would suggest first of all that the moment of writing in Nietzsche, as in
Hegel's reference to the "utterance" of "this," puts in question the *presence* of
consciousness ("We do not present before our mind in saying so the universal
this"), but unlike Hegel, we are not led to another presence—that of the
realm of universality (which will ultimately be subsumed under that general
self-presence of Spirit). For Nietzsche, the moment of writing is the moment
of undecidability—undecidability, that is, and as always, within the framework that insists on a clear answer to the question: one or many?

I would recall here Nietzsche's suggestion that eternal recurrence is
"the being of becoming," or "the closest approximation of being in becoming." Abstractly, it might suggest (and one might conclude) that eternal
recurrence is the way the force of becoming is finally *betrayed*. Becoming, it
might be said, is the one thing that never changes, and hence embodies the
value of *permanence*, which is the hallmark of Being.

Is there not something very strange about such a formulation? "Becoming is the one thing that never changes" seems like a perfectly intelligible sentence. But what if becoming was the greatest threat to all thinking involving an "is"? The "approximation" of the becoming of eternal recurrence to being should be read neither as Nietzsche's confession of a new metaphysics, nor as a naive admission of that fact.

Rather, the approximation of the becoming of eternal recurrence to being should be seen as a disruptive *substitution* of an *undecidable* term into metaphysical discourse. "Becoming" in Nietzsche functions io the way in which, for example, "writing" functions in Derrida.[39] Becoming is no more in simple opposition to Being than *writing* (in Derrida's new sense) can be simply *opposed* to speech.[40] The function of eternal recurrence is to allow the reinscription of becoming within the discourse of metaphysics in a way that undermines that discourse.

The point is that becoming demands a quite "different" logic of identity to that of being.

It is now possible, I think, to distinguish more clearly three quite different functions of eternal recurrence. The first, as I have suggested, undermines the privilege of seriality by introducing, via exact cycles, a vertical depth. The second is to parody the values of a philosophy of being, by providing for every particular a kind of universality. But there is a third function, and that will require yet another thinking through of the meaning of eternal recurrences.[41]

I have already addressed what Nietzsche has to say about the state of rapture as an affirmative projectivity.The Dionysian poet could be said to "go out of himself,"[42] and when Nietzsche discusses the will to power, he says, "it must will something *higher than any reconciliation.*" In each case the basic structure is one of a self-exceeding that is not appropriated, but that, precisely, risks the self, and does not aim at a higher reconciliation. This, I would argue, is the fundamental structure of the *moment* for Nietzsche. And corresponding to this futural element of risk is the importance of forgetting; both are aspects of nonappropriative thought.

How is this idea of a self-exceeding that is not appropriated compatible with talk of the eternal recurrence of the same?[43] Does not the reference to sameness imply, if not identity, at least a continuity? And is that not precisely what is put in question by an exceeding that does not "return"?

There is a way of reading this reference to the eternal return of the "same" that does make sense. It can be treated as a description not of repeated *contents* of experience, but of the dynamic structure of experience—the rhythm, the pulse of excitement and fatigue, of arousal and consummation, of exhilaration and passivity, of the rising and setting of the passions . . . It is this *movement*, the movement of becoming, that is repeated eternally. Or, with Deleuze,[44] one can say that it is *the returning that returns*. But what of my reference to a nonappropriating exceeding, a nonreconciliatory will? Surely the point is that sameness is not what is repeated, not what returns, but, again, is constituted by that return. The "return to self" operates without a self, *is* the self.[45] Sameness is this eternal recurrence. Eternal recurrence is the condition of and not merely the extension or prolongation of sameness.

I now want to return to my opening question—whether Nietzsche's thought is another kind of philosophy of presence or whether he might be said to have *exceeded* the parameters of that framework. Without claiming to have mastered the complexities of Heidegger's reading of Nietzsche, let us remind ourselves that Heidegger does claim that Nietzsche's philosophy is in this way metaphysical. The will that wills the past, that affirms all that has been and is then able to will "the eternal return of the same," is, he says

> The supreme triumph of the metaphysics of the will that eternally wills its own willing.[46]

This will to power is interpreted as a form—the highest form—of subjectivity, of self-presence. But surely everything could turn on how one thinks of the moment?

If Nietzsche's account of the moment renders its relation to the question "one or many" undecidable, and if the ecstatic moment can be treated not as an *exception*, but simply as the highest possibility of intensification of experience, and if the realization of that possibility is, as I believe it is for Nietzsche, the essence (i.e. nonessence) of time, there surely is a case for saying that Nietzsche's thought here at least aims beyond presence and self-presence. Everything hangs on our being able to accept the idea that a general description can be given of Nietzsche's various accounts of the intensities of a moment, and that this description is of a willing/thinking/affirming beyond that does *not* aim at its own preservation, but *risks* itself perpetually, a going out that, even as it anticipates a *return*, puts in question what it is that will be returned to.

If this is not Nietzsche's thought, Nietzsche must be far less important than he is supposed. As Derrida might say, however, deconstructive theses coexist in Nietzsche with those that remain inscribed within metaphysics.

READING HEIDEGGER READING NIETZSCHE: AN INTERIM REPORT

Clearly a challenge is being posed here to Heidegger's reading of Nietzsche, and I will devote this last section to an all too perfunctory elaboration of that challenge. I continue to draw on the work of Deleuze to that end. Unlike Heidegger, he sees the concepts of Will to Power and Eternal Recurrence as successfully deconstructing the matrix of metaphysical conceptuality. And it must be with Heidegger in mind that he writes:

> We misinterpret the expression "eternal return" if we understand it as "return of the same."[47]

To seaders of Derrida, Deleuze's argument will be familiar, although the direction it gives to the thought of the eternal recurrence is new. He writes:

> The synthetic relation of the moment to itself as present, past and future grounds its relation to other moments. The eternal return is thus an answer to the problem of passage.[48]

I shall return to discuss this passage (together with "The Vision and the Riddle") shortly. He continues:

> And in this sense it must not be interpreted as the return of something that is, that is "one," or the "same." We misinterpret the expression "eternal return" if we understand it as "return of the same."[49]

But what then follows is an inversion:

> It is not being that returns but rather the returning itself that constitutes being insofar as it is affirmed of becoming, and of that which passes. It is not some one thing which returns but rather returning itself is the one thing which is affirmed of diversity or multiplicity. In other words, *identity in eternal return does not describe the nature of that which returns but*, on the contrary, *the fact of returning for that which differs*. This is why the eternal return must be thought of as a synthesis, a synthesis of time and its dimensions, a synthesis of diversity, and its reproduction . . . [emphasis added][50]

Derrida's position seems very similar here. He wrote:

> And on the basis of this unfolding of the same as *différance* we see announced the sameness of *différance* and repetition in the eternal return.[51]

For both Deleuze and Derrida, the key underlying idea is that identity is not a fixed point one needs to presuppose for differences to be possible; matters are rather the other way round. And the possibility that a thing can appear again and again at different times is what *gives it* an identity; it is not dependent on its having a prior atemporal identity. Time, then, is not only *constitutive* of identity, rather than a mere medium in which things unfold, but is *itself* constituted by its role in supporting identities and differences. But even if one cannot in any simple way say "what" it is that returns, independently of its returning, there are still questions that need answering. Perhaps I can put my disquiet like this: When Deleuze talks of "the returning itself that constitutes being," is he talking here in fact of Being or beings? Is he referring to Time itself, or to things in time? I take it that Nietzsche fairly plainly talks about things in time, or if not things, at least events, configurations of forces. And yet if we take seriously these remarks of Deleuze, eternal return is being interpreted as the ground of "time itself." It may or may not be possible to square this with any account of the return of particular (especially nonhuman) beings, but it would certainly suggest that, yet again, eternal recurrence is functioning as a device for the deconstruction of time—here, time seen as the locus of identity.

Here Deleuze raises explicitly the question with which we began, that of presence. I suggested at the outset that Nietzsche might perhaps have offered an account of the present, and indeed of time based on the present, that was not subject to Heidegger's (or to Derrida's) criticisms. The vital question will undoubtedly be the status of *becoming* in Nietzsche. I shall now begin to open up this question.

Heidegger's verdict on Nietzsche is rather different from that of either Deleuze or Derrida.

First, Heidegger quotes Nietzsche's remark "that *everything recurs* is the closest approximation of a world of Becoming to one of Being—peak of meditation,"[52] and comments:

> With his doctrine of eternal return, Nietzsche in his way thinks nothing else than the thought that pervades the whole of western philosophy.[53]

Why? Because he thinks Being as Time without thinking it as the *question* of Being.

> Eternity, not as a static "now," nor as a sequence of "nows" rolling off into the infinite, but as the "now" that bends back into itself: what is that if not the concealed essence of Time? Thinking Being, Will to Power, as eternal return, thinking the most difficult thought of philosophy means thinking Being as Time.[54]

So Nietzsche does not think of the *question* of Being (and Time). But might one not justly respond that the *thought* of eternal return is a continuous questioning, that to use such an idea as an explication of time as Becoming is to lodge a question as deep as possible into the heart of time? It may be that when understood as "the mere bending back" of the "now," the eternal return no longer had that disturbing undecidability that we have consistently noted, but perhaps that is a deficiency in Heidegger's reading. Might not Deleuze be right to query Heidegger's reading of eternal return as (always) eternal return of the same?

Our second source is *What Is Called Thinking?*[55]

> ... the answer Aristotle gave to the question of the essential nature of time still governs Nietzsche's idea of time.

I have already alluded to his argument: that Nietzsche's use of a transvaluing will to affirm the past betrays a traditional valuation of Being (including the "Being" of time) as present. Our response, we may recall, was to say that he took no account of Nietzsche's reference to a will that did not seek "reconciliation"—that Heidegger was *refusing* the radicality of Nietzsche's affirmative willing.

Finally, I would like to suggest a way of reading Heidegger on Nietzsche, one that anticipates some of the ambiguities in the notion of authenticity that are noted in our long discussion of Heidegger's *Being and Time* (see pp. 166–214). In the discussion of authenticity will be discovered tendencies toward closure (for example, the idea that one's "ownmost possibilities" could ever be anything more than a question) and tendencies that would preclude such a closure, such as references to anxiety, the abyss, and thrownness.[56] I would like to suggest that this tension between these two motifs is not just found generally in *Being and Time*, but is found *specifically* connected with the question of that ongoing rupturing of selfhood that I have associated with the Nietzschean moment. Even more interestingly, Heidegger offers us, within the space of a few lines—though without posing it *as* a problem—the very question that is most pressing: How to understand this "rupturing" in terms of Being? Finally, he does this at one of the very few places at which he invokes the name (in brackets!) of Nietzsche. These are the sentences in question:

> Anticipation discloses to existence that its uttermost possibility lies in *giving itself up*, and thus it shatters all one's tenaciousness to whatever existence one has reached.[57]

What is this "giving itself up" ("Selbstaufgabe")? Is it just death in the narrow sense, or is it not precisely the risking of all one is and has known? How does Heidegger continue? Doesn't he temper the radicality of the suggestion he has just made?

> In anticipation, Dasein guards itself against falling back behind itself and behind the potentiality for Being which it has understood. It guards itself against "becoming too old for its victories" (Nietzsche).[58]

The important thing here is how one understands "falling back behind itself" and the "potentiality for Being" (which it has understood).

These remarks can be given a direct Nietzschean interpretation—the "understanding of Being" that it has understood is of course not a self to which one clings, but, I would suggest, a grasp of the sense and responsibility of the "intensity" of experience. The problem about "not falling back" is the same problem as that of selectivity. Only what can will its own return can/should return.

The clear Nietzschean influence here suggests what will seem obvious when stated: that we would be wise not to divorce Heidegger's reading of Nietzsche from Heidegger's continuing attempts at a self-interpretation.[59]

I would like to have shown that if we suppose that the model of Dionysian excess provides a standard by which to measure the intensity of the moment, and if that excess is a nonrecuperable rupture with all "presence," then Nietzsche's "moment," so far from being the reworking of the metaphysical value of presence, is the scene of its explosion. But is that what one should conclude about Nietzsche? Does he achieve the magical result of a nonmetaphysical philosophy of the present? Perhaps matters are not quite so clear-cut. What he does do, I believe, is force us to make a distinction between two levels at which one can understand the meaning of "presence" as a metaphysical value. The first might in modern terms be called foundationalist, and the second appropriative. To each corresponds a different stratum of that mode of textual inscription that Derrida insists makes for metaphysics. By foundationalist, I mean a kind of thinking that reduces to, or derives from, one fundamental point the entire developed structure of some theoretical field. Arguably, Nietzsche does this at least in a formal, and perhaps only strategic, way, if I am right in giving the moment the status I have. Nonetheless, one could say, in Levinasian language, that for Nietzsche, time, in the shape of this "moment," is an opening onto the other, onto otherness, onto what may never be appropriated, made identical, brought back. Here Nietzsche does break with the second characterization I have given of the metaphysical value associated with presence. Nietzsche, on this reading, would be a nonappropriative foundationalist.

This characterization brings back the question raised when discussing whether Nietzsche's was a special or a general theory. And for all the value of seeing the ecstatic moment as an ideal, it surely does not *actually* capture the general structure of time. It is precisely because it does not that it can function as an ideal. And what that suggests is that one may learn more about the possibilities of exceeding metaphysics from the nonappropriative stratum of his thinking than from its foundationalism.

Whatever one decides about the adequacy of Nietzsche's accounts of eternal recurrence as strategies in the service of what has been called his transvaluation of time, it is worth drawing from them a connection that will return throughout this book, and that constitutes one of its central subplots—the connection between the question of language and representation, on the one hand, and that of one's "critical" strategy in thinking about time, on the other. Nietzsche has not only a profound distrust of the adequacy of descriptive language, but an equally profound respect for the possibilities opened up by innovative styles and strategies of writing. And despite his calling eternal recurrence his "most scientific hypothesis," we see the various accounts of eternal recurrence as witness to the power of strategic writing. For it is precisely the descriptive inadequacy of his accounts that pushes us to other ways of understanding him. Equally,while Nietzsche seems to want to address the question of time through the question of its fundamental unit—the moment—the upshot of this is an account that undermines its very capacity to function as a theoretical building block.

Nietzsche's work opens up two different general strategies for the deconstruction of time, which in Part 4 are pursued in parallel: First, the "theoretical" displacement or questioning of the fundamental element (the present moment) of traditional theories of time. Phenomenology thought this could be done "directly," by a reconstructive phenomenological analysis, but discovers that the problems of strategy, of language, of representation return again at the very end of the analysis. Second, moving in the opposite direction to the explosion of the Moment, Nietzsche's doubts about the metaphysical tendencies of language can lead us to a radical pluralization and deontologizing of temporal discourse. Moving in this direction restores, though with a very different philosophical justification, something of the analytical concern with the structure of experience that Husserl opened up to us. In the detailed discussion that now follows, I will press for an opening up of that account to the pressure of the plural, and for the overcoming of the metaphysics of consciousness. The radical differences between Nietzsche and Husserl can, I believe, be creatively synthesized.

Part 2
Husserl's Phenomenology of Time

1 *The Intuitional Foundations of Husserl's Phenomenology and the Requirement of an Original (Preobjective) Temporality*

The powerful and enormously productive thrust of Husserl's phenomenology is the affirmation of a value that in his eyes all previous philosophy had lacked. Contract, immediacy, fullness, primordiality, intuition all cluster round this central value, which could be called "presence."[1] Along with a number of other thinkers (such as Marx and Freud) whose thought in this respect shares the same structure, Husserl's affirmation of this value sustains a repeated opposition between a public, completed, "objective," derived, *inauthentic* account of things, and the directly graspable, subjectively intuitable, original, *authentic* account of things.[2] As Marx had elaborated an ideology-critique and Freud the practice of psychoanalysis, Husserl too developed a special method for prising apart the merely taken-for-granted from the intuitively graspable, and for describing delicately and in detail the region of intuitive transparency that this distinction opens up. This begins with a stage that is variously called an *epoché*, bracketing out, suspension, and phenomenological reduction. The operation that takes place at this stage is crucial for the whole subsequent course of phenomenology. Husserl draws both on the distinction just made, between the lived experience transparently grasped as such, and the taken-for-granted or "natural" attitude that we unphilosophically take up, but also on the distinction between reality as such, and our grasp of it, however primordial or derived this latter might be. Husserl wants to set aside, bracket out, *both* "reality as such" and our "natural attitude" toward it: " . . . we are concerned with reality only insofar

39

as it is intended, represented, intuited, or conceptually thought."[3]

Our "natural attitude" gets "put aside" not because it is false (indeed, he will not make that claim) but because part of what is involved in this "attitude" is its lack of concern with the ground of its truth, its reflexive complacency. And "reality as such" is "put aside" because we can know or say anything about it only via an account of our own constitutive meaning-giving activity, not as it is "in itself." What phenomenology seeks, then, is access to those meaning-giving acts on which our grasp of the real depends, and that will illuminate the knowledge we already possess by exposing its grounding in such intentional acts.

It might fairly be said, and Husserl was well aware of this, that the possibility of phenomenology rests on negotiating a delicate passage between psychologism on the one hand and some sort of linguistic philosophy on the other. Husserl's concern lest phenomenology be thought to be nothing other than a specialized branch of psychology dealing with mental activity was well founded. The logical principles with which he first concerned himself could not be the subject matter of an empirical science at the same time as being essentially presupposed by any science. Clearly there *could* be (for there is) some psychological interest in logical thinking—when it arises, how it develops through childhood, under what conditions it is abandoned and so on—but psychology could never tell us *what* logical principles were, nor why they were valid. While logical principles can be embodied in or be the object of thought, nothing guarantees that thinking will be logical, and if we are concerned with the common characteristics of all thought (such as predication), there is nothing *essential* that psychology—which deals, as does any science, with variations—can tell us about it. Husserl's insight was to have realized that the most general cognitive framework of science (that is, any rigorous/disciplined thinking) could not itself be the subject matter of an empirical science, let alone be justified by such an approach. His solution was not to abandon any attempt to ground thinking absolutely, nor to focus in a consequentialist manner on where different conceptual systems lead us, but to develop an analytic method for describing what he would call the *ideal* structures of our conscious life. And this should not be taken as an exercise in the imagination that would idealize its subject matter with flattering artistry, but as a method that *discerns* the actually operative ideal in experience. Phenomenology, then, is an analytical method devoted to describing the qualitative constants of human experience. Insofar as that experience has more or less primitive aspects, and is an *activity*, not merely a receptive surface, phenomenology is also concerned with bringing out the constructive, or constitutive depth of experience.

It is through this concept of ideality (of "essence") that both the method and subject matter of phenomenology can be kept distinct from that of any empirical science, including psychology. But it is equally important, if we are to retain Husserl's sense of his own enterprise, to steer clear of another false alignment.

It would be easy to suppose that the ideality Husserl discerns in experience is none other than that of the ideality of language. Is it not through language that meanings get fixed, that ideality is born?[4] But to follow this path would naturally lead to the abandonment of any interest in experience or conscious life as such, because there would always be a more direct route through some sort of philosophy of language. This is not the path Husserl followed. The reasons for this are historically momentous. If Frege's 1894 critical review of Husserl's *Philosophy of Arithmetic* as a piece of psychologism had converted Husserl to his own views, we would not have had the great divide in the twentieth century between linguistic philosophy and phenomenology. Quite how far Frege's review influenced the course of Husserl's thought is a matter of debate.[5] What is less a matter of debate is that phenomenology claims to provide a *distinct* solution to the problem of psychologism, one that not only cannot lean on, or be reduced to, a covert philosophy of language,[6] but would itself account for precisely that ideality which language exhibits.[7]

For example, when discussing the difficulties that ordinary language— often ambiguous and vague—presents for the development of a scientific method, Husserl clearly gives to language the task of adequately representing what has *already* been intuitively clarified. In *Ideas*, section 66, he talks of

> the requirement that the same words and propositions shall be unam-
> biguously correlated with certain essences that can be intuitively ap-
> prehended and constitute their completed "meaning."

And even when he describes the countervailing tendency to associative ambiguity and dispersal of meaning, the same dependency of language on intuition is maintained. He talks of "cancelling" "other meanings which under certain circumstances thrust themselves forward through the force of habit." (ibid.) For Husserl, "essences" discovered by reflection on our conscious life precede language and make it possible. Ideality is a prelinguistic phenomenon.

There are a number of ways to criticize Husserl's phenomenological approach. The subsequent history of twentieth-century philosophy is littered with positions that either explicitly take issue with it (Marxism, structuralism, *existential* phenomenology, to name but a few) or offer alternatives based on premises that phenomenology would dispute (such as logical atomism, logical positivism, ordinary language philosophy, pragmatism). One could put in question its apparently foundational model of justification.[8] One could doubt the very possibility of the pre-predicative description that seems to be entailed.[9] Or one could dispute the adequacy, possibility, and viability of the particular kind of "foundation" Husserl constantly demands. This will be discussed later when we look at Derrida's critique of "presence."

But before taking up such external positions, it is always worth pursuing a philosopher's own discussion of, or confrontation with, the *limits* of his thought. Before considering Husserl in this light, drawing on a number of sections of *Ideas* 1, my underlying strategy deserves an explicit comment.

I hold the value of philosophical criticism (and indeed the claim can be extended to other theoretical disciplines, including the natural sciences) to lie not so often in the *refutation* or *confirmation* of a philosophical position (or theory or conceptual framework) but in the demonstration or exploration of the *limits* of its scope of productive application. A philosophical position and the like formulated or held in the absence of any reflective grasp of the limits of its application could be said to be formulated or held metaphysically.[10] From the point of view of intellectual productivity this can be, at different periods in the life of a position, a "good" thing, or a "bad" thing, "necessary" or "remediable," and so forth. It can be important that a position be initially formulated in an exaggeratedly universalistic way to seem to be a plausible successor to a current position similarly formulated.

One danger with such talk of limits is the ultimately complacent assumption that all successfully "limited" philosophical positions could be compatibly ranged or articulated in a single conceptual space. On this view, criticism would be analogous to the proceedings of an international boundary commission, in that it would suppose that the nature of this "space" was not itself a matter of philosophical debate. It always is. Furthermore, and to prevent any misunderstanding, there is no denying that philosophical positions *can* be plain incoherent, inconsistent, based on false premises, and in various other ways inadequate, and that criticism can in these cases be healthily destructive. There is no proposal to establish a rest home for lame ducks.

I now turn to a discussion of the limits of Husserlian phenomenology, which will prepare the way for a critique of his attempt at a radically antimetaphysical "theory of time."[11]

One of the key areas in which Husserl gives expression to the limits of phenomenology (without himself conceiving these limits negatively) is in his discussion of phenomenological method. This is hardly surprising for two reasons: (1) It would seem quite generally true that a philosophical method is something that is always already at work (or in play) as soon as one writes. If one subsequently returns to question it, that questioning would constitute a further elaboration of one's method. Does not that resultant complex strategy ultimately just stand or fall without further justification, and this constitute a "limit" to any philosophical position? (2) Such a *general* argument seems to be especially important for phenomenology because its claims for philosophical renewal rest on adopting a radically new method.

I shall focus first on a remark Husserl makes in *Ideas*, section 3, entitled "Essential Insight and Individual Intuition." He writes:

> At first "essence" indicated that which in the intimate self-being of an individual thing discloses to us "what" it is. But every such What can be "set out as Idea." Empirical or individual intuition can be transformed into essential insight (ideation)—*a possibility which is itself not to be understood as empirical but as essential possibility*. The object of such insight is then the corresponding pure essence, whether it be the highest category or one of its specializations, right down to the fully concrete. (p. 54) (emphasis added)

Husserl is explaining here how essences can be generated as objects of a distinctive kind of insight, by a certain transformation of our understanding of the essential nature of particular "things." We shift from thinking of a quality as instantiated in a thing to thinking of it in its own right. He will proceed shortly to some of the implications of this possibility, but before following him I would like to focus here on the apparent circularity of the words, "a possibility which is itself not to be understood as empirical but as essential possibility." To see that this transformational possibility is "an *essential* possibility" (and not merely a move that, as it happens, we can perform), we have to be able, presumably, to make it itself the object of essential intuition. This involves seeing that the "essential" properties of individuals have a built-in independence of any particular individuals in which they happen to inhere ("Whatever belongs to the essence of the individual can also belong to another individual," section 2), and that it is consequently possible to have a grasp of such properties independently of their concrete inherence. This grasp would involve "essential insight."

But if there is a circularity here, it is not vicious. Essential intuition does not depend for its validity on being a transformation of individual intuition, and "judgments of essence" do not depend for their truth on their being transformable from and into "judgments of essential generality." There is rather a quite fundamental assumption as to the existence and importance of a distinctive kind of intuition that Husserl calls essential. What appears as a circularity is in fact just the recursive use of what is deemed a privileged mode of cognition. When he laments Külpe's critique of his "theory of categorial intuition" (section 3, n. 1), he immediately goes on to refer to it as a "simple and quite fundamental *insight*." So the *theory* of categorial intuition is an insight, that is, an intuition. And this confirms the suggestion that what we are dealing with in Husserl's work is a repeated appeal to a privileged source of a privileged kind of knowledge that precedes and sustains any

subsequent "theory." That is, we will not (without transforming this valuation of intuition into one) find a basic premise, or set of premises, underlying
Husserlian phenomenology, but rather a certain (alleged) cognitive possibility.

Now it is of course open to question whether we do in fact possess the
capacity for "essential intuition" as he describes it, and whether it has the
epistemological status he claims for it. (He claims it is the basis of a whole
range of "eidetic sciences.") It might be argued, for instance, that Husserl is
not justified in supposing that just because an essential characteristic of a
thing is independent of its instantiation in any particular individual item, it
can be grasped independently of any instantiation at all. But Husserl's
position here is quite subtle. He insists that "no essential intuition is possible
without the free possibility of directing one's glance to an individual *counterpart* and of shaping an illustration." In fact, while there is a difference in
principle between essential and individual intuition, each depends on being
convertible into the other. Husserl is not advocating a complete separation of
essential intuition from concrete instances. And interestingly, such instances
may as well be imagined as real.

The first tentative conclusion to be drawn is that phenomenology rests
on a primitive valuation of intuition, and this will constitute—not necessarily
in a negative way—the first *limit*.

The second focus will be on an important chapter in *Ideas*, "Preliminary
Considerations of Method," and certain immediately preceding remarks.

In "The Suspending of the Material-Eidetic Disciplines," part of the
previous chapter on the phenomenological reduction, Husserl is explaining
the method of reduction by which phenomenology brackets out of consideration all that "transcends consciousness," so that it can fulfill its own
methodological standard, and claim "nothing that we cannot make essentially transparent to ourselves by reference to Consciousness and on purely
immanental lines." The question he poses at the start of section 60 is whether
there are any limits to this reductive procedure. He claims there are:

> As regards the eidetic fields of study on their material side, one of these
> is of such outstanding significance for us that the impossibility of
> disconnexion can be taken for granted: that is, the essential domain of
> the phenomenologically purified consciousness itself. . . we could not
> dispense with the a priori consciousness. A science of fact cannot
> alienate the right of making use of the essential truths which relate to the
> individual objectivities of *its own* domain. Now it is our direct
> intention . . . to establish phenomenology itself as an eidetic science, as
> the theory of the essential nature of transcendentally purified conscious-
> ness. (p. 177)

Husserl is saying that there are some "essences" that are instantiated in things that transcend consciousness, and others that are instantiated within the "immanent formations of consciousness" itself. An essential grasp of the concepts involved in an abstract "chronology" or "topology" or "mereology" . . . for example, could be bracketed out, but (I take him to be saying) the distinctions (see sec. 1, ch. 1) between *fact* and *essence* or between *transcendent* and *immanent*, for example, could not. There is an obvious reason for this. The very value and possibility of undertaking these reductions is explicable only using these distinctions. If disconnection, suspension, bracketing out, could be glossed as "making no use of" certain concepts, then the limit reached would not merely be one beyond which the intelligibility or importance or interest in the reductive procedure might be put in doubt, but a logical limit. One simply *cannot*, in principle, successfully employ a procedure that involves applying certain concepts so as to suspend the use of such concepts. If one were to "succeed," one would retroactively annul the procedure by which that success had been brought about, and so annul the success itself. Briefly, reduction is an achievement dependent on the concomitant satisfaction of procedural norms.

But to say it is a logical limit is just not simply to confirm Husserl's claim about the "right" of any science of fact to utilize its own essential truths; it turns one's attention toward the very methodological and conceptual framework within which such limits arise. In our previous discussion we concluded that phenomenology rested on a primitive valuation of "intuition" (that is, the transparent grasp of the essence of some conscious formation, a close cousin of Descartes's "clearness and distinctness"). And that would suggest that critical distance to Husserl could only be taken up by scrutinizing this notion, or the founding value he gives it.[12]

Curiously, perhaps, in his chapter on method, he gives a somewhat different solution to the problems of reflexivity, circularity, and limits in phenomenology. In "The Reference of Phenomenology Back to Its Own Self" he writes

> It might be a stumbling block to someone that whereas from the phenomenological standpoint we direct our mental glance towards this or that pure experience with a view to studying it, the experiences of this inquiry itself, of this adoption of a standpoint, and this direction of the mental glance, taken in their phenomenological purity, should belong to the domain of what is to be studied. (pp. 189–90)

Husserl gives three different justifications of this reflexivity. First, he begins by claiming that it is not problematic because it is no less true of psychology than of logic ("the thinking of the psychologist is itself something psychological, the thinking of the logician someting logical . . ."). But this is surely an

unconvincing response for (a) it is hard to accept a justification of phenomenological reflexivity that involves drawing parallels with empirical or purely
formal disciplines from which Husserl has at all other times been at such
pains to distinguish it; and (b) while it is undoubtedly true that "the thinking
of the psychologist is itself psychological etc.," what is disturbing (at least
potentially) about the self-reference of phenomenology is that *unlike* psychology or logic (at least according to phenomenology), phenomenology is
concerned with its own ultimate grounding, as an ordinary factual or formal
science need not be. The question is, does this self-reference have any bearing
on phenomenology's claim to have achieved, or to be able to achieve, this
grounding?

Husserl seems to deny that it does:

> This back reference upon myself would be a matter of concern only if
> upon the phenomenological, psychological and logical knowledge of this
> momentary thinking of this momentary thinker depended the know-
> ledge of all other matters in the relevant spheres of study, which is, as
> anyone can see, an absurd presupposition. (ibid., p. 190)

But this implausible exaggeration of the original problem is not a satisfactory
response. That problem is: What effect does it have on the *status* of phenomenological scrutiny that this scrutiny (and all its accessory features) should
itself belong to the field (presumbly of immanent formations) that is itself to
be studied phenomenologically? If one were to pretend that such scrutiny, at
whatever level it occurs, had as its object the awarding of "epistemological
value," this might seem an innocent enough fact. But the self-reference in
question would precisely have the effect of confirming that the method of
phenomenology did indeed conform to the standards it itself embodied, and
thus give the appearance of an epistemological status that would block or at
least deter subjecting these standards to an independent assessment. And
such an assessment might be thought desirable.

Indeed, having begun by saying that the self-reference does not present
a difficulty, Husserl goes on to admit that there is a sort of difficulty, one that
phenomenology shares with other sciences, and this leads him into his second
justification for reflexivity. Husserl in effect offers an account of the development of formal sciences (and of the formal structure of natural sciences) in
which the self-reflection of phenomenology is no mere logical possibility, or
difficulty with which we must somehow come to terms, but the operation by
which a procedure ("directing our mental glance towards a certain experience"), which is first carried out more or less innocently, is raised to a more
"scientific" (that is refined, precise, self-transparent) level, as reflection
would do to any *practice*. In other words, the character of the second "glance"

is different from the first; it is specifically concerned to transform the "unsophisticated form" of the procedure into something more rigorously formulated. This is not merely a useful means of clarification, it is an essential phase in the development of any science: "Without preliminary and preparatory consideration both of matter and method, the sketch of a new science could never be outlined."[13] (ibid., p. 190) So Husserl is claiming that the reflection involved is no mere logical possibility but the means by which phenomenology develops itself as a science. Self-reflection is defended as a necessity, even a virtue, rather than as presenting any sort of difficulty. Husserl's third justification is that the intuitively compelling and yet modest method that phenomenology makes use of precludes the possibility of difficulties arising:

> If it figures as a science *within the limits of mere immediate intuition*, a pure *"descriptive"* science of essential Being, the general nature of its procedure is *given in advance as* something that needs no further explanation. (last emphasis added) (ibid.)

This modest self-limitation is surely an attempt to play down the "epistemological validation" side of phenomenology by referring to its descriptive method and its restriction to intuitions. If it does not explain or justify and only says what is so clearly the case as to be indisputable, no problem could possibly arise. It is interesting to note the phrase "given in advance as," for, whether or not Husserl likes it, this raises yet again all the questions about how far phenomenology is pure description. It is quite true that "immediate intuition" and "description" suggest there is no need for further explanation. But this way of thinking might be wholly illusory. Something can be "given" in a certain way, but not actually be that way. This is quite obvious when one thinks of the relation between spatial aspects and actual shape. But equally, Husserl's "given in advance as" suggests the kind of rationally deducible future (insofar as it concerns the development of phenomenology) that might lead one to deduce the completeness of arithmetic from the possibility of providing it with formal axioms.[14] And how many more opportunities there are for gaps to develop when one is dealing with such qualities as "need for explanation." Such needs can develop from unsuspected quarters even when not originally apparent. And one obvious way in which such a need can appear is through a reflective grasp of phenomenology's limits, in which internal self-satisfaction (for instance, about the adequacy of an intuitive ground) is subjected to external critical appraisal.

I would like now to suggest what conclusions we can draw about the limits of phenomenology from these probings of the points at which Husserl himself has appeared sensitive to the issue. I shall then suggest certain lines of criticism, and then show that Husserl's analysis of internal time-consciousness was intended as an answer to at least one of these lines.

The most conspicuously reiterated conclusion to be drawn is that intuitive transparency is a standard that is so fundamental to phenomenology that it not only characterizes the everyday products of phenomenological method, but also the results of phenomenological self-reflection, reflection on that very method. There is nothing necessarily corrupt about this. If another standard did arise at the level of reflection, it could undermine the operation of the original standard at the level of ordinary phenomenological description. No critique of phenomenology, then, could deny that phenomenology was self-critical. But what might still be scrutinized is the value accorded to this pervasive standard of "intuitive transparency."

I would like now to propose three different critical perspectives on phenomenology, from which, singly and collectively, the importance to Husserl of a theory of internal time-consciousness will become evident:

THE PROBLEM OF RECONCILING THE DESCRIPTIVE AND LEGITIMATING ASPECTS OF INTUITION.

The appeal of phenomenology (and perhaps any positive philosophical valuation of intuition) is that it (uneasily) combines (1) the idea of a purely descriptive method, that, after certain purifying procedures, is licensed to tell it how it is, and (2) the idea that it is only through an intuitive apprehension that the concepts and the principles that organize our scientific knowledge can be legitimated.[15] Now it is easy enough, perhaps too easy, to argue that one cannot derive legitimations from descriptions unless those descriptions covertly include valuations that would deprive them of their status as pure descriptions. As strict adherence to Hume's position can easily lead one to being unable to understand any valuations as anything other than irrational preferences, one might be excused for thinking that the matter was not so simple, and perhaps even that this uneasy combination of fact and value in "intuition" might indeed secretly hold the solution to this problem. This does not seem likely, at least in the form that Husserl presents it. What it seems to preclude is the undermining of the status of an original intuition in the light of the consequences than can be drawn from it. Moral intuitions, for example, are notoriously vulnerable to subversion by the deduction of unacceptable consequences from them, despite the utter clarity and transparency of the original insight. And yet it might be said that it is not the *value* itself that is undermined, but the scope of its application. It ceases to be thought of as *absolute*; it loses its metaphysical status.[16] The recognition that there might be circumstances in which one would be prepared to kill does not refute a no-killing principle, it merely limits it, by demonstrating its potential conflict with other values.

Nonetheless, the possibility of questioning the offerings of moral intuition raises the question of whether something parallel is equally possible in the formal and empirical sciences. And of course it is. Even the self-evidence of certain logical laws has been questioned. (Does the law of the Excluded Middle apply to statements about the future, or to statements embodying false presuppositions? Have I or have I not sold my submarine?) And the danger for Husserl is that intuition would be reduced to being a source of *candidates* for logical primitiveness that would be judged by other standards (for example, hypothesis-generating capacity, practical utility, compatibility with the conceptual frameworks of other theories, and so forth).

NOTE: We do not mean to suggest that Husserl was not aware of many of these questions. He was well aware, for example, that intuition was not a foolproof basis for knowledge, or at least he became aware of this. In his "Phenomenology of the Reason" he brings these issues to the fore.[17] One of the most important moves he makes, having pinpointed self-evidence as the feature of intuitive apprehension that gives it its status, is to distinguish between *adequate* and *inadequate* self-evidence (section 138) in which, as will be clear, "self-evidence" is used in an "extended sense." *Adequate* self-evidence is that found in, for example, simple arithmetical propositions ($2 + 2 = 4$). Its chief characteristics are that subsequent experience cannot strengthen or weaken it, and that it lacks "the graded differences of a weight." Yet, "a thing in the real world . . . can within the finite limits of appearance appear only 'inadequately,'" for all our intuitive grasp of worldly things is in principle subject to confirmation (harmonious "filling-out") or cancellation, fragmentation into a multitude of hypotheses, and so forth. A façade, for example, may lead us to posit a house that closer inspection will reveal not to be there.

Three points should be carried forward from this note for consideration:
1. Husserl supposes that the various modes of confirmation, cancellation, fragmentation, and so forth of "inadequate" self-evidence can be given a rational organization, that they conform to various standard types.
2. He takes for granted (that is, presumably as "adequately self-evident") the distinction between *adequate* and *inadequate* and its corresponding application to domains of objects distinguished as *immanent* and *transcendent* (as I understand it). This would seem to leave somewhat undecided the status of "complex rational objects," if I might so label complicated proofs, difficult mathematical equations, theories, and so on. Descartes faced just this problem, and his solution—speed reviewing of the simple parts and their relations so as to bring them all within a single momentary gaze[18]—shows exactly how the problem of time arises for Husserl.
3. For it is clear that the difference between adequate and inadequate self-evidence is nothing but a *reflection* of the distinction between what can or cannot be exhaustively grasped *in a single moment*, in the present. What

is inadequately self-evident is in effect only a convincing pointer to a fullness that one *at that moment merely posits*. And yet the idea of anything at all taking place in an instant (an infinitely small temporal interval) is quite implausible. What is called for, and what, in effect, Husserl supplies, is a phenomenology of the present.

Having concluded this note to our first critical perspective on Husserl, I turn now to the second.

THE PROBLEM OF PERFORMING (AND INDEED OF MAKING SENSE OF) THAT DETACHMENT FROM THE WORLD ON WHICH PHENOMENOLOGICAL SELF-SCRUTINY DEPENDS

The systematic use of intuition as the basis for organized scientific knowledge requires, for phenomenology, the successful completion of an act of cognitive purification—a reduction of our conscious life to the life of consciousness, and then, as we have indicated, a focusing on the immanent structures of that consciousness.[19] Husserl was quite aware of the possibility of skepticism about the possibility of such detachment and indeed attempts to deal with it (in section 64, "The Self-Suspension of the Phenomenologist"). The fact that existentialists make such skepticism the backbone of their objection to Husserlian phenomenology suggests that this idea of detachment was not wholly convincing. In Heidegger's existential ontology, for example, knowing is just a mode of being-in-the-world, which radically undercuts the status Husserl gives to the reductions. Merleau-Ponty thought of the reductions much more as a way of focusing on the structures of existence, an operation that can never be completed, and not as capable of bringing about any sort of ontological self-transformation. Obviously one question that will concern us is how far Husserl's account of time-consciousness is dependent on his belief in the possibility of existential detachment.

THE PROBLEM OF GIVING PUBLIC EXPRESSION (IN LANGUAGE) TO PRIVATE INTUITION WITHOUT EPISTEMOLOGICAL CORRUPTION.

The last critical perspective one ought at least to adumbrate is one for which we have already laid the groundwork in discussing Husserl's charting his course between the Scylla of psychologism and the Charybdis of offering merely a disguised philosophy of language. In his *Rules*, Descartes claims at one point that while intuition is unchallengeable, as soon as one proceeds to *judgments* about such intuitions, fallibility immediately appears.[20] The problem is how one retains the virtues of prepredicative insight at the level of linguistically articulated expression. For the relation of "adequation" or "fulfillment" or "correspondence" between what it is one has grasped and the conditions of presentation that make its public appearance possible is extended to the point of fracture. If phenomenology is not to be forever plagued by the problem of the linguistic form of its insights, it must hold one of two "theories" of language: (1) that at some level, whether it be the level of the primitive *elements* of language, such as morphemes or words, or that of such articulated complexes as sentences, language must be capable of adequately capturing the intuitive contents of phenomenologically purified experience; or (2), that even if language in its normal, imprecise form cannot do this, that it can be sharpened up, used in careful accordance with definitions, rules, and the like, so as to achieve this result.

In its broadest form, the difficulty with either view is that even if there were some sense to be given to "language capturing its objects," it is hard to see how the ideal of such a precise relationship could survive unscathed the material conditions of language use—the permanent sway of a *pragmatic* dimension (context, local conventions of use, range of particular intentions in play). If speaking/writing/understanding language is an interminable process (both, in principle, for each item, and because of the infinite number of such items) rather than the simple emitting and receiving of a message, then the prospect of phenomenology as an apodeictic[21] science would be dim. This line of argument is not weakened by the actual existence of empirical sciences, for they do not claim the apodeictic qualities of certainty, necessity, and clarity that phenomenology makes its own. The proper comparison would be with formal sciences such as geometry or mathematics. Surely they exist, and are successful? And yet Husserl claims that to grasp what geometry is about we have not merely to be able to do geometrical proofs, but to be able to reactivate the original insights that fundamentally sustain geometrical thinking. But if at the level of mere symbolic manipulation, the internal coherence of (a particular) geometry may be quite unchallengeable, the introduction of another standard—that of the adequacy of one's inner understanding—puts in question all that purely terminological and operational precision. So, even if the ideal of an apodeictic science were a plausible one for phenomenology to aim at, even if comparison with geometry was appropriate, one would still be faced with the difficulty that the apodeicticity in question remains a property of basic insights and does not extend to the relation between those insights and their linguistic form.

The question that persists, and in its more complex shape can usefully be divided into two parts, is this: First, even if one allows "within its own limits" the possibility of intuitions grounded in the transparent immediacy of present experience, does not the requirement to embody these intuitions in a public language—subject to all the vagaries of historical, social, scientific, even political pressures—explode the privileged interiority of intuition, and deprive the phenomenologist's public version of the value he aimed at?[22]

This could be called the problem of the clothing of intuition.[23] Second, more radically, one may come to question the alleged independence of this intuitional layer from the structure of signification found in language. This move, which can be attributed to Derrida, has (for us) the important corollary that the whole ideal of a realm of (evidential) presence, and of an internal time-consciousness that would sustain it, is undermined. Derrida's denial of an alternative nonmetaphysical temporality is clearly dependent on his view that the epistemological function served by Husserl's account of time-consciousness both determines the nature of that account in advance and teaches a wider lesson about the philosophical status of theories of time. (Derrida's critique of Husserl is discussed at length in chapter 3.) The immediate task is to explain how Husserl's theory of internal time-consciousness reflects, depends on, and is limited by the theoretical tasks for which it was invented.

2 *Husserl's Analysis of Time-Consciousness*

> The conclusions we advance in regard to these matters, because of the
> difficulty and obscurity of Husserl's analyses, will involve a considerable
> degree of speculation and should be taken as tentative.
>
> John Brough

In this chapter I try to show how Husserl's quest for an "authentic time"
leads him to a point at which the very idea of time is itself put in question.
This startling conclusion sheds an entirely unexpected (and hitherto unnoticed) light on Derrida's insistence that time is an essentially metaphysical
concept. I begin, however, by trying to answer certain fairly analytical
questions: How far is Husserl's phenomenology of time-consciousness a
complete and successful philosophy of time? If there is *more* to be said, does
this mean an expanded phenomenology, or a going beyond phenomenology?
What scope does Husserl claim for it? How far does his *Phenomenology of
Internal Time-Consciousness* (hereinafter referred to as *PITC*)[1] rest on
questionable/unexamined/optional assumptions? What schemas, conceptual
frameworks does Husserl make use of?

Answering these (and allied) questions should determine whether Husserl's phenomenology is indeed "the highest point of the metaphysical
tradition" (Derrida), and in particular the nature of the relationship between
Husserl and Derrida, which is so often misunderstood. I argue in the next
chapter that Derrida is perhaps even closer to Husserl than he knew. The
position of the *PITC* is critical in this regard, as Husserl (albeit in his own
terms) fully understood. If one of the key characteristics of metaphysics is the
privileging of presence, then a work that actually deals thematically with
both the temporal and the evidential present, as *PITC* does, will not be a
metaphysical work by any failure to reflect, or by its temporary commitment
to certain themes or procedures. Its metaphysical status will have been
achieved carefully, deliberately and wholeheartedly. This is possible because
of what Heidegger calls the "unthought" in any thinking, which does not
refer to shadowy parts of an otherwise illuminated field, to questions one just

happens not to think about, but to what is structurally hidden by one's very method of inquiry. (How, for example, could a questioning of reflection as a philosophical method that also made use of reflection as its mode of scrutiny do other than repeat and hence endorse, at least at some level, that very reflection?)

QUESTIONS OF SCOPE

It is a natural, or at least a common, assumption, whether or not we accept Husserl's account of time-consciousness, that it is a candidate for acceptance as a philosophy of time in general—indeed the correct one. That is to say, it is often supposed not merely that it offers an account of experiential time, but that such an account either says all there is to be said about time, or that anything else that could be said would rest on or be derivable from such an account, and that Husserl's own account of experiential time is the correct one (over and against those of Bergson, Brentano, James, and the like). This assumption will be progressively questioned in the course of this chapter, but it will be instructive first to consider Husserl's own position on this matter.

In keeping with phenomenology's general procedure, Husserl begins *PITC* with a double move. He both distinguishes the phenomenological treatment of internal time-consciousness from other treatments of time (sections 1 and 2) and claims a privileged status for the phenomenological treatment. So he does not think that a phenomenology of time-consciousness says all there is to say about time, but given that what he distinguishes it from are psychological and other scientific studies, the clear suggestion is that there is no other (satisfactory) philosophical treatment of time. I now discuss this double move in detail.

A phenomenological analysis of time-consciousness, insofar as it is phenomenological, involves (section 1) "a complete exclusion of every assumption, stipulation, or conviction concerning objective time." The (perfectly natural) assumption that there is a single all-embracing objective time would be set aside. More important, certain ways of studying time, based on naturalistic assumptions such as the psychophysical correlation of objective and subjective measurement of time intervals, are also excluded: "Psychological apperception, which views lived experiences as psychical states of empirical persons . . . is something wholly other than the phenomenological." It is consistent with Husserl's view of the philosophical radicality of the phenomenological approach that he does not specifically mention other *philosophical* treatments of time, with the exception of a passing reference to St. Augustine, and then, at greater length, Brentano. The implication, I believe, is that it is only by the phenomenological approach that philosophy (let alone, for example, a particular empirical science of psychology) can avoid making naturalistic and hence psychologistic assumptions about our relationship to it. If the case of Brentano can be taken as exemplary, it becomes clear that finally succumbing to some form of psychologism is a risk that even the subtlest of philosophers can fall prey to. But again, if we take the study of psychophysical temporal correlations to be exemplary, then it is clear that Husserl believes there can be other ways of studying time, but simply that they are not properly philosophical in the sense of dealing with the *essence* of time. When they do not pretend to be philosophical, they can simply be excluded; when they do, they have to be criticized.

And yet at the same time as Husserl is distinguishing the phenomenological from the naturalistic/psychologistic approach, he is also claiming a privilege for the former. What the latter approach takes for granted (objectivity, objective time), phenomenology offers an account of. It is this move that leads to " . . . the most extraordinary difficulties, contradictions and entanglements" that motivates the strictness of the phenomenological method. In giving an account of what it distinguishes itself from, phenomenology establishes a priority. "Phenomenologically speaking, objectivity is . . . constituted . . . through characters of apprehension and the regularities which pertain to the essence of these characters. It is precisely the business of the philosophy of cognition to grasp this fully and to make it completely intelligible." So a phenomenology of time both distinguishes itself from any account that takes the objectivity of time (or indeed of anything) for granted and goes on to give an account of how that is constituted. But while these two moves can be separated analytically, they are one and the same move in Husserl's writing. For the distinctivenes of the phenomenological approach in fact consists in its dealing with what precedes any objective formations, and such formations could have no other basis than these immanent acts that phenomenology uncovers.

However, if one asks what the relationship (of constitution) comes to, it must be concluded that it can itself be understood only phenomenologically. Immanent acts do not literally constitute objects in the way that cakes or candles have ingredients. The language of "act" and "constitution" cannot be understood in any ordinary sense. There is no question of events that occur "in time," nor indeed any temporal features at all. A closer but not precise parallel would be with a logical or conceptual relationship. So if a phenomenology of time, whether or not the only possible philosophy of time, deals with the most primitive forms of time that any other philosophy of time would have to rest on, this privilege is one that it itself explains and justifies. This would seem to suggest that the final answer to the scope of a phenomenology of time remains undetermined, for all that is available is a phenomenological answer, an answer, that is, that rests on the phenomenological concept of constitution. How far is this true? Might not phenomenology justifiably claim to be simply bearing the mantle of philosophy here? Are doubts being raised about the general approach that tries to deal with time in its fundamental features, rather than with each and every concrete chronological phenomenon? Does not the concept of constitution simply give a phenomenological name to the relationship between these fundamental features and derived forms? Is it not simply a consequence of its being a philosophical approach that phenomenology assesses its own scope, and no sort of vicious circularity? Again, this is precisely *phenomenology's* position— but is it compelling? Thus:

> From the point of view of theory of knowledge, the question of the possibility of experience (which at the same time is the question of the essence of experience) necessitates a return to the phenomenological data of which all that is experienced consists, phenomenologically. (*PITC* sec. 2, p. 27)

and again

> The *question of the origin* is oriented towards the primitive forms of the consciousness of time in which the primitive differences of the temporal are constituted intuitively and authentically as the originary sources of all certainties relative to time. (Ibid., p. 28)

Husserl is claiming that phenomenology is required by the theory of knowledge, and his conviction that phenomenology is not just a more or less desirable philosophical option can be seen in his phrase "phenomenological data," which, echoing his slogan "to the things themselves," suggests that phenomenology is not simply a good method, but is in some way an approach *required* by what it has to deal with. This same suggestion appears even more strongly when Husserl refers to "the primitive forms of the consciousness of time . . ." What is at stake here is not, it seems, a hypothesis, but a fact (of sorts).

Now, it would be an immense task to try to adjudicate phenomenology's claim to be "first philosophy," and one that would divert me unjustifiably from my own questions, but it is clear that it was not Husserl who invented the idea of a theory of knowledge and the quest for primitive elements and forms of experiences; the belief in the possibility of focusing in on "lived experience" is common ground particularly for that philosophical tradition we call empiricism. Phenomenology retains sufficient of the goals and orientations of philosophy to be radically *philosophical*. And certainly its abandonment of a largely passive and receptive model of experience is a great gain. So it would not be unreasonable for us to take seriously phenomenology's claim about the scope of the phenomenology of time. But one phrase, "as the originary sources of all certainties relative to time," gives us pause for thought. At the beginning of section 7, Husserl asks two questions: "How the apprehension of transcendent temporal objects which extend over a duration is to be understood?" and "How, in addition to 'temporal objects,' both immanent and transcendent, is time itself, the duration and succession of Objects constituted?" (p. 42)

These two questions fall on either side of another question about the relationship between the analysis of time and the theory of knowledge, on which this phrase takes sides. The role of time or time-consciousness in making possible our grasp of temporally extended objects is one in which it is contributing to the theory of knowledge. And without an understanding of the role of retention, protention, and memory in experience, no account of knowledge could be possible. On the other hand, Husserl is also concerned with time as an object of knowledge in its own right. He asks how time *itself* is constituted, and, as in the phrase being discussed, he is looking to account for our "certainties relative to time," and a return to "the *primitive* forms of the consciousness of time" is the direction he takes.

Does this link between epistemological issues and the analysis of time indicate anything about the scope of Husserl's project here? Both a weak and a strong interpretation are possible. On the weak interpretation, Husserl is quite right to point out the importance of a theory of time to our knowledge of objects. Without temporal extension there would be no objects, and without an understanding of the possible modes of temporal extension our knowledge would be fundamentally deficient. (Consider the differences between, say, objects extended in time, objects temporally structured, and objects that structure their own time.) And the concern to get to the bottom of our common-sense understanding of time itself is nothing more than an attempt to bring relief to those who, like St. Augustine, find themselves at sea when they try to get a reflective grip on time. Husserl's answer to obscurity, difficulty, and uncertainty is always, and understandably, to return to those most fundamental layers of experience in which primitive clarity has not yet been obscured.

The strong (and potentially critical) interpretation would be that this two-way involvement with epistemological considerations does constitute a real limitation to the scope of Husserl's inquiry. The simplest statement of how this limitation can arise would go something like this: There is no necessary connection between *truth* and *certainty of apprehension*. Suppose for the sake of argument that there are indeed two kinds of knowledge that yield certainty—that which concerns mathematical and logical truths, and that which concerns what is *immediately apprehensible*. Many would deny the latter, and some the former. But even if they were to be accepted, it would require an enormous act of faith to suppose that all truths can be grasped by reduction to one or other of these kinds of knowledge. Logical atomism, for example, elegantly orchestrated these two modes, treating complex propositions as logical constructions from simple ones that could be grasped with certainty. But it requires either faith or some sort of stipulative act to suppose that all knowledge can be dealt with in this way. And neither *faith* nor *stipulation* are consistent with the standards of certainty, clarity, and groundedness that such programs affirm.

We need not take on board all his accompanying intellectual baggage, but Hegel was surely right to wonder whether "the fear of error (might not be) . . . the initial error."[2] And the way that Hegel and other philosophers such as Dilthey developed this, by an appeal to "understanding" rather than to "knowledge," still has some relevance today. For what traditional theory of knowledge takes for granted is that knowledge is a relationship between a clearly distinguished subject and object, and that every putative item of knowledge will appear, as such, as an object of some cognitive act (such as "apprehension"). Only under this kind of spotlight can there be anything like certainty, or that clarity and distinctness Descartes demanded as a standard. Now whatever one thinks of the possibility in principle of isolating objects of knowledge from the subject grasping them as such, there would seem to be clear cases in which such a separation could be expected to prove difficult, if not impossible. For example, widely different theorists such as Foucault, Gadamer, and Kuhn take this impossibility to lie behind the structural impossibility of historical reflection on one's own present period, on the grounds that historical detachment cannot be purchased by greater reflective effort. One has to wait for the present to become past for it to be capable of becoming an object. This is a negative example. Positively, the hermeneutic tradition does suppose that understanding may be possible where knowledge in the sense already discussed (a relationship between an independently identifiable subject and object) falters. Understanding (*verstehen*) proceeds by explication of what is being "lived through." It is appropriate for self-understanding, for existential reflection, for historical understanding (at one level), and for exploring the wealth of our taken-for-granted world.[3] But it does not set itself certainty as a standard. That is not to say that it treats accuracy and correctness with disdain, but rather that it recognizes the inapplicability of the subject/object schema to which certainty as a standard belongs.

Now the question as to the scope of Husserl's *PITC* can be posed anew if it is linked, as it seems to be, to "knowledge." If, in other words, there are aspects or features, of "time" that are neither immediately available to phenomenological scrutiny, nor derivable from those (the primitive features) that are available, then Husserl's *PITC* will for all its virtues not be able to claim to be a philosophy of time, but *only* of time-consciousness. I hope to be able to show that what Heidegger and Derrida each offer are different ways of limiting Husserl's phenomenological treatment, by developing accounts of what it cannot adequately deal with. The enormously problematic nature of each of these efforts can be seen in the fact that the later Heidegger abandons talk of "time" proper, and "substitutes" a more primitive "time-space," while Derrida explicitly claims the metaphysical nature of any concept of time. To treat these "results" as the logical consequences of abandoning the link between time and epistemology would cast a new light on Husserl's enterprise, although, as with all light, a new shadow will be cast too. For if Husserl is seen as having taken *to the limit* the working out of the metaphysical concept of time, it might be judged either a finite project successfully completed, or, with a slight shift in one's angle of reading, as an enterprise doomed from the start by its metaphysical blinkers. In fact neither of these extreme judgments will quite fit.

One of the interesting features of Husserl's analysis, as will become clear, is that he cannot *in fact* confine time to the status of an *object* of knowledge precisely because of the second strand of his original double claim about its involvement in epistemology—that time enters into the constitution of objects of knowledge. It does so both at the level of the "object" and at the level of the "subject," and once this latter is established one can begin more easily to ask skeptical questions about the possibility of an epistemological delimitation of the subject, questions that lead in the direction of Heidegger's existential problematic in *Being and Time*.[4] And if in Heidegger's work there proved to be something called existential time that was not amenable to the kind of scrutiny that Husserl's phenomenology provides, then it would constitute a positive limit to the scope of a phenomenology of internal time-consciousness. If there were a distinctive temporality of the unconscious, that too might escape the limits of a phenomenology of time.[5]

HUSSERL'S ANALYSIS OF TIME-CONSCIOUSNESS

To demonstrate the pervasive and yet variable penetration of Husserl's account of time-consciousness[6] by epistemological concerns, it is necessary to trace out in detail the path of Husserl's thinking. This discussion will be based on the lectures Husserl gave at Göttingen, especially in 1904–5 and from time to time up until 1910, which were edited by Heidegger and published in 1928.[7]

Husserl, then, begins with an exclusion and an apparent self-limitation by which his phenomenological approach gains its specificity and privilege. But the exclusion of "objective time" can be understood in two different ways, and as so much is often at stake in the early moves in a philosophical text, I shall take the opportunity of elucidating this exclusion by trying to adjudicate between the two ways. Husserl could be thought to be affirming the existence of "world time, real time, [and] the time of nature," simply adding that he instead will deal with "temporal experiences," because "we deal with reality only insofar as it is meant, presented, looked at, or conceptually thought of." Alternatively, it may be thought that he leaves that question open, putting the external reality of time in question from the very beginning. Sokolowski supposes that Husserl is committed in these lectures to the first position, which affirms but ignores "world time" and so forth, so that by the time of his *Ideas* (1913), there is a definite change. But this is not really convincing. It is not at all difficult to read Husserl as writing in scare-quotes when he uses expressions like world time and real time, and even in the sentence Sokolowski quotes as most convincing[8] in which Husserl finds it "eine interessante Untersuchung" to relate subjective time-consciousness to real objective time, he can quite easily be read as referring to the way psychophysicists conceived of their project. The crucial move by which the passage is made from a dualistic position to one that is neutral with respect to ontological commitment is to be found in the claim that terms like "real" and "objective," and indeed "time" (in phrases like "real time" and "objective time") derive their sense from our constitutive acts. (Husserl's strong and notorious version of this, which is said to commit him to "idealism," can be found in *Cartesian Meditations*, Section 62.) And this, it is supposed, compromises the idea that such phrases could *refer* to anything that as such had an independent existence because without that constitutive activity there would be no "as such." Versions of this move are to be found in various places in Husserl's two introductory sections. For instance:

> . . . we make the attempt to account for time-consciousness, to put Objective time and subjective time-consciousness into the right relation and thus gain an understanding of how temporal Objectivity . . . can be constituted in subjective time-consciousness . . . (sec. 1, pp. 21–22)
>
> . . . Objectivity belongs to "experience," that is to the unity of experience, to the lawfully experienced context of nature. Phenomenologically speaking, Objectivity is . . . constituted through characters of apprehension and the regularities which pertain to the essence of these characters. (ibid., p. 27)
>
> . . . that these lived experiences themselves are temporally determined in an objective sense . . . does not concern us . . . On the other hand, it does interest us that "Objective-temporal" data are *intended* in these lived experiences. (ibid., p. 29)

The point of bringing into prominence these remarks is to demonstrate that Husserl from the very beginning talks of Objectivity and Objective time as being *constituted* by our lived experiences. And this seems to count against Sokolowski's attributing dualism to Husserl at this stage. On this reading, remarks that would lead one in that direction are better taken as remarks made in the course of a transition from everyday language to a phenomeno-logically displaced language, and should not be taken as suggesting a theoretical commitment to dualism on Husserl's part. There is then much less need to reconcile Husserl's views here with those in *Ideas*. The *exclusion* is of "what is called" reality, transcendent objects, world time, and so forth. Husserl is already ontologically noncommittal while programmatically anti-cipating a relationship of *constitution*. The relationship of constitution cannot hold between independently existing things, and so the more he writes about constitution, the less plausible is Sokolowski's dualistic reading at this juncture.

Husserl's initial move, then, is one that distinguishes a subject matter ("lived experiences of time") that excludes, as preventing a solution to the problem, any consideration of temporal transcendencies, that claims to be able to answer the question of the "essence" of time by a phenomenological analysis of such experiences, and that already outlines the kind of generative power (in the theory of *acts* and *constitution*) that such an analysis would have. As a forceful example of the latter, for instance, he suggests the possibility of explicating the *a priori features* of temporal order (infinite, two-dimensional, transitive, and so on) by reference to time-consciousness.[9] But what precisely is meant by time-consciousness? By "lived-experiences"? The language of "acts" and "characters of apprehension" and "constitution" correctly sug-gests that Husserl's positive contribution will not merely be to have cham-pioned subjective time over against objective time, but to have established a rather subtle relationship between them, and to have done so with an account of "subjectivity" distinct from that of St. Augustine, James, Bergson, and Brentano, each of whom might superficially be thought to have cham-pioned subjective time in the same way. The particular position Husserl held can best be grasped by following his analysis and critique of Brentano's theory of time (section 3). With this analysis we shift from his introductory remarks to an account of "lived experience," albeit one needing correction. While the aim of this move has been explained, it is also important to make explicit Husserl's justification for it. It runs like this:

> . . . time and duration . . . are absolute data which it would be senseless to call into question. To be sure, we also assume an existing time: this, however, is not the time of the world of experience but the *immanent time* of the flow of consciousness. (sec. 1, p. 23)

and he goes on:

The evidence that consciousness of a tonal process, a melody, exhibits a succession even as I hear it is such as to make every doubt or denial appear senseless. (section 1, p. 23)

For a (broadly speaking) foundationalist epistemology, these sentences constitute a justification for focusing on immanent or inner time rather than world time. We are directly aware of inner time, and its principal feature, *succession*, is indisputable. Even if we have the most unlikely dreams, the wildest hallucinations, the severest of illusions, the fact of temporal succession within and among these false images will not itself be in doubt. (However teleologically askew, succession is even found in those dreams that, as Freud reminds us, involve an inversion of the proper order of events.) But these sentences are much more than a justification for dealing with a specific subject matter (inner time). They offer the precise form of the underlying question that guides Husserl's whole analysis: What must temporal succession be like in order for our evidence for its occurrence to be beyond all possible doubt? Husserl's difficulty can be posed in this way: Is not "being beyond all possible doubt" a feature reserved for what is *immediately* grasped? And does not any grasp of succession involve grasping a relationship between what is *immediately available* (a current sensation) and what is now past (a previous sensation)? Yet how can a relationship between something immediately grasped and something no longer immediately grasped be itself immediately grasped? Is not "succession" in fact a *judgment* we make about our experience that, however unlikely it might seem, could be mistaken, just like any other judgment? Husserl's view, quite plausibly, is, I think, that succession is not a relation built up out of parts and a relation between them, but something primitive, and moreover that it does not breach the conditions that "certainty" demands.

HUSSERL'S TREATMENT OF BRENTANO

Husserl's discussion of Brentano's account of time must now be pursued, with this question in mind. Brentano clearly shared with Husserl the idea that the question as to the "origin" of time (that is, its fundamental "essence") was to be answered by isolating its *primitive* element. Brentano's name for this was "the primordial sensations." These it was that would account for the ideas of duration and succession. The part of Brentano's theory (and here I am paraphrasing Husserl; the adequacy of this presentation will be discused later) that deals with these primordial associations can be summarized in the claimed "genesis of the immediate presentations of memory which, according to a law that admits no exceptions, are joined to particular presentations of perception without mediation." Consider what it is like to listen to a tune. What happens to each note as I hear it? If it simply disappears, I would never hear the tune but merely a succession of isolated notes, and I would never even know it was a succession. If it stays around unchanged with nothing to distinguish it temporally from notes that followed, I would just have a simultaneous plurality of sound. The solution must lie in a third position. Certainly what seems to happen is that notes after first being perceived come to seem "temporally shoved back," without entirely disappearing without trace. How precisely does this happen according to Brentano, asks Husserl? It must be the case

> . . . that that peculiar modification occurs, that every aural sensation, after the stimulus which begets it has disappeared, awakes from within itself a similar presentation provided with a temporal determination, and that this determination is continually varied, [only then] can we have the presentation of a melody in which the individual notes have their definite place and their definite measure of time. (p. 30)

> It is a universal law, therefore, that to each presentation is naturally joined a continuous series of presentations each of which reproduces the content of the preceding, but in such a way that the moment of the past is always attached to the new. (p. 30–31)

Crucial, then, to the third alternative is the idea that instead of remaining unchanged, or just disappearing, sensations do not *themselves* linger on, or slowly fade, but generate "a phantasie-idea like, or nearly like, itself, with regard to content and enriched by a temporal character."

This idea will also generate a modification, and so on. The point of this, it seems, is that the arrival of a new sensation and the phantasy modification of a previous sensation will occur at the same time, giving rise to a "primordial association." One consequence of this, of course, is that one does not actually perceive succession on Brentano's view. It simply offers an account of how one seems to see it.

But if this supplies the basis for understanding temporal succession, more is required to understand the idea of time as an infinite series. For this we need to consider the role of phantasy in creatively generating the idea of the future "from the appearance of momentary recollections." Husserl's account of Brentano's position is quite inadequate at this point. It is not good enough to compare the generation of the idea of the future from the past with transposing a melody into a different key. I can only suppose that a fuller account of how "phantasy forms ideas of the future" "on the basis of momentary recollections" would refer to an imaginary self-displacement in which one comes to see that the possibility of these being recollections is based on those original past experiences being remembered at some future date, relative to their present, namely, "now." If the past has one of its futures realized in the present, then phantasy can suppose that this present, when past, will too. This is the most plausible expansion of Husserl's account of Brentano that I can construct; it is odd that contrary to his usual habit, Husserl offers such a condensed account.

Finally, Husserl explains Brentano's insight that the temporal modes "past" and "future" modify rather than define the ideas they are attached to, as do predicates like "possible," or "imaginary," while "current" or "present" or "occurring now" are defining characteristics. However difficult it may be to swallow, it follows "that non-real temporal determinations (past, future) can belong in a continuous series with a unique, actual, real determinateness to which they are joined by infinitesimal differences." So the succession of sensations or ideas should be seen as a perpetual shifting from nonreal to real temporal determinations.

Husserl's critique of Brentano's theory of time is an attempt to distinguish in that theory those elements that are genuinely phenomenological and can be built on, and those in which naturalistic, psychologistic presuppositions linger on. These presuppositions are surely at work, for example, in Brentano's talk of "stimuli" and "objects" "producing" sensations in us and the like. One can, however, focus instead on the properly epistemological (and hence phenomenological) aspects of this theory. The crucial first point of scrutiny is Brentano's analysis of (the appearance of) succession. Is he correct to claim that a "now" and a "past" can be unified in a succession only by the "past" appearing in phantasy? Husserl's doubts run like this: it is possible to distinguish two levels of Brentano's analysis—the original intuition of time (which I perceive as temporal succession) and the secondary grasping of the idea of infinite time. These two are contrasted, Husserl claims, by the former being "authentic"—that is, involving direct intuition, while the latter is not. Now there clearly *is* a distinction here, but for Brentano, both are the work of phantasy, and

> if the original intuition of time is indeed a creation of phantasy, what then distinguishes this phantasy of the temporal from that in which we are aware of a past temporal thing, a thing therefore that does not belong in the sphere of primordial association and that is not closed up together in one consciousness with perception of the momentary, but was once with a past perception? (sec. 6, pp. 36–37)

Husserl's point is that the distinctiveness of Brentano's "primordial association" is *threatened* by the role of phantasy in its explication, for it is just the same mode that is involved in the extension of the original intuition to that of infinite time. Brentano fails to grasp that as a merely psychological connection phantasy cannot serve to constitute an original intuition.

Husserl proceeds to demonstrate the further inadequacies of Brentano's position from the phenomenological point of view. The main difficulty arises from Brentano's ignoring of the "act"-aspect of consciousness. Basically, Husserl claims that Brentano treats time, or temporality, as a special feature of the *contents* of consciousness, while he himself wants to credit specific *acts* of consciousness with responsibility. But Husserl's claims here seem to waver into inconsistency. He begins by claiming that Brentano does not make the act/content distinction:

> Brentano did not distinguish between act and content, or between act, content of apprehension, and the object apprehended. We ourselves must be clear, however, as to where to place the temporal element. (ibid., p. 37)

But later he says:

> . . . even if [Brentano] . . . was the first to recognize the radical separation between primary content and character of acts, still his theory of time shows that he did not take into consideration the act-characters which are decisive for this theory. (ibid., p. 40)

So, on the one hand Brentano fails to make the act/content distinction, and on the other he makes it but fails to employ it at the right point. Perhaps closer attention to his criticisms will resolve this apparent shift of ground.

Husserl's question is how temporality can arise from a mere association of ideas, taking into account only the contents of those ideas. How can a temporal difference appear through a concatenation of ideas differing only in "richness and intensity of content"? Husserl shows that for all the ingenuity of Brentano's account, the attempt to make a particular *content* of consciousness both be present (a product of phantasy) and also bear the quality "past," or signify "past," must fail in the absence of any reference to conscious acts, that is, without reference to the intentional dimension of time-consciousness. Husserl's critical position can be rendered consistent: he is saying that Brentano had elsewhere made the act/content distinction, but had failed to apply it to his theory of time.

SOME DOUBTS ABOUT HUSSERL'S TREATMENT OF BRENTANO

Husserl also makes a remark which, if taken quite strictly, gives the reader pause for thought.

> It is most extraordinary that in his theory of the intuition of time, Brentano did not take into consideration the difference between the

perception of time and the phantasy of time, for the difference, here obtrusive, is one that he could not possibly have overlooked. (*PITC* p. 36)

Husserl is making an even more pointed version of the kind of remark discussed earlier concerning Brentano's failure to apply to time the act/ content distinction that he, Brentano, was the first to draw in this area. Now Husserl is claiming that he *must* have seen the difference between perception and phantasy, but made nothing of it. Rather than supposing Brentano to be lacking in intellectual acumen, we must suppose that Husserl is accentuating the limitations of Brentano's putatively psychological framework, which allowed him to grasp at one level, the psychological level, what he could not make use of at another, the epistemological. But there is another way of responding to Husserl—that indeed Brentano could not, and hence *did not* overlook the distinction, and that consequently Husserl is misrepresenting him.

The interests of scholarly justice require here a brief reference to a rather critical, indeed at times polemical, assessment of the way Husserl deals with Brentano.[10] Oskar Kraus, who published his paper in 1930, two years after Husserl's lectures on the phenomenology of internal time-consciousness had belatedly been made public, vigorously defended Brentano against Husserl's criticisms. The main points he makes are these:

1. That either Husserl himself, or Heidegger as his editor, ought to have pointed out that Brentano had abandoned the theory of time Husserl critically attributes to him by 1895, that is, not only before the 1928 publication, but a full ten years even before Husserl's original lectures. Husserl's critique of Brentano was based on notes he had taken at Brentano's lectures (I take it during Husserl's visit to Vienna in 1884–86). Certainly by 1911, with the publication of Brentano's *Von der Klassification der psychischen Phänomene* (Leipzig: Duncker and Humbolt), a copy of which Brentano sent to Husserl, Husserl knew about Brentano's new position and should have pointed this out to his editor. (Or was this perhaps one of those points on which it is said he was less than pleased with Heidegger's editing?)

2. Husserl's criticisms of Brentano's attempt to found time-consciousness on temporal variations of the object were predated by Brentano's own autocritique.

3. Husserl's new position, which makes use of "modes of consciousness" rather than differences in objects, had already been anticipated by Brentano by March 1895. Brentano wrote of his new belief in a letter to Anton Marty, which can be found in Kraus's *Toward a Phenomenognosy of Time Consciousness*:

that every sensation is bound up with an apprehension of that which is sensed. (p. 228)

... before ... we had a limited continual series of objects in the proteraestiesis, we now have a limited continual series of modes of apprehension of the same object. (p. 228)

And he went on

[Perhaps] ... now some things are conceivable that were inconceivable before ... (pp. 228–29)

There is much else of great interest in Kraus's paper that it would be out of place to pursue here. With the exception of one or two small points, Kraus is largely putting the record straight, criticizing Husserl's misrepresentation of his own originality. But on the whole the value and interest of his published lectures—whoever's views they originally were—is not contested. The real value of Kraus's piece today, half a century later, is that it shows us that the frame of reference within which internal time-consciousness appears as a problem was quite widely shared. Kraus's concern for accuracy here is I think justified, especially in the light of the importance of Brentano's reflections on time for the phenomenological movement, and indeed for Brentano himself.

HUSSERL'S TWO MAIN QUESTIONS

Husserl's analysis of time-consciousness involves him, then, in answering, at least initially, two distinct questions, the first concerning temporal objects, and the second, time itself. He claims a priority of the first question over the second, "a phenomenological analysis of time cannot explain the constitution of time without reference to the constitution of the temporal object." (p. 43) (By "temporal object" he means unified temporally extended object.) This priority claim is a little complicated by the fact that he also claims that "all these questions are closely interrelated so that one cannot be answered without the others." (p. 42) But certainly the approach he makes to his central schema of time in section 11 is in contrast with two further deficient theories of our understanding of temporal objects, each of which attempts to solve the same problem that Husserl has set himself. The comprehension of a temporal object requires an act of cognition in which the temporally separate parts are brought into some sort of unity or synthesis. The two views that Husserl criticizes differ at this point, over the temporality of that act itself. The Herbart-Lotze view (which influenced Brentano) is that this unifying act must itself take place at an instant in time. For the grasp of the whole involves transcending those temporal differences into which experience has dissolved it. If this consciousness of the whole were not momentary, but itself took time, then it would perhaps itself require a further act of unification. On the other view, that of the "specious present," which Husserl attributes to Stern,[11] there are said to be (at least) some cases in which the apprehension of the unity of a temporally extended object develops alongside it, and is completed with it. Listening to a song would be a good example.

Husserl is not satisfied by either version, for they each fail (as Brentano had done) to make certain fundamental distinctions. In other words, their presuppositions about the nature of experience were, phenomenologically speaking, naive.

His two main objections seem to be these: that the question of how such temporally transcendent objects are constituted (for example, as enduring unaltered, as constantly changing) is left undiscussed and unexplained. And how, on the momentary apprehension view, is it possible to understand the apprehension of time itself? (Surely a momentary apprehension of flux would *have to* distort it.) Against the Herbart-Lotze view, Husserl says it is perfectly obvious that the "perception of duration itself presupposes the duration of perception"; perception does not occur at an instant. Hearing a melody is a case of perception, for instance. But when trying to explain how this occurs, it is easy to fall back into talking of a cognitive act that mixes current perception with memory and expectation. Even to focus on the very shortest phases of individual tones keeps us within the kind of framework Husserl criticized in Brentano.

In all his remarks so far, Husserl has merely been clearing the way for, demonstrating the need for, a phenomenological description of the most primitive temporal phenomena.

AN ANALYTICAL PREVIEW

Husserl's account of internal time-consciousness has something like the following structure, by which the dependence of Husserl's account of time on its role in constituting temporal objects will be apparent.

The originary constitution of temporal objects	1. A description of the of modes of appearance of immanent temporal objects
	2. A description of our consciousness of those appearances
The consciousness of time given via temporal objects	3. An attempt to explicate the nature of the continuity of our experience of temporal objects through commentary on a diagram—"the diagram of time"

Despite numerous detours and picking of flowers on the way, our discussion follows the same path as Husserl's text.

THE IMMANENT TEMPORAL OBJECT

Husserl's persistent critical stance is that previous accounts of time-consciousness have focused on the *contents* or *objects* of experience at the expense of the ways in which we experience them. His repeated criticisms are designed to show that any such account will inevitably leave vital questions unanswered, especially about the way temporal objects are constituted, and about the constitution of time itself. His phenomenological elucidation begins with two parallel descriptions of our experience of "immanent temporal objects," first as they present themselves to us and, second, the possibility of using the results of the second description as the basis for a general account of our consciousness of time that is free from the difficulties Husserl previously found.

Husserl takes his example of an immanent temporal object from the sphere of sound. Initially, he considers using a melody as his focus. But the question of how different notes, or tones, are unified into a cohesive whole already takes for granted the temporal unification of discrete tones, each of which endures. So he takes the single briefly enduring tone as his example of an immanent temporal object. A tone is chosen, one supposes, because it allows the basic structure of time-consciousness to emerge with minimum sensory content. Sounds, even treated as material phenomena, are essentially evanescent, essentially temporal. Arguably, persistent tones are to time what lines are to space, and it is, accordingly, no accident that Husserl resorts to a geometrical schema to illustrate the basic structure of time. And his concern to establish the nature of the temporal *continuum* completes the parallel. Husserl's moves here would suggest that it is not the attempt to understand time on the model of space that has prevented an adequate grasp of time heretofore, but the particular way that comparison has been drawn. At its simplest, if the usual mode makes out time to be a line with a direction,

Husserl is adding a "depth" to that line. But as shall shortly be seen, the nature of that "depth" is far from simple.

Husserl provides a pair of descriptions. First, that of immanent temporal objects and their modes of appearance in a continuous flux.

The first description is of the various ways I am (temporally speaking) conscious of an immanent object "as" . . . Thus, "the sound is given: that is, I am conscious of it as now, and I am so conscious of it 'as long as' I am conscious of any of its phases as now." (sec. 8, p. 44) Most of the ways the object presents itself "as" are marked in the text by scare quotes: "before," "during," "for a while," "expired." He describes quite uncontentiously the way I become aware of a sound as beginning, as ending, as coming to an end, as having just finished, and being past. He also introduces the key concept of retention which will later play such an important role: the way in which we still hold on to sounds that have ceased to be impressions and have "sunk back," as he puts it.

"What we have described here," he says, "is the manner in which the immanent temporal object "appears" in a continuous flux, that is, how it is given." And he concludes by carefully drawing attention to the fact that his description of the manner of the object's appearance is not in fact a description of its temporal duration, for that duration is part of the immanent object (the same sound) itself and is presupposed by the whole description. (p. 45) Through the various modes of its appearance the same duration is lived, "expired," or remembered.

The second, parallel description (section 9) concerns "the way in which we are 'conscious of' all differences in the 'appearing' of immanent sounds and their content of duration." It is this shift that will inaugurate the discussion of the *constitution* of temporal objects through consciousness.

The key distinction he makes in this passage centers on the application of the term "perception," a term he will discuss thematically in sections 16 and 17, which is the occasion of a great deal of careful footwork throughout. Here, Husserl wants to claim that while the sound is perceived we can say of the period of its duration only that it is perceived moment by moment.

> We speak here with regard to the perception of the duration of the sound which extends into the actual now, and say that the sound, which endures, is perceived, and that of the interval of duration of the sound, only the point of duration characterized as now is *veritably* perceived. (emphasis added) (p. 46)

The significance of the term "veritable" (*eigentlich*) will only appear during his later distinction of the two senses (restricted and extended) of perception in section 17. For now, it need only be pointed out that Husserl is setting up the problem of temporal constitution. For he is claiming that the sound as a whole, the enduring sound, the temporally extended sound is *perceived*, and yet the interval of its duration is only perceived at each now. Strangely making no reference to future-oriented constituting modes of consciousness (later called "protentional"), his first stab at an account of temporal constitution, at aligning the two levels of perception, at constituting a whole out of temporal parts, involves the deployment of the idea of retention, and that of degrees of clarity and distinctness in the objects of "retentional consciousness." The retentional mode of consciousness operates at two different levels—both as functioning within the duration of (for example) a sound, in which to every now there attaches a retentional "tail"—and with respect to "the running-off of the entire duration." And coupled with each of these levels is a graded loss of clarity and distinctness as the distance from the "now" point increases, which again allows Husserl to draw the spatial analogy.

On the basis of these brief remarks, Husserl reiterates and clarifies the basic distinction he is making, one by which his difference from Brentano will be made explicit. The "object in the mode of running-off" is involved in constant change and as such is "always something other." (And yet, "we still say that the object and every part of its time and this time itself are one and the same.") But, this is not a form of consciousness, any more than a spatial appearance would be, even if, as is just as clear, there must be for Husserl an analytical interdependence between "the object-as" and the mode of consciousness.

This distinction is essential to understand what precisely is going on in Husserl's "diagram of time."

THE DIAGRAM OF TIME

Husserl entitles section 10 "The Continuum of Running-Off Phenomena— the Diagram of Time," and for all his insistence on the distinction between "the object in the mode of running-off" and the consciousness for which that occurs, there is very little if any need to refer to this latter at all in understanding the diagram, which, whatever it is supposed to be, is actually a representation of the temporal constitution of an immanent object. Modes of consciousness such as running-off are exhausted, from a diagrammatic point of view, by what they "intend." Perhaps this should be put another way: the *structure* of Husserl's analysis here, of the double continuity of the modes of running-off of the object, seems to us compatible with various interpretations of the status of this *Ablaufsphänomene*, so it is not obvious to me at least that the particular status Husserl wishes to claim for them is *demonstrated* by this analysis or even specifically presupposed by it. "Running-off phenomena" are, for Husserl, synonymous with "modes of temporal orientation," which are modes of consciousness. And yet the particular concept of consciousness required by this analysis is at best a passive, registering one, rather than one involved in, say, some of the higher order complexities of constitution. There is for example no reference to protention, no reference to the epistemological mechanism by which an object is identified as *this* or *that* *x* through time, and so on.

Husserl's real achievement is to have supplied an answer to the problem of continuity through time, although, as we have already suggested, it is in no way complete. For the reader's ease of reference Husserl's diagrams and his explanations are here reproduced.

HUSSERL'S DIAGRAMS

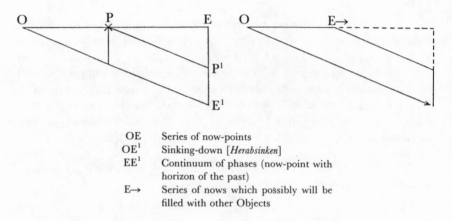

OE	Series of now-points
OE^1	Sinking-down [*Herabsinken*]
EE^1	Continuum of phases (now-point with horizon of the past)
E→	Series of nows which possibly will be filled with other Objects

From *The Phenomenology of Internal Time—Consciousness*, trans. J.S. Churchill (Bloomington: Indiana University Press, 1964), p. 49.

To these explanations we ought to add that P-P^1 is an arbitrarily chosen perception or experience or phase of the object somewhere between the beginning (O) and the end (E). The first diagram concerns the ongoing temporal constitution of an enduring immanent object (for example, a sound), and the second places such a completed object of consciousness in the framework of the past and undecided future.

Two key terms Husserl uses, *Ablaufsphänomene*, (modes of) running-off, and *Herabsinken*, sinking down, call for comment. When we hear an enduring tone, a sound that lasts, each analytically distinguished phase of each note begins with a "now" and then reappears as something "past." These expressions name "running-off characters." *Ablaufsphänomene*, "running-off" itself, however, involves understanding these as reflecting "modes of temporal orientation." Presumably, then, this is meant to illuminate the fact that while at one moment we are aware of something happening, at the next, we are aware of it as having happened. Certain of Husserl's formulations suggest that *each* of these is in-itself a kind of running-off, but I take it that the word "mode" points to the essential unity of each of these phases, and that the "running-off" itself is the flux from temporal mode to temporal mode.

Herabsinken, "sinking down," adds the idea that each impression is retained in sequence, in a kind of depth of sedimented layers that attaches itself to any later "now." So essentially the vertical axis just repeats the horizontal axis, but having converted something sequential into a sedimented depth, it is a repetition that transforms. The most important consequence of this transformation is that it generates a continuum, and it is the purpose of this section to explain more carefully the nature of this continuum.

Husserl does not say so explicitly, but as I understand him, a temporal continuum is required for us to successfully "intend" objective duration. That is, we know or suppose that the sound we hear is a "single enduring sound." We may in fact not hear it *in* an unbroken sequence, while we are nevertheless conscious of it *as* unbroken, *as* a single sound. The continuity of our own experience of it forms the basis of this. We may then surmise that this continuity will equally play a role in allowing us to understand the unity of time itself. So how is this continuity explained?

His first explanation is something close to an assertion, coupled with an explanation of why it might not *seem* to be a continuum. He writes:

> With regard to the running-off phenomenon, we know that it is a continuity of constant transformations which form an inseparable unit, not severable into parts which could be by themselves nor divisible into phases, points of the continuity which could be by themselves. (sec. 10, p. 48)

In other words, he is claiming from the start that any talk of bits or parts of time (moments, instants, and so forth) as actually existing is a false abstraction from a natural flux. This is quite understandable, and indeed something of a commonplace among such theories. But it is quite another thing to explain the structure of this confusion. It is Husserl's striking claim that there is a *double* continuum, or rather continuity operating at two distinct levels, to which the two diagrams correspond. His most concise account of the position is this:

> . . . the continuity of running-off of an enduring object is a continuum whose phases are the continua of the modes of running-off of the different temporal points of tie duration of the object. (sec. 10, p. 50)

Or again:

> Since a new now is always presenting itself, each now is changed into a
> past, and thus the entire continuity of the running-off of the parts of the
> preceding points moves uniformly "downward" into the depths of the
> past. (ibid.)

What I take it he is claiming is this: that if each moment of experience is
coupled to a continuum of past phases, any subsequent moment will have the
previous moment *with its attached continuum* as one of its own phases. If we
allow brackets to enclose attached continua of past phases, then we could
represent this complex continuity thus:

abstracted point moment (starting with *a* as the original moment)	structure of continuum of phases
Series 1	Series 2
a	a
b	b(a)
c	c(b(a))
d	d(c(b(a)))

etc.

We could already call this a "double continuity," but I do not think this is
what Husserl means. We call series 1 a continuum because of the merely
abstract nature of the "now"-points. Or, to put it better, what this series *a*, *b*,
c, *d*, and so on *abstracts from* is a continuum. Series 2 is a continuum by virtue
of its constant inclusions of itself. But perhaps Husserl has in mind a different
conjunction, between series 2 (or the unity of series 1 and 2), and a continuity
established among successive past phases of a completed sound. This, I take
it, is what he means when he says (and here he is commenting on the second
version of the diagram) that

> [a] series of modes of running-off begins which no longer contains a now
> (of this duration). The duration is no longer actual, but past, and
> constantly sinks deeper into the past. (ibid.)

This conjunction of continuities would then be established between the
constitutive continuity of our consciousness of duration of an immanent

object, *as it unfolds*, and the continuity among our retentions or recollections of it, once it had unfolded. And the initially constitutive continuum of temporal phases is enclosed as completed, within the continuity established by our memorial consciousness in its successive phases.

NIETZSCHEAN DOUBTS ABOUT HUSSERL'S CLAIMS

How ought these claims to be assessed? Husserl's method, it is worth recalling, is purely *descriptive*. He is not offering a model of how things might be (that might be verified or otherwise), or how things ought to be (whether they are or not), but of how things are, and how it is perfectly clear that they are on reflection. Confronted with a concept of time as merely flat linear extension, it is certainly an achievement to have begun, at least, to shake that picture, to have offered an account of what the continuity of lived, experienced time *could look like*. But is it more than that? Or is Husserl, in effect, offering us a geometry of time that idealizes rather than describes our actual experience? To pursue this line of criticism, consider the claimed (double) continuity of our experience of an immanent temporal object. If the succession of sinkings-down (which he will soon call *retentions*) were a natural process, like that of the sedimentation of silt in rivers, then the order of succession would indeed be perfectly followed. But if, as he claims, retention is an *intentional* consciousness (or an aspect of such), are there not going to be other considerations, interests, motives, than the simple recording of the actual order of incoming impressions?

An example drawn from Nietzsche is apposite here:

> *The error of imaginary causes.*—To start from the dream: on to a certain sensation, the result for example of a distant cannon-shot, a cause is subsequently foisted (often a whole little novel in which precisely the dreamer is the chief character). The sensation, meanwhile continues to persist, as a kind of resonance: it waits, as it were, until the cause-creating drive permits it to step into the foreground—now no longer as a chance occurrence but as meaning.[12]

What this example suggests is that if consciousness is thought of as having a central "meaning-bestowing" function, then not only is there no guarantee that this will ensure the faithful reproduction in experience, of the real order of events, but there is every reason to believe that this does not always happen. But this conclusion must not be drawn quite so quickly. Two difficulties stand in the way of considering Nietzsche's example as an objection to Husserl. First, it could be said, he works within a clearly prephenomenological dualism, in which there are then problematic alignments of objective and subjective times. Second, Nietzsche is arguably utilizing a notion of the "Unconscious" that would be quite alien to Husserl.

Husserl explains early on why he is not interested in experimental comparisons of inner and outer time. But there is surely a difference between experiments demonstrating lengthening and foreshortening of perceived time with respect to "real time," on the one hand, and demonstrations that between the two there are changes of order, structural transformations. These would have an *essential* (and not merely qualitative) significance. If so, however, it would seem to cast doubt on the practice of suspending all consideration of "objective time" (assuming it to be *possible*) as Husserl recommends. He takes for granted, I would suggest, that the structural transformations that occur between the two are limited to those that ultimately preserve the original order.

In fact, we can read Nietzsche's suggestion in two ways: as suggesting a break between objective and subjective time, and as suggesting a break between the order of sensation and the order of experience.

If it is objected that the former considerations are inadmissible within Husserl's framework, we are surely entitled to consider the latter, particularly given Husserl's own reference to sensations (*hyle*).

What is so interesting about Nietzsche's example is that like Husserl he makes use of a concept of retention ("the sensation, meanwhile, continues to persist, as a kind of resonance") and like Husserl, his "retention," while being a phase of an intentional act, is not in itself an act (see p. 79ff.). And yet Nietzsche is not, I believe, claiming that his retention is merely a physical or empirical psychological phenomenon. He is, I would say, claiming an unconscious retentionality—something that Husserl could not envisage. But for all that, Nietzsche's is an account of the necessary *meaningfulness* of consciousness. Here this necessity demands that the sound be experienced not merely as an external interruption of a dream sequence but, suitably interpreted, as playing a significant part in it. The sound that actually *was* a cannon shot is incorporated into the dream (in, say, a dream involving a house) as the slamming of the front door.

The justification of taking such an example seriously within a Husserlian framework is this: (1) that if we consider the facts of *sensation* rather than those of "real events," we move on terrain Husserl (rightly or wrongly, from his point of view) has already legitimated, and (2) that even within Husserl's framework we could say that the hyletic level is an unconscious level, in that it does not, by itself, constitute consciousness.

But the gap between Nietzsche and Husserl remains. For Husserl, consciousness begins, as I understand it, with the transition from the impressional consciousness of a sensation of sound to a retentional consciousness of the same. For Nietzsche, we remain at the level of sensation, even with retention added on, until the sound is taken up into consciousness by being given a meaning.

What, one wonders, would Husserl say about such an example? It clearly demands a response. (Freud quite independently discusses the way in which dreams utilize inversions of natural order to create their effects.) He *could* suggest that dreams are not the proper basis on which to construct an account of ordinary experience. And yet it is Nietzsche's (and Freud's) belief that it is impossible to restrict the scope of these mechanisms (for example, of delay, of inversion) to specially contained regions of "unconscious life." The demand that our experience exhibit a certain sort of intelligible structure underlies our belief in the Self, our insistence on causal connections, belief in God, and so on. And, to maintain a parallel in conscious life with the specific dream example, if we *do* come across an event of unknown provenance, we assume it has causal antecedents sufficient to explain it, and will usually assume it to be of this or that sort. And if Kuhnian accounts of experimentation in periods of "normal" science are to be believed, it is much harder to even *notice* things happening for which we have no explanation. This would suggest that a similar sequence of *delayed registration of sensation → successful scanning for interpretive schema → conscious registration of sensation*, operates in ordinary experience. Differential delay in the second stage can cause inversions of the sort that Nietzsche has described. This would mean that the singularity of dreams would not stand up as an objection. Needless to say, even if one were to concede the idiosyncratic character of dreams, Husserl would still have to account for such experiences, and it would seem that his conceptual framework is far from ideally suited to handling them.

So far the focus has been on the concept of retention. In essence, all of Husserl's arguments are designed to show that it is a mode of access to the past that is "originary," not to be confused with any of the ordinary everyday empirical acts that it itself must play a part in constituting, in making possible. It is by this claim that Husserl distances himself from any kind of psychologism, which is always presented as a *confusion of levels*. Retention is an ineliminable and irreducibly temporal phase of consciousness. Husserl's discussion, as thus far presented, has focused on its role in constituting immanent temporal objects, objects whose identity is inseparable from their temporal extension and whose existence as such is inseparable from our experience of them. However successful this account, a further step is required to explain how this helps in understanding how immanent *time itself* is constituted. This is one of the next tasks. And beyond that we have to ask whether the immanent time grasped through the consideration of temporal objects is not perhaps a limited conception of time, and whether any more original temporality could be unearthed. This opens onto a consideration of

Husserl's description of Absolute Flux. But first it must be realized that an account of immanent time cannot be developed without going beyond primary meaning, retention, to consider the positive role of reproductive, secondary memory. This is an enormously rich and valuable part of Husserl's work, for he is showing that even if retention is *essential* for the constitution of time, it is not sufficient.

RETENTION AND REPRODUCTIVE CONSCIOUSNESS

In discussing the relationship between retentional and reproductive consciousness, reflection on the expression "retentional consciousness" itself is a useful way to proceed. The intimacy that holds between primal impression and retentional consciousness is founded on two conditions: (1) that between the *contents* of each there is a continual passing over from impression to retention, and (2) that both impression and retention are modes of *consciousness*. Impressional consciousness is (now) directed toward what is sensed, retentional consciousness is (now) directed toward what has just been sensed. Both are immediate, nonreproductive modes of consciousness.

The language of "peeling off," "passing over," "running off," "sinking down" is all carefully designed to suggest a mere change in "temporal place" of the same content, with only a new angle being required for the intentional ray directed toward it. Two further points must be borne in mind: first, what I have called the *succession of inclusion* involved in retention, and second, that all these accounts in which reference is made to this or that particular retention must be understood in the light of the fact that individual retentions are only abstract moments of a continuous flux. Husserl's reference (in section 11) is to a "continuous modification" (p. 50) and, later, to retentions as mere *phases* of a flux. (p. 54) The relation between a phase and that of which it is a phase is vital to Husserl's characterization of retentional consciousness. Phases are not independent parts of wholes. They are, as Husserl wants to claim for retention, *logically* connected to their fluxes, to the structure of any such flux—namely, that it always *begins* with an impression, a now[13] source. A retention is always a retention *of*. But the uniqueness of this relationship is that it is not a reproductive relationship, it does not involve images, or signs. Husserl's position stands or falls with the claim that it is an originary mode of consciousness. It is no more an imagistic, reproductive consciousness than it is a mere quasi-physical echo or reverberation.

THE DEPENDENCE OF RECOLLECTION ON RETENTION

Here Husserl is not just making analytical distinctions, nor "merely" describing phenomena as they present themselves to us; he is doing phenomenological epistemology. In section 11 he has, in effect, an argument to the effect that, for example, symbolic, imagistic, and representational relationships to the past must at some point as a matter of principle depend on our having an "intuition of the past" (p. 53), which he calls an "originary consciousness." It is only because we have a *retentional* grasp of the past that other (secondary) forms of memory can be *understood* to be "of the past." The supposition is that only what we *can* be acquainted with directly can we give sense to. Without retention, we would, I take it, be unable to distinguish— even in principle—memory from imagination. Of course we cannot in practice always do this, but the access to the past that retention (in a limited way) provides[14] gives sense to the distinction that we may or may not on any particular occasion be able to make accurately.

Husserl here seems to be discounting the possibility that we might be able to make sense out of the various ways in which acquaintance, direct grasping, confronts its *limits*, which it cannot penetrate. To be specific, he presupposes that our grasp of the past has to be a positive rather than a negative one. But he would respond, I take it, that to talk of "the various ways in which acquaintance, direct grasping, confronts its *limits*," begs the question, for it does not explain how these "ways" are in fact distinct. Both memory and imagination deal with what is "not actual"; the difference between them is that the former deals with what "has been" the case, and the latter with what "might be/have been" the case. The "beyond" that memory deals with is distinguished from the "beyond" of imagination (say) by the meaning of "has been." And our only source for that is primary (retentional) memory.

Of course the reply provided for Husserl here could itself be challenged. Surely we do have other ways of distinguishing memory from imagination, ways that rest on the fact that memory of states of affairs that still persist will give us knowledge we can check. I may "imagine" that I have a tail, and "remember" that I do not (and perhaps, qua human being, could not). The difference between these instances of memory and imagination, determining which is which, will be established by feeling for my tail. Of course such tests are not in themselves conclusive. Not all that one remembers is still the case, and some of what one imagines is actually true. That one can easily construct scenarios in which such tests would mislead is not the point. The very fact that one could broadly distinguish a class of experiences that mislead and another class of these that, making due allowances for the ways things can change, does not is what is significant. However, unless one at some point begs the question it is hard to see how such a distinction could do more than provide a way of guessing more or less accurately, assuming one had no other way of telling, which experiences were memories and which phantasy. But it would not quite so obviously provide us with the *sense* of these terms. They might be thought synonymous with: more predictive/less predictive! Perhaps Husserl has a point. But what would he say to the further suggestion that precisely because retention is a mode of self-evidence, or direct awareness, that it cannot convey *pastness* to us, for the past is essentially what is no longer *in any sense* present? This question will be touched on again in later sections on Perception (see p. 93–7). The beginning of an answer might go like this: that the force of this objection is such as to focus one's attention on the importance of supplementing retention by a consideration of reproductive memory. It might be, in other words, that while reproductive memory was dependent on retention, the possibility of forming the basis for a reproduction was just as essential for retention to be the primitive form of our access to the past.

While considering alternatives to Husserl's use of retention to provide us with a grasp of the past, it is worth reflecting that from the human point of view, the most striking feature of the past is not at all that it "has been" but that it now *must be*, that it *cannot now be changed*. What is done, is done. There is no use lamenting what is past. It could be said, however, that this could not serve as a defining feature of the past because the present cannot be changed either, and perhaps some aspect of the future are similarly immutable. As regards the future, however, it would be only a kind of folly (it would not be absurd) to try to prevent the sun rising tomorrow. Our inability to cut a cake into three halves is another form of determination quite different from the fixedness of the past. One can of course often restore affairs to their original state—retying a shoelace, rebuilding a wall, washing the dishes—but that does not repeal the past. One does not thereby wipe out the fact that the laces came undone, that the wall fell down, that the dishes were used, even if one wipes out all evidence of this fact. Again, it could be argued that the past, or at least the *future past*, can indeed be changed by my current and future actions. I can, today or tomorow, bring it about that I am (at some future date) remembered as a generous benefactor. Nonetheless I cannot bring it about that I *did* do something yesterday if I did not. It is the indeterminacy of the future that prevents symmetrical remarks being made about what I *will* do from having the same force.

How about the claim that the present, at least, can no more be changed than the past, and so, again, immutability cannot serve as a distinguishing characteristic of the past? It is difficult to respond neatly to this. *If* the present is understood as the instant of sensate consciousness, then it is true that if I am sensing x at t^1, there is no changing that fact, except by moving on to t^2, to *another* present. With such a view of the present, one would have to add "and not at this instant being sensed" as a rider to the unchangeability of the past. But if one thinks of "the present" as having some duration (or indeed as being conscious duration) as we do in a loose colloquial way, then it is clearly a locus for change, indeed perhaps the only one. If I look around to see what is making a purring sound at my ear, I change the focus of my attention. I can only *act* in the present, and in doing so, of course, I take time, and so change the present! It would seem clear that the present as commonly understood cannot be thought of as unchangeable, for it is precisely the locus of change.

This however suggests a further objection—that the future cannot be changed not because it is fixed, but because it is not fixed. If it is not fixed, it cannot be changed, for there is nothing for it to be changed from. Perhaps we should say this—that the distinctiveness of *a* past consists in the fact that our current actions can have no effect on it, while they can affect what will be the case.

If this general contention can be sustained, does it not provide an alternative candidate for "the meaning of the past" to that provided by retention? Clearly these are fundamental issues involved here, for a pragmatic, active, existential dimension would be being introduced into a contemplative, epistemological framework, and somewhat vulgarly at that. But the possibility that it might be in just such a dimension that the past derives its "meaning" cannot be ruled out.

There is, however, one final qualification I would like to make to the suggestion that the unchangeability of the past might be its defining feature. It would seem to confine our understanding of the past to a collection of events, the "limits" of which are fixed at the time at which they occur. But there are two ways at least in which, by abandoning that presupposition, it would be possible to change the past. We can act in such a way as to change the meaning of a feature of the past. If I make many sacrifices toward the attaining of a supreme goal, and I am not at the point of being able to achieve it, I can act in such a way as to make sense of those sacrifices, or to make them, in retrospect, futile. A more interesting example can be found in the act of forgiveness, as Levinas discusses it. The act forgiven cannot be undone, but forgiveness can effect a kind of moral wiping out of the past. These sorts of questions are returned to later.

THE DIFFERENCE BETWEEN RETENTION AND REPRODUCTION

> . . . we characterised primary remembrance or retention as a comet's tail which is joined to actual perception. Secondary remembrance or recollection is completely different. (sec. 14, p. 57)

How is this difference described?

As usual Husserl offers us first a simple story, which he attributes to Brentano, one that seems plausible but that does not survive critical scrutiny:

... actual perception is constituted on the basis of phantasies as representation, as presentification[15] ... just as immediate presentifications appear without being joined to perception. Such are secondary remembrances. (ibid.)

In other words, on this account both retention and recollection are representational phantasies, the difference being that the former are not "joined to perceptions."

The flaw in this story, put very simply, is that it assimilates memory to phantasy,[16] but more important, even if (as section 19 suggests) that assimilation is permitted, to treat retention, primary memory as rooted in phantasy, is to make it impossible to understand how secondary remembrance can *reproduce* what was once "experienced," that is, *perceived*. *Reproductive* memory must refer back to an original self-given, nonreproductive grasp of an immanent object, and that cannot be provided by an essentially reproductive "phantasy."

It would be a misleading portrayal of Husserl's position, however, to focus too singularly on the nonrepresentational nature of primary, retentive consciousness. This *is* a vital component of his account of time, but it might suggest that reproductive modes of consciousness are merely reproductive, with no special contribution to make to our understanding of time. And it is surely one of the great strengths and indeed virtues of his position that he recognizes the vital positive constitutive role(s) played by reproductive consciousness (see especially sections 18 and 32). The argument has the structure of what Derrida will call "supplementarity" (see chapter 3). This will be pursued after a fuller discussion of the difference between retention and reproduction, primary and secondary memory.

Reproduction is not all of the same kind. Husserl makes a distinction between at least two different forms of reproductive memory, which we could call (a) simple recalling and (b) recapitulative memory. The first affords us largely unanalyzed glimpses of the past; the second, a systematic reconstitution of the past experience. Husserl seems typically to focus on the latter.

He also mentions a spontaneous form of reflection on, say, some just completed phases of a melody, which is neither simply itself retention, nor a reproduction of the sequence. We simply have the ability to glance back at what we have just experienced. What is puzzling about this is the status it is meant to have. While Husserl does not say so, it could be argued that this capacity for spontaneous retro-glancing is quite *vital* for understanding our ability either to experience immanent temporal objects of any significant duration, or to recall them in their extended unity of duration. For even if we do not (as psychologists have variously done for short-term memory) give any precise quantified limits to retention, it is hard to see how one could

actually hold even a short story in one's retentive grasp, let alone a sym-
phony, without some intermediary form of partial and successive retro-
glancing that would allow the process of constitution of an immanent
temporal object to work at different levels.[17] I read these brief remarks as
Husserl's attempt to solve the problem of the limitation of the range of
retention. But it is nonetheless theoretically problematic. It is not a "mode of
accomplishment of reproduction" because it is not reproductive. Nor is it
simple retention. Husserl's words surely belie a certain unease ("Es scheint
also, das wir sagen können: . . . sec. 15, p. 59)—as if we seem to be able to say
it, but he does not quite know how. His justification is of course that of one
who is faithfully describing what he sees, whether or not it is theoretically
convenient. Its place in this section on the varieties of reproduction rests on
what it shares with reproductive modes of consciousness—that "this givenness certainly refers back to another 'primordial' (ursprüngliche) one." (sec.
15, p. 60) But instead of immediately pursuing the consequences of trying to
assimilate this rearward glancing in some theoretical way, I want to draw
attention to a phrase Husserl employs at the end of the first paragraph of
section 15, for it opens up a more general question about how we should
understand the phenomenological status of Husserl's claim in particular
about reproduction. After briefly describing recapitulative memory (distinguishing it from what we have called simple recalling), he remarks, "However, everything [here] has the index of reproductive modification." But
what *is* it to have the index of reproductive modification? In particular, does
this mean that reproduction bears (or bares) its status as reproduction on its
phenomenological sleeve? Can one always or usually tell "at the time" just
by spontaneous reflection on the experience itself whether it is, say, primary
or secondary memory or phantasy or "living experience"? Or is he in fact
saying that it is the analysis or description of reproductive memory that "has
the index of reproductive modification"?

IMMANENT DESCRIPTION AND ANALYTICAL REFLECTION: HUSSERLIAN AMBIGUITIES

The thrust of these questions must be clear. Similar questions could be asked
by skeptics of various hues. But the aim here is not to cast doubt on the
concept of recollection. I am interested, however, in marking the point at
which Husserl moves from *immanent description* to what we could call reflective
(analytical) description, and for two reasons. First, Husserl himself does not
always pay attention to this shift, although from his *own* (phenomenological)
point of view, it is quite crucial. And second, one would suspect that it is at
the reflective, analytical level that not only an intensification of theoretical
insight, but also the quiet accumulation of dogma, would be found.

To appreciate the importance of this distinction between descriptive and analytical phenomenology, consider the following possibility: much of what Husserl wants to say about the distinctive features of "recollection" cannot be said to fall under the heading of "immanent description." Indeed, in section 19, Husserl himself, having already devoted a number of sections to the distinctive features of recollection, goes so far as to point out, as if for the first time, that between "representifying memory and primary remembrance" there exists "a great phenomenological difference," which "is revealed by a careful comparison of the lived experience involved in both." (p. 68) And what he means by a "great phenomenological difference" is, it seems, "radical differences in content." (p. 69) There remains, however, this possibility: that however much it may be crucial for Husserl's general program, and indeed his particular analysis of time-consciousness, that he go beyond Brentano's interest (however short-lived) in the *contents* of consciousness, it is to just such differences in content that Husserl returns to bring out the "great phenomenological difference" between retention and recollection. Even if it would be unfairly to exaggerate this remark to suppose that he was claiming the only difference was one of content, why does he use the word "phenomenological" just at the point at which he wants to talk about the contents of lived experience? (With one trivial exception, it is the first time he has used the word "phenomenological" for twenty-five pages.) Might it be that differences at the level of acts are not phenomenologically accessible if one understands by phenomenology the immanent description of lived experience? Certainly many of the claims Husserl makes are simple deductions from certain basic principles he has established about experience (for example, the necessity for all experiences to be either self-given or related to one that is, or was, self-given). Whatever ought to be concluded here, I hope now at least to show that many of the differences between retention and reproduction owe more to reflection, to what must be the case, than to immanent description. And it makes the relationship between the two levels of description problematic in principle, as it certainly was in practice for Husserl himself.

Consider section 14, "Reproduction of Temporal Objects—Secondary Remembrance." Clearly taking reproductive memory as his model, one of Husserl's main points is that from the point of view of the experiencer, "the entire phenomenon of memory has, mutatis mutandis, exactly the same constitution as the perception of the melody" (p. 57). "Everything thus resembles perception and primary remembrance and yet is *not* itself perception and primary resemblance." What separates the two cases are *facts about them* (which we are not, in memory, actually hearing) and all that flows from these facts about the epistemological status of recollection ("presentified but not perceived"). In section 19 Husserl will attempt to distinguish these two, phenomenologically (that is, in terms of content, as we have noted), but his preparedness to follow through a very simple model of reproduction without much thought already calls for comment.

By his "very simple model" I mean the idea that reproduction is indeed just that—the reliving of the original experience just as one originally experienced it. It is important to remember that Husserl takes this to be the standard case, not an ideal, realized perhaps under special conditions. The implausibility of this model of perfect reproduction can be seen by reflecting on the following sentence, in which he is describing how in reproduction the operation of primary remembrance remains intact, the seal unbroken:

> with the apprehension of the sound appearing now, heard as if now, primary remembrance blends in the sounds heard as if just previously and the expectation (protention) of the sound to come. (ibid., p. 58)

It is extraordinary to have had to wait so long for a reference to protention, but here surely it is fatal for his account. For a recollection that *ignored* the question of whether or not the protentions in the original experience had or had not been fulfilled, a recollection that abjured the benefits of hindsight would surely be a very odd and unusual thing. If the memory is of our first listening to a piece of music, our knowledge of how it goes will interfere with the protentional structure of the original experience on recall. Where once we did not know what to expect, we now do. . . . To be fair, Husserl discusses some of these points in section 24, but section 14 exhibits no anticipatory awareness of the coming destruction of its model of recollection.

If Husserl has wanted a critical difference between primary and secondary remembrance, one that spans the space between what I have called analytical and immanent description, a discussion of the interference patterns produced by the shape of subsequent protentional fulfillment would surely have been appropriate. What does Husserl offer us in the way of a "phenomenological difference"? His analysis in section 19 of the "hearing of two or three sounds" is intended to provide the answer. Unfortunately, his discussion is at best confusing and at worst itself confused. (pp. 68–70) His attempt to demonstrate a phenomenological difference between the two (that is, between primary and secondary remembrance) seems to involve both the identification of phantasy and presentification and yet importantly to distinguish between the two, and it is altogether unclear what the final upshot is. When a clearer claim does emerge, it seems, as often happens, to be an analytical distinction posing as an immanent descriptive one. Husserl claims that there is a difference between the modification of consciousness involved in the transition from "an originary now into one that is *reproduced*," on uhe one hand, and "the modification which changes the now . . . into the past." But after having first suggested that the difference lies in the way the latter involves a "continuous shading-off," he then states that "reproduction . . . requires exactly the same gradations although only reproductively modified." And unless what is meant by "reproductively modified" can be made phenomenologically clear (that is, clear in immanent description), we are dealing again with another analytical distinction that does not make itself experientially apparent.[18]

What is the upshot of the claim that Husserl's references to experience are often overtaken by his analytical method? It is (surely) that the reference to experience can in part be interpreted as a strategy. Phenomenology claims to be descriptive of experience both because of the wealth of what is opened up, and because of the legitimacy conferred on such accounts by the neutrality that "description" suggests.

But the analytical tendencies of phenomenology are important to recognize not because some particular theoretical description (say, of time) is smuggled in, but because the theoretical itself is confirmed in its capacity to illuminate time. This is problematic because of the temporal commitments of "theory" which could be summed up as the neutralization of time in the interests of generalization. The very categories that the traditional concept of time is bound up with, and that contribute to its metaphysicality, reemerge as the working oppositions of analytical phenomenology. This suggests that the value of presence, the commitment to logocentrism, appears in phenomenology not just in the commitment to the immediacy of certain fundamental experiences, but in the concomitant commitment to a set of categories (indeed to the analytical/theoretical use of categories) by which the play of differences is organized, stabilized, identified, and represented as a world of enduring things.

THE FREEDOM OF REPRODUCTION

In his discussion of the "freedom" of reproduction there is, however, a plausible candidate for an experientially distinguishing feature, even though there is a certain exaggeration involved. Husserl claims that there is a basic difference between the passive way in which we can merely observe the sinking-down of original impressions into their retentional phases, and the free activity involved in reproducing (presentifying) the successive phases of a temporal interval.

> We can carry out the presentification "more quickly," or "more slowly," clearly and explicitly or in a confused manner, quick as lightning at a stroke, or in articulated steps, and so on. (sec. 20, p. 71)

This is surely an important claim. And it raises the most interesting questions in its own right. I shall consider three. The first two concern the status of the term "can."

In the section title, the word "freedom" is in scare quotes, as if Husserl was not sure it was quite appropriate. In fact Husserl is deeply committed to the idea of freedom, even if it does not play for him (as it does for, say, Sartre) a central role. For example, when endorsing (while modifying) Descartes's universal doubt (in *Ideas*, section 31) he writes, "the attempt to doubt everything has its place in the realm of our perfect freedom." There, too, the power of the "can" devolves from our capacity as conscious beings to modify our mode of consciousness to "transform" our cognitive attitudes. There is an essential element of voluntarism operating here; the "can" does not just refer to possibilities (as in, say, "beetles *can* come in many shapes and sizes"). But it is a little difficult to know how, for example, it is within our freedom to modify the level of clarity or confusion in our reproductions. It would seem to presuppose that the "base" for this freedom in reproduction is perfectly clear access to the past, which we then deliberately may allow to become confused. It is hard to see how one could reconstitute clarity from confusion. But it is in fact just as difficult to see how we can *introduce* confusion into clarity while at the same time (in some sense) having access to the (clear) version being confused. In fact the assumption of perfectly clear access to the past is itself problematic for two reasons: (1) it would itself have to involve reproductive memory, and hence pose the same difficulty again, or be a kind of access not yet discovered, and (2) while we *may* have such clarity on odd occasions, it is hardly something we can take for granted, and when we lack it, it does not seem to be remediable by an exercise of our freedom. The second point to be made about this "can" is not a criticism, rather an observation: it itself involves a *temporality* that requires a complex use of Husserl's categories to be explained. One stab at an analysis of its specific temporality would be this: it is the reflective assertion (which transcends but includes "expectation") of a generalized protentional possibility. If anything like this is true, it would at least call for a reassessment of any suggestion that the difference between retentional passivity and reproductive activity is one open to simple immanent observation, for the sense of "(I) can" must surely be seen either as a reflective addition to the bare activity of reproduction, or as introducing reflection into the very experience of reproduction.

THE PRESERVATION OF ORDER

Finally, it is interesting to note the privilege accorded to the *order* of events or phases being reproduced, even though it is permissible to break the continuity into "articulated steps," which would seem to set the scene for transformations of order too. But why not extend the exercise of our freedom to the order as well? The simple answer is that it couldn't be a reproduction of the original experience if the order of its parts was tampered with. Variations of speed or clarity, and a kind of phase parsing seem to be acceptable, but not changes of order. The intuition behind this would seem to be that a change of order could only take the form of *error*, an intrinsic failure of reproduction, like adding an element that had never occurred. Clearly one can make mistakes about order. (Indeed, this possibility is discussed in section 22.) But not all transformations of order need be mistakes. If reproduction cannot avoid the destruction of the innocence of the protentional unfulfillments of its original experience, it might well be that the reproduction of an original experience could begin with a glimpse at the outcome or the end. (Think of trying to run over the construction of a plot in a novel one has just read, or the course of an important conversation, or a trial, or a joke.) There might be resistance to calling this reproduction, but (a) it is clearly not the same thing as error, for there need be no deception involved, no supposition that the end phase actually *did come first*, (b) nor is it a case of phantasy, because there is genuine reference to a past that actually occurred; and (c) in the light of the effects wrought by knowledge of protentional fulfillments, one might conclude that there is nothing *radically* different in reproductive representations involving, at some level, complex splicings of past and future, rather than the simple maintenance of the original "order."

Husserl would resist allowing transformations of order to be included in the freedom of reproduction. He could nonetheless allow that they exhibit a wider freedom. But the reason, I suspect, is paradoxically that he readily acknowledges the complex temporal overlappings and interweavings on which we have predicated such transformations. But he wants to preserve a distinction between the order of what is experienced, and that of the intentional nexus that accompanies it in (for example) reproduction. It is in the way he begins to explore this that section 24, "Protentions in Recollection," is so promising. What has to be taken account of is that "every act of memory contains intentions of expectation whose fulfillment leads to the present." (p. 76)

THE PRODUCTIVE ROLE OF REPRODUCTIVE MEMORY

This gives a hint of the *productive* role of reproductive memory. In reviewing the past in the light of outcomes, one can learn from it and give it meaning. There is an interesting illusion possible here, however, that it is worth pointing out, just so that it becomes clear that it does not *follow* from a phenomenological analysis. It is that the *only real meaning* of past events is to be determined on the basis of the fulfillment or otherwise of their protentional horizons—how things worked out. First, Husserl's concern at this point is with memory, but it would be a mistake to suppose that this frame of reference is definitive. Possibilities that were not realized, hopes that were dashed, dangers that did not materialize are in their own way every bit as "real" as those that were "fulfilled." Indeed, retrospection need not merely serve to consolidate, or to draw the lessons of experience. It can just as well discover forgotten possibilities, find honor or courage in actions that resulted in failure, and so forth. And although such judgments may often arise only retrospectively, the horizon they celebrate is not the past but the future. Courage attaches to an act, whether successful or not, because of the horizons of danger and uncertainty in which it is carried out. It is only an accident that the recognition of courage (with medals, for example) is so often a consolation for practical failure.

Husserl's discussion of protentions in memory makes it clear that one of the great virtues of a phenomenological analysis of time is its ability to register and delicately disentangle the multiple strands of temporal intentionality. It does not necessarily mean that these strands are wholly susceptible to phenomenological separation; that remains to be seen. But it does mean that phenomenology offers a powerful way of broaching the question of the depth structure of time.

THE PLACE OF PROTENTION

We have already mentioned the paucity of references to protention. Husserl's discussion in section 24 is certainly some compensation. But is it not significant that it should first receive thematic treatment in the context of memory? What this seems to suggest—though this may be unfair—is that the fundamental importance of protention for Husserl is the way it anticipates possibilities of objectifying reflection. Protention is the name for the way the adventure of the future—its fundamental openness—is closed off by anticipation. The future is seen as a reservoir for generating an ever tidier past, insofar as it is (subjectively) filled with anticipations of completions of as yet only partial objectivities. The interesting question this raises is whether any truth there is in these remarks should be traced to the epistemological context of Husserl's writing, or whether it does not perhaps offer us a way of coming to understand the fundamentally temporal project of epistemology itself.

Husserl, of course, also concerns himself with expectation as well as with protention (though some of his remarks would seem to refer to either, particularly section 26, where he compares "memory and expectation)." The possibility of clear sight of the future and imperfect memory makes "determinateness" an inadequate distinguishing feature. But there is an essential asymmetry, which is particularly interesting from our point of view, for it is of immediate phenomenological interest. Our expectations are fulfilled, or otherwise, perceptually, that is, by a future experience, the occurrence of which will end the expectation. Whereas the only *subjective* basis for confirmation of memories is internal coherence, or, as Husserl puts it (p. 80), "the establishment of nexuses of intuitive reproductions."

I would like to make two comments on this, one that would add further complexity to the analysis, and the other, more important, which will draw us into a dimension that we will take up only much later.

The first is this: there is a strange inverse mirror relation between memory and expectation. In terms of fulfillment, as we have seen, expectation alone can appeal to perception. And yet in terms of *intention*, of course, it is memory that has a special relationship to perception. We remember only what we have experienced and so what has actually happened. And memory posits such an occurrence. Expectation can posit what it likes, but it may not happen. Memory has a special relationship to actual perception; expectation relates to *possible* perception. Husserl's asymmetry rests on the function of coherence in allowing us to test meanings, and perception those of expectation. But it ought to be pointed out that expectations too can be tested by considerations of coherence ("All this must . . . dovetail into a context of similar intuitions up to the now"). But such considerations are surely equally applicable to expectations. Some expectations—for instance, those inconsistent singly or severally—can be discounted in advance. That the moon will be found to be made of green cheese, that my dead grandparents will turn up for dinner, that my book will write itself, and so on. In this respect, the contrast with memory is less marked than one might think.

EXPECTATION AS AN INADEQUATE PARADIGM FOR CONSCIOUSNESS OF THE FUTURE: AN ALTERNATIVE

My second comment may perhaps seem more tangential. Husserl is taking expectation as something of a paradigm for our relationship with the future. And yet does it not exhibit a strange unconcerned neutrality? Expectations treat the future as a matter of detached curiosity, unless "expectation" is treated as a mere analytical moment in some more interested or involved relation. It is surely somewhat rarely that we can relate to the future in a purely cognitive manner, and for two reasons. First, insofar as we are active beings (and not simply observers), the future is not something independent of our own actions. Second, even when we are not in a position to affect the future, we may be no less *concerned* about it. In the case of active involvement, it may be said that it both presupposes an expectational nexus involving all sorts of hypothetical assumptions about "what would happen if . . ." and it results in certain expectations about what after one's intervention *will* happen. This is all quite true, but in each case, expectation has become (if ever it was anything different) an analytical component of a more complex relationship to the future. I shall suppose for the moment that a plausible name for a better paradigm of our relationship to the future is "desire," and briefly suggest ways in which the analysis of desire would have to depart considerably from that of a mere analytical component— expectation.

Whether one adopts a Platonic or a neo-Nietzschean conception of desire—whether, that is, one thinks of desire as the expression of ontological lack (Plato, Hegel, Lacan, Sartre), or whether one thinks of it as an expression of the way we each exceed and overflow ourselves into the future (Nietzsche, Deleuze), there is a sense in which desires may not be fulfilled as expectations ideally are. An expectation is a cognitive distance from a perception, and the perception (ideally) totally fulfills the expectation. The satiation of a desire, however, is different. For the satisfaction of desire only sustains more desiring. Indeed, it is the occasion for the discovery—certainly on the Platonic view—of the impossibility of satisfying what desire truly seeks—possession, completion, unity with the Other (or the other object), the end of striving, and so forth. In other words, desire has a metaphysical dimension to it that expectation per se lacks. If, however, expectation is always conjoined with some conative element, then we cannot exclude that, albeit by association, the same would be true of expectation. But only because there is no *mere* expectation. Expectation, pure as it may always be, is an index of our status as temporal beings. Particular expectations themselves occur, as we have said in a different context, within nexuses of other expectations, all drawing on previous experience, and as these new expectations are (or are not) fulfilled, this body of experience will be itself enlarged, generating a new range of expectations. . . . But although the fulfillment of expectations could in this limited sense be said to generate new expectations, it would be more accurate to say that it modifies the framework of understanding within which new expectations are generated. The generation is indirect. But are matters so different with desire? A proper account of this question will have to await a fuller analysis of desire. Perhaps the clue to the fundamental difference is precisely in the apparent disinterestedness of expectation, and the overwhelming interest involved in desire. The possibility of metaphysical illusion is not confined to the idea of disinterestedness (as the hollowness of at least some desires has suggested, for example, to Epicurus, and to Buddhism).

HUSSERL'S EPISTEMOLOGICAL FOCUS: A LIMITATION?

How far has this discussion of Husserl got? I do not claim yet to have *proved* anything, by comparing expectation to desire, about the intrinsic limitations of phenomenology. But I do hope to have at least suggested that the move from thinking about expectation to thinking about desire is a move that brings the scope of Husserl's treatment of time-consciousness into a new perspective.

That is, it shows again the limited *epistemological* concerns on which Husserl's treatment is predicated. If this is a fair claim, a number of possibilities arise:

1. That there may be a good reason for this limitation—that time is either itself an epistemological concept, or that epistemological and temporal concepts have some common root, or common field, of operation.[19]
2. That if this is not true of time, it is nonetheless true of time-consciousness, and this latter is fundamental.
3. That we need to expand our frames of reference to accommodate other concepts of time, leaving Husserl's intact, but as having finite scope, limited application.
4. That we need, in addition, to place epistemology itself within a wider context, within which Husserl's concept of time may or may not have its place.

These alternatives are not all incompatible. Arguably, the second is a refinement of the first, and the fourth refines the third. The relationship between these two pairs is complex, indeed one of the key issues of this book will be whether moving the consciousness of time out of the framework offered by Husserl would mean its expansion, or its transcendence/displacement/destruction/deconstruction . . .

THE SCOPE OF PERCEPTION

The distinctiveness of Husserl's account of internal time-consciousness rests on the particular way he treats retention. Not only is it *not* understood as a change in the contents of consciousness (for example, by their acquiring a new quality—that of pastness), it is, positively speaking, treated as a nondependent or "originary" mode of consciousness. Retention in particular is not a reproductive or representational mode of consciousness. Husserl calls it a kind of "perception," and allows that we perceive the past. And yet there seems to be a difference between our awareness of what we are immediately sensing *now*, and our retentive grasp of the "immediate" past. Can we really call both "perception" in the same sense? And if we talk about perceiving what is in one way or another immediately available to us, can we really talk of *perceiving* a melody that endures through time? Husserl *does* so because he wants to stress that the unity of the melody is *constituted* by intuitive acts, and is *not* the result of mere associations. How can he reconcile these three senses (or types) of "perception," especially considering the kind of epistemological *privilege* that such a term conveys?

TWO LEVELS OF PERCEPTION

What, in fact, Husserl provides is an explanation of the different motives for the double application of "perception." It has already been suggested that the idea of the present and of presence involves a fusion of a temporal and an evidential sense, and this ambivalence plays an important part in Husserl's discussion of perception.

Consider what it is to listen to a melody, and the problem becomes apparent. On the one hand, distinctions are called for between individual tones actually being perceived, and those that have gone by, that are no longer perceived. And on the other hand, it is clear that melody is perceived as a whole.

Husserl offers two ways of explaining the latter. First, the whole melody is said to be perceived because each part of it is, one after the other, perceived. But this seems insufficient because the unity of the whole melody as an immanent temporal object is not accounted for. A randomly organized sequence of the same tones would do quite as well. Husserl seems to admit this when he writes:

> The whole melody . . . appears as present so long as it still sounds, so long as the notes *belonging to it*, intended in the *one* nexus of apprehension, still sound. (emphasis added) (sec. 16, p. 61)

This reference to "belonging," to "oneness," to the "nexus of apprehension" brings out the second condition, that the individual tones be perceived as *part of a whole*. What he calls "adequate perception of the temporal object" requires that.

> the unity of retentional consciousness still "holds" the expired tones themselves in consciousness and continuously establishes the unity of consciousness with regard to the homogeneous temporal object, ie. the melody. (ibid. p. 60)

Husserl seems to say that the elasticity of the scope of "perception" results from the variability of the intentional focus. One legitimately says one "perceives" whatever is intended in experience and is therein constituted as such through time, as "one," as a unity.

After this, Husserl follows up two seemingly inconsistent lines of

thought, which can be explained in the following way: what counts as
perceived depends on whether we are making a distinction within the
domain of what is "originally self-given," or whether we are distinguishing
that field itself from what is not self-given at all (but is only presentified). The
internal distinction gives rise to a distinction, one that emphasizes the
temporal sense of "presence," between perception (of the now) and retention
(of the just gone), while the external distinction emphasizes the evidential
sense of presence and calls perception whatever is self-given, whether or not
it is so given in the immediate temporal present. In brief, he distinguishes
perception versus retention from perception versus presentification (for
example, recollection).

His successive consideration of what we have called the *internal* and
external distinctions leads him to interestingly different conclusions.

If the distinction is made internally, our apprehension of an immanent
temporal object could be described as a product of mixing perception and
nonperception. At any time, there will be impressions (perceived) and
retentions (not-perceived). And yet when it is recalled that it is only the
phases of a "continuum which is constantly being modified," it becomes clear
that the restriction of "perception" to the now is only an ideal limit, never
actually realized. He goes on to say that "even this ideal now is not
something toto caelo different from the not-now, but continually accommodates itself thereto." This is a momentous remark. Husserl is first of all
saying that once it is admitted that the now is only a phase in a continuum, it
has only an ideal existence, and *then* he sees that it would be wrong to restrict
ideality to the function of making static abstractions from a continuum. The
true ideality of the now is itself a *dynamic ideality*, a continuous accommodation of the now to the not-now.

What this surely implies is that the internal distinction between perception and non-perception breaks down, when it is clear that the distinction
between the now and the no-longer-now itself breaks down, and not only "in
practice," but ideally too. What is *actually* perceived is not the now, but *the
now as it passes away into retention*. And the now has no existence independent of
its own becoming not-now.

The "external" distinction—between perception and recollection—
involves treating the "nowness" of narrowly conceived perception as merely
an instance of originary self-givenness. If this latter is allowed to be the
criterion for perception, then retention can be a mode of perception and as
such is quite distinct from recollection, which is only presentification. The
extraction of "originary self-givenness" from the apprehension of the "now"
is in effect the extraction of evidential presence from a blend of temporal and
evidential senses of presence, claiming that the evidential sense extends

beyond the temporal, and is sufficient for "perception" to occur. If this move is made, it is possible to "perceive" the past.

"THERE HAS NEVER BEEN ANY 'PERCEPTION'" (DERRIDA)

What Husserl has to say about perception in sections 16 and 17 leads in a strange direction. The structure of his epistemological thinking is such as to treat our immediate awareness of the "now" as the model of originary givenness, with which retention and recollection will later be compared. While retention does not give us the present, it is a mode of direct access to what is self-given, and if *this* feature is considered to be what is epistemologically central to our awareness of the now, then retention is thought of as a form of perception. But Husserl is quite aware that the distinction between impressional and retentional consciousness is hewn from what is actually a continuum, and a dynamic continuum at that. And as a continuum, it is characterized by infinite divisibility, so that at any level one chooses, the sinking-down into retentional consciousness will already be apparent. And if, as he claims, the now is only an ideal limit, it is tempting to treat retention as the primal phenomenon, with impressional consciousness as extrapolated or derived from it, or perhaps as only one of its abstract phases. Husserl has, of course, earlier vigorously resisted anything like this: "We teach the a priori necessity of the precedence of a perception or primal impression over the corresponding retention." (sec. 13, p. 55) And such a conclusion, as has been suggested, would make it somewhat difficult to credit retention with the epistemological status of "now"–perception in the way that was done before, because, at the very least, they are inseparable. Put more strongly, however, *there is no now-perception* as such, and so nothing with which retention could be compared.

Is there not a parallel here with Hegel's discovery at the beginning of the *Phenomenology of Spirit* of the difficulties involved in the idea of simple sense-certainty? I am not in fact convinced that the indexical terms "this," "that," "now," or "here" can be thought to bear the seeds of that universality that threatens the very possibility of simple sense-certainty. Nonetheless, it is arguable that "this," "here," and "now" not only have no meaning, they have no referential value either until they are filled out, or qualified in some way ("this phrase," "here in Chicago," "as I speak to you now," and so on). The argument then is that any attempt to specify the particularity or individuality of an item (for example, a sensation) automatically makes use of some sort of conceptuality, simply to delimit the item being indicated.

Husserl's willingness to *extend* the scope of "perception" to immanent and *enduring* temporal objects could be seen to constitute the basis in an extended now-perception from which he proceeded. For it begins to seem that *any* now one might choose has its own inner retentionality. (The parallel with Hegel would be complete only if one were able to relate the role of *concepts* in Hegel to the primitive origins of constitution in Husserl—the structure of retention. This could perhaps be accomplished by seeing Kant's *schemata* as supplying the temporal underpinnings of Hegel's concepts, and Husserl's retentionality as supplying the basic unit from which schemata could be constructed.)

Husserl's repeated example of the *melody* (or a tone) may seem to simplify the question of temporal constitution, but at the same time it leaves open the way in which concepts actually do supply the unifying grounds for other immanent objects. Husserl's answer, of course, is his theory of constitution. But it is far from clear that the operations involved in constitution can all be derived from retention. And if the claimed epistemological privileges of "perception" can be extended to retentional consciousness, and indeed to the experien*cing* of temporally extended objects, it is hard to see how it can extend to the products of extended constitution.

CAN "PERCEPTION" BE EXTENDED TO ANY RETENTIONAL CONTINUUM?

Husserl has offered us a model of how this can happen, by progressive retentional inclusion:

> . . . in each of these retentions is included a continuity of retentional modifications . . . each retention is in itself a continuous modification which . . . bears in itself the heritage of the past in the form of a series of shadings. (sec. 11)

This is both a simple and a generatively powerful model, and does indeed offer a superficially plausible way in which we might have some sort of direct access to the past via a kind of retentional backlog. But there are certain difficulties with it.

First, precisely because the principle of retentional inclusion can be recursively applied without limit, there should be no limit to what can be perceived. But it is perfectly clear that even if we accept this terminology such extended retentions shade into recollection (and indeed into forgetfulness, of which more later) without warning. And yet this whole analysis rests on there not being shades of self-givenness. Retentional inclusion is a neat but a priori solution, but it cannot explain why it ever comes to an end, or falters.

Second, and more important still, there seems to be no scope, on this model, for understanding the role of other schematic devices for organizing time. A very simple example will suffice. If I listen to four bars of a song, retentional inclusion has either to ignore the intentional structure of the song (that is, minimally, that it is divided into four parts) or to take this into account. If it ignores it, it is blind; if it accommodates itself to it, it transcends the simplicity of its structure of progressive inclusion toward a recognition of the temporal structure of objects themselves, or of ways of *representing* their temporal structure that are *inseparable* from the ordinary ways of experiencing them, ways that could not be eliminated for being "representational" (and hence secondary, and so on) without eliminating the object itself.

Consider, for example, listening to a (regularly) repeated sensation, such as a heartbeat, a ticking clock, a throbbing pain. In this experience the regularity, the equality of the intervals between the "beats," which is a fact *about* the series of sensations, is equally part of the experience itself. It could be replied that, of course, retention does not deal with representational elements in a temporal sequence. My claim is that the kind of extended retentional constitutions to which Husserl alludes always require the scaffolding of types of temporal structuring that transcend retentional inclusion.

Husserl is confronted with an unpalatable choice: (a) either face the prospect of "perception" being narrowed down to a very restricted band of experiences, or possibly even becoming an unrealizable ideal; or (b) abandon the epistemological status of "perception" as, for example, offering certainty ("what I am retentionally aware of, we say, is absolutely certain"). (sec. 22)

If there are doubts about how much we could be said to "perceive" by retentional inclusion, it should not be thought that Husserl restricts our experience to this *perceptual* paradigm. Not at all. Recollection, even though it is not itself an original, self-given kind of experience, makes a particular positive contribution to the constitution of time. This must now be explored.

THE POSITIVE CONTRIBUTION OF RECOLLECTION

Reproductive consciousness, secondary remembrance, representational memory . . . in each case the question is simple: What account can be given of the relationship to primary, original consciousness of the past (in retention) such that recollection can play a role in the constitution of "time"? Why is this relationship problematic at all? Unlike retention, reproduction can offer no guarantees of accuracy. The order of succession may be misrepresented. Elements may be added or lost. In fact, in section 22, Husserl allows the possibility of transmitting the certainty of retention through to reproduction only under very restrictive conditions, namely, when we inwardly repeat what is still "fresh" or "vivid" in our memory. In this way, the "reproductive flow" can coincide with "a retentional one." This would seem to apply to a rather narrow range of cases. It might properly describe the strategies by which one tries to keep in mind a telephone number until pen and paper are to hand, but does it really apply any more widely? According to Husserl, it does. In section 14 it provides the basis for his account of the role of recollection in constituting the consciousness of duration and succession. His argument can be summarized as follows, focusing on his account of the constitution of succession.

Take the simple succession of two tones, *A* and *B*.

First, in addition to the experience of *A* and the experience of *B* (and their respective retentional trains), there is the experience of the succession *A,B*; this succession we *perceive* in an originary way (via *Retention*). Second, it is always possible to ("I can") repeat my consciousness of this succession, "presentify" it; that is an a priori fact about my freedom. Third, I can do this indefinitely, generating a succession of consciousnesses of succession, and I can even make any such sequence into one of the units of a similar sequence. Finally, in this way succession is constituted.

Husserl's presentation contains many subtleties not here reproduced. But I shall try now to clarify further the central line of the argument, remembering that it is the role of *recollection* in the constitution of succession that is in question.

Husserl begins, as we have seen, by claiming that consciousness of the succession of two tones (*A* and *B*) is "an original dator consciousness." I take it that is achieved in retention, or is a kind of "retentional consciousness." His question, then, if I understand him, is how is "succession as such" (and not this particular succession) constituted from this particular example, and his answer is that we have the capacity to reproduce the first constituted succession at will. His claim seems to be that the series of memories of this particular succession (and memories of memories of succession) generated by this capacity gives us succession freed from its particular original content (*A* and *B*). Again interpreting his remarks, he seems to be claiming that reproductive consciousness, by generating a succession of successions, allows us to focus on the form of succession itself, presented in the most appropriate way—successively.

Husserl's conclusion is most important for understanding the fundamental status of succession in the constitution of an enduring object.

> In the succession of like objects (identical as to content) which we are given only in succession, and never as coexisting, we have a peculiar coincidence in the unity of one consciousness. (p. 67)
> We have an interrelatedness which is not constituted in a relational mode of observation, and which is prior to all "comparison" and all

"thinking" as the necessary condition for all intuition of likeness and difference. (pp. 67–8)

Could this be read as saying that repetition precedes any consideration of identity or similarity?

RECOLLECTION AND THE PAST

As far as the "just past" is concerned, we have originary access to it by virtue of retention. But this does not suffice to accounu for the more remote past or the past in general. And it is here that recollection plays a vital constitutive role. The fact that recollection posits (as "having been" in this way or that) the now it reproduces, serves to distinguish it from fantasy. Recollection also "gives it a position with regard to the actual now and the sphere of the originary temporal field to which the recollection itself belongs." (sec. 23, p. 74)

Recollection can play a constitutive role in this regard because it itself takes the form of a flux of presentification, which is "a flux of phases of lived experiences constructed exactly like every other temporally constitutive flux." In other words, the phases of primal retention and experience of the now occur just as freely with presentifications as their content. So this flux constitutes itself as a unity, but it also constitutes as its object—its intentional object—"the unity of the remembered," the sphere of memory itself.

I have already discussed one of the most powerful ways in which this sphere of memory is constituted, via the reworking of the protentional horizons latent in what is recollected, in the light of what happened. This of course establishes nexuses within which particular past durations can be placed. And as these nexuses have as their ultimate context the whole of my conscious life, both past and still in flux, that sphere of memory is itself subject to modification. "Everything new reacts on the old . . ." (sec. 25, p. 77)

This account is of the utmost importance. It should be noted in particular that (1) it is in the multiphased flux of lived experience that constitution occurs; (2) this can as easily take reproductive modifications as its "content" as anything else; (3) this permits all manner of nonoriginary modes of consciousness to play constitutive roles, and is the vital way in which the limitations of retentional consciousness are superseded; (4) it is precisely the interaction between originary protentions as recollected, and the continual process of recollection that brings about the progressive knitting together of the unity of time-consciousness; and (5) there is a fifth claim which, while clearly important, is open to two somewhat different interpretations. Husserl's way of stressing the fact that the chain of reproduced objects is not a mere succession of associated intentions, is to talk of "the past as reproduced bear[ing] . . . an indeterminate intention toward a certain state of affairs in regard to the now . . . an intention which in itself is an intention toward the series of possible fulfillments." (sec. 25, p. 78)

Husserl is saying, I take it, that the past as actually reproduced at any one time is to be understood against a background of greater fulfillment, which is steadily realized. On this account his large-scale intention would be the ideal fulfillment (or otherwise) of all the protentional gaps in our memory, and hence the maximizing of the internal intentional integration of our past.

But his subsequent remarks leave us less sure. He writes:

> . . . this intention is a non-intuitive, an "empty" intention and its objectivity is the objective temporal series of events, this series being the dim surroundings of what is actually recollected. (sec. 25, p. 78)

and shortly thereafter:

> The component "unauthentic perception" which belongs to every transcendent perception as an essential element is a "complex" intention which can be fulfilled in nexuses of a definite kind, in nexuses of data." (ibid.).

This seems to be a rather different claim—namely, that what recollection actually intends is not just what I experienced, but what *happened*. That serves as the background against which what I can actually recall stands out. We posit "the past," which always in fact transcends what I will ever be able to recall or understand about it, but which could ideally be presented as "data."

If this is what Husserl is claiming here, it is an important instance of one of the transcendencies excluded in Section 1 reappearing as the object of an overarching intention.

RECOLLECTION AND THE CONSTITUTION OF TEMPORAL OBJECTS

This discussion of the positive constitutive contributions of recollection will be completed by an explanation of how Husserl treats its involvement in the constitution of temporal objects and objective time. This is dealt with in sections 30 to 32, but with enormously greater precision and clarity in appendix 4. I deal first with the constitution of temporal objects.

Husserl's claim is lucidity itself:

> The identity of temporal objects is a constitutive product of unity of certain possible coincidences of identifications of recollection. (p. 144)

Recollection has a feature that merely perceiving again lacks—the temporal object is quite identical on each occasion of recollection. Temporal objectivity can be established in subjective time-consciousness by virtue of the reidentification that recollection makes possible.

Recollection also has a feature that perception lacks—that while present time exhibits flux, and an always new now point, for recollection

> . . . every point exhibits an Objective temporal point which can be objectified again and again, and the interval of time is formed from purely Objective points and is itself identifiable again and again. (p. 144)

Husserl seems to claim that while the *unity* of a temporal object rests on "the series of primal impressions and continuous modifications" through which it is generated—either as something unchanging, or as a unity involving change—its *identity* rests on the possibility of reidentification, in which the same "enduring unity" simply changes in its "mode of [temporal] givenness."

RECOLLECTION AND THE CONSTITUTION OF TIME

But what is the relationship between the constitution of Objectivity in time, and the constitution of time itself? Husserl says they are not the same and that "the possibility of identification belongs to the constitution of time." (p. 145) But what further contribution does recollection make to this constitution? Husserl is clearest on this point in appendix 4. There he argues that the possibility of freely returning, via recollection, to an identical portion of time, over and over again, which applies in principle to every part of my past experience, *constitutes* that experience as occurring within an objective temporal field. The analogy is drawn with the parallel feature of an objective spatial field—one can return to parts one has been to before. Recollection functions as the condition of accessibility of the temporal field, and hence constitutes it *as* a field.

For a grasp of how Objective time is constituted as such—how, that is, we come to be able to think and have experiences for which we know a place on a single temporal continuum will always be assured—recollection is not enough. The idea of Objective time is inseparable from this a priori assurance of the locatability of any particular temporal object or duration. And in this concluding section 33 of the second part of the book, Husserl spells out some of the further intuitive certainties that are involved, each of which is immediately comprehensible to us when we think about temporal position. This section fulfills the promise of section 2. He discusses simultaneity, transitivity, the homogeneity of absolute time, and so forth.

But how is this account of the constitution of Objective time to be assessed? I would like to make two related observations.

TIME AND IDENTITY

Subtly, and somewhat without announcement, Husserl has moved from discussing internal time-consciousness to explaining the constitution of what we think of, or experience, as objective time. The key role in this shift has been played by the apparently idealizing capacities of recollection, and the transitional, if lengthy, emphasis on the constitution of temporal objects. These each sharpen a certain orientation toward the interpretation of time in the light of objects in time (including essentially temporally extended objects), that is, discrete unities in time, subject to identity—establishing recollection. His discussion of time is in effect the discussion of *the establishing and preservation of identity in and by time*, by which the phenomenology of time contributes to the project of "fundamental ontology" (the term is Heidegger's). Recollection deals with completed durations; immanent temporal objects endure within distinct temporal limits, are nameable, reidentifiable, and the bearers of predicates.[20] Husserl does of course mention both protention and expectation but, as I have already suggested, his most important discussion of protention is in the context of *recollection*.

But is it really fair to suggest that Husserl's account is limited by its concern with identity? Surely it is simply a result of trying to deal with the most obvious and difficult question time poses—How *can* there be stable identities given the flux of time? It is not a theory that ignores the fact of flux, it is precisely built on that fact. What can be said, however, is that there is little sense here of what might be called the negative side of time, the threat it poses to all constitution of identity, even as it sustains it. The concentration on the past is for its positive epistemological value. "Wesen ist was gewesen ist" as Hegel put it. This seems to leave for another occasion a serious discussion of the future, and those attitudes, emotions, orientations, and practices that require its openness.

THE IDEALIZATION OF TIME

What is to be said about the idealization involved in Husserl's descriptions? The first thing to say is that Husserl himself acknowledges it. In section 41 he writes,

> ... we operate with descriptions which already are in some respects idealizing fictions. It is a fiction for example that a sound endures completely unaltered ... (p. 113)

There is of course a *general* justification for idealization in any theoretical discipline, namely, that it brings out what is essential. Where would Euclidean geometry be without the fiction of straight, infinitely thin lines? But this analogy, so helpful in some ways, also sharpens doubt. The dropping of the assumption that all lines are straight, for instance, allows the development of the geometry of curved surfaces, for example. What if the same sorts of restrictions are still operating with time? Husserl, for example, seems to treat as legitimate idealization that everything we have perceived can be recalled at will, indefinitely, and without alteration. But this is not a description of experience, but the elucidation of a *model* of experience to which our actual experience *ideally* corresponds.

It should be possible to formulate a general rule to capture those cases in which this difference matters and those in which it does not. It would go something like this: idealization is permissible when those factors that are being excluded are of no significance to the matter at hand. So a map of the British Isles could, for most demographic purposes, square off the coastline quite a bit without affecting its capacity to communicate accurate information. But for an inshore yachting map it would be useless.

IDEALIZATION AS DISTORTION: A SPECIAL PRINCIPLE

A very general description can be given of a type of case in which idealization may be dangerous: whenever the phenomenon being dealt with is itself the product of a certain (variable) failure or success at idealization. An idealization that ignored this struggle, compromise, and so on would be critically misleading. Suppose I draw this O. I can say:

"I have drawn a circle." This is an idealization of my achievement. In many circumstances it would not be misleading. But suppose I had been trying to draw a more or less *perfect* circle. Then it would be more accurate to say: "I have *tried* to draw a circle." An idealizing act must ensure that it does not take as its subject matter something that has an intrinsic relationship to its own ideality, for it then risks covering up the importance of the gap between actuality and ideality. Idealization would make every attempt into a success.

How does this apply to time?

Husserl gives recollection a vital role to play in the constitution of objective time. But it is an *idealized* recollection, one that does not know the meaning of failure. Can it do more than constitute an ideal time? It is particularly curious that the form this idealization takes ("I can" [always return]) actually *presupposes* the very ideal objective temporality it constitutes. Does not the "I can" protend indefinite participation in the same future, connected to the past? Interestingly enough, though not noted by Husserl, this would make the past and indeed Objective time dependent on an idealization of the future. To be explicit, recollection presupposes a certain model of the temporal dimension which could be called transparency, accessibility, mobility, and preservation. Nothing is lost, everything is still available. But if that were not so? If *forgetting* was quite as common, and identical recollections always involved some element of fantasy? Would it not at the very least be necessary to consider the status of "Objective time," or at least Husserl's account of its constitution? Might it not (just) be a projective idealization?

The power of Husserl's account is that it is in many ways intuitively persuasive. No doubt the objectivity of "time" has something to do with our power to freely order, revert to, and reproduce past temporal points. But is it any more than an imaginary infinitizing of that power?

Husserl's orientation toward the "completed," the way his road to "objective time" takes him through temporal objects, has been noted. But it would be misleading to leave it there. For Husserl is led to posit not just "objective time" but an Absolute Flux, in which time, as a preobjective, preconstituted ground of any constitution, is discovered. I will consider this idea, and where it leads Husserl, shortly. It is the main concern of the third and last part of his book and of a number of appendices (6 and 13, for example).

THE PATH TO ABSOLUTE FLUX

The idea of the absolute flux of consciousness was not obvious to Husserl at the beginning of his lectures on time-consciousness. John Brough pinpoints its emergence in late 1906 or early 1907.[21] Husserl's position is presented by means of a contrast that had not previously been put to much work, that between the constituted and the constituting. And although he proceeds in a descriptive manner it is clear that an argument does underlie the move. It is that the time-consciousness and "objective time" I have previously been discussing have been the constituted products of a more basic constitutive level not yet unearthed. In section 32 he will call it "the absolute, temporally constitutive flux of consciousness." What is so interesting about it is that it differs so much from what I described as his previous orientation toward "completeness," "unity," and "identity," for it is understood as the constitutive ground of all such "products," and is not to be confused with them.

> . . . any object which is altered is lacking here, and inasmuch as in every process "something" proceeds, it is not a question here of process. There

is nothing here which is altered, and therefore it makes no sense to speak here of something that endures. (sec. 35, p. 99)

So, neither the categories of object or of process apply. As it is a flux, we want to talk about change, in some sense, but how?

> . . . we find necessarily and essentially a flux of continuous "alteration," and this alteration has *the absurd property* that it flows exactly as it flows and can flow neither "more swiftly" nor "more slowly." (emphasis added) (ibid.)

What is *absurd* is that one wants both to talk of time as a flux or flow (*Fluß*) and yet not talk about the *rate* of flow. But it would of course be even more absurd (if that is possible) if one were to talk of it flowing quickly or slowly.

"FOR ALL THIS, NAMES ARE LACKING"

Section 36, "The Temporally Constitutive Flux as Absolute Subjectivity," ends with the sentence, "For all this, names are lacking." Husserl has drawn the ultimate (and obvious) consequence from the priority of this flux to any constituted objectivity (coupled with the thesis that language and meaning are also *products* of such constitutive acts)—that the description of this flux puts a fundamental strain on language itself. The word "flux" itself must be treated as a metaphor. What is in question here is absolute subjectivity. He wants still to speak of "the lived experience of actuality," "primal source-point," a "continuity of moments of reverberation," but language is clearly a struggle.

> . . . temporally constitutive phenomena . . . are not individual objects . . . not individual processes, and terms which can be predicted of such processes cannot be meaningfully ascribed to them. Therefore, it can also make no sense to say of them (and with the same conceptual meaning) that they are in the now and have been previously, that they succeed one another temporally, or are simultaneous with respect to one another etc. (p. 100)

In other words, the language of temporal predicates belongs to the realm of the constituted and cannot properly apply at all to the constitutive ground itself. He writes that "we can only say that this flux is something which we name in conformity with what is constituted, but it is nothing temporally objective." I take this as saying that we can attribute to this flux such "properties" as are proper to it by virtue of its relation to what it constitutes, whatever they may turn out to be.

It is worth at this point explaining why Husserl calls this flux "absolute." What he means (see, for example, appendix 6, p. 153) is that the shape of the flux—the constant sinking-down from impression to retention—is not something that any change in circumstances could affect; it is not contingent. It could not suddenly be the case that the flux got stuck in an impressional phase, for example. "The variety of its phases can never cease and pass over into a self-continuing of ever-like phases." (p. 153)

But one has to ask what the relationship is between this absolute flux and the original discussion that centered on the diagram of time, of the various modes of temporal awareness. Has something quite new been discovered here, a new temporal depth hitherto obscured by the discussion of objective time? In fact, Husserl's thematization of absolute flux is a taking up again of the discussion of primal[22] impression and retention, the fundamental modes of temporal awareness, this time in the light of his attempt to delineate the constituted states of "immanent," that is, objective time (see section 39). With that behind him, the question of the temporality of the consciousness responsible for that constituting activity is posed again. Merely to talk about different temporal phases is not enough. Two problems in particular arise: If temporal predicates properly apply to constituted (immanent, objective) time, what can we say about the temporality of the flux? And surely every flux must be constituted a unity—but how?

What Husserl tries to do is to derive features of the flux from a discussion of its relation to and difference from immanent time. This most explicitly begins in sections 37 and 38.

ABSOLUTE FLUX AND IMMANENT TIME

We know (or at least we suppose we know!) that all objects and processes and events can be located within the framework of one immanent time. We know that within that time we can give sense to "simultaneity" for example. Does this concept, just to take an example, apply to the absolute flux?

Impressional consciousness is at any moment surely a plural affair. We are aware of many different sensory contents, and this awareness is equally "plural." Husserl refuses to talk here of simultaneity but instead of *Zusammen, Zugleich* (altogether, all at once):

> the primal sensations themselves are not simultaneous, and we cannot call the phases of the fluxional before-all-at-once [*Vorzugleich*] simultaneous phases of consciousness any more than we can call the succession of consciousness a temporal succession. (sec. 38, p. 104)

Husserl is making a distinction between different *levels* of predication. Standard temporal predicates, to repeat, are being restricted to constitu*ted* time. And yet we do want to talk about the/this/any flux as possessing unity of some sort. How could such unity arise?

The illusion that would treat this flux as a temporal series like any other is an illusion born of adopting a retrospective viewpoint on it. It then appears constituted like any other object of reflection. But if that were the case, one would still be left with the problem of the temporality of the *other* level at which it was constituted, and so on . . . unless, of course, some way could be found of avoiding this infinite regress. Husserl makes a formally traditional move, and claims, to Sartre's subsequent satisfaction, that the Absolute Flux is *self-constituting*.[23]

In section 24 he had invoked the idea of double intentionality (somewhat by analogy with the axes of two-dimensional space) of recollection. Here he uses the same term in relation to retention. The retentional phase both serves to constitute immanent temporal objects (for example, melodies) and thence objective time, via the "I can"; it is also self-constitutive in that these very retentive phases that make it up each refer "back" to a previous mode of consciousness (ultimately impressional) and "forward" to the next retentional one.

How should we describe this flux? Husserl talks of

> the *quasi-temporal* disposition of the phases of the flux, which ever and necessarily has the following now-point, the phase of actuality and the series of pre-actual and post-actual (of the not yet actual) phases. This *pre-phenomenal, pre-immanent* temporality is constituted intentionally as the form of temporally constitutive consciousness and in the latter itself. (p. 109)

As far as one can tell, through the intricate forest of his descriptive subtleties, Husserl is arguing that the modes of consciousness (impressional, retentional, retentions of retentions, and so forth) in themselves, and without considering the "objects" they intend, possess a kind of longitudinal or inner transverse intentionality by which they constitute themselves as a flux, indeed as an absolute flux, as they are not the work of any more fundamental agency. At a later point (appendix 8) Husserl writes of this unity as one of "lived experience" given to us by a shaft or ray (*Strahl*): of attention. He adds

> That this identifying is possible, that here an object is constituted, lies in the structure of lived experience, namely that every phase of the stream changes into retention "of," this again, and so on. (p. 158)

And:

> The flowing consists in the transition of every phase of the primordial field (therefore, of a linear continuum) through a retentional modifica-

tion of the same, only just past. (p. 158)[24]

The remark I would like to focus on in these sections is one I have not yet quoted. In section 38 he writes: "We can no longer speak of a time of the final constitutive consciousness" (p. 104) and in appendix 6 he writes, "Subjective time is constituted in an absolute timelessness" (p. 150)[25]

He does talk of *quasi-time* in one of his appendices. He calls it in section 39 "a one-dimensional quasi-temporal order," and a "pre-phenomenal, pre-immanent temporality." Standard temporal predicates apply only at the risk of confusion of levels. The word *Fluß* itself is an acknowledged metaphor.

There is something very obvious and also very strange about this discussion of absolute flux. That consciousness is a flux is both important and undeniable. What is special about Husserl's position is that he does not think of it, as many will, as a flux of contents (ideas and impressions running through one's head) but as a flux of *modes of consciousness*. He has drawn on the basic structure of intentionality: consciousness of/object I am conscious of, and to the level of the temporal object there corresponds an immanent time, while to that of consciousness itself corresponds the absolute flux. So one is not just dealing with a flux theory of consciousness, but one that reflects a particular model of conscious life (and of human life itself). It is not easy to know how to deal with this claim, but I shall venture the following comments.

Husserl claims that the idea of "unconscious contents" is an absurdity. Yet surely the idea of logic implied by such a notion of absurdity is itself one that operates at the level of the constituted. Does not the operation of its standards here perhaps involve a confusion of levels? (Did he not himself endorse the "absurdity" of "the flux flowing as it flows"?) If concepts like "succession" and "simultaneity" are inapplicable, is it not misleading even to talk of its unity? Is the claim that it is self-constituted simply a logical solution, or does it rest on phenomenological evidence? My concerns about the categorial commitments of analytical phenomenology surface again here.

Husserl's use of attentional rays as ways of revealing the longitudinal intentionality seems misguided. Surely what is essential to flux is its prelogical and (preunified) status, one that does not require such a "device" as self-constitution. (Is there any real difference between self-constituted and not constituted?)

The point of all this is that what the idea of Absolute Flux opens up is the possibility that the logical, the rational, that which conforms to the categories of what is objectively constituted (taking these to occupy a common field) might rest on something that exceeds these categories even more radically than Husserl allows. Has not Husserl perhaps peered over the edge of the brink and drawn back?

Husserl's discussion of Absolute Flux would suggest that if one pursues the concept of time to its "origin," it ceases, in important ways, to exhibit temporal properties. This conclusion is interesting enough in itself; what makes it all the more fascinating here is that one can find the same movement of thought, over a longer biographical period in Heidegger's work too, in the transition from his account of existential time (1927) to what he calls time-space (1962).[26] Even more tantalizing is the way in which Husserl's position here can be aligned with that of Derrida, who in denying "perception" is claiming, as was said at the beginning, that time is an essentially metaphysical concept, and who seems to distance himself so much from Husserl. It may be wondered whether the distance between Husserl and Derrida is not in part mirage.

3 Derrida's Reading of Husserl

In 1967 Derrida published a book entitled *La voix et le phénomène*[1] that some years later, and after various more ambitious and eccentric writings, he could still refer to as the work of which he was, from a philosophical point of view, most proud. Husserl scholars have not reacted too favorably to it, but it has had an enormous impact on the wider perception of the limits and indeed the very possibility of phenomenology. It is subtitled "Introduction to the Problem of Signs in Husserl's Phenomenology," and the "problem of signs" is the other face of what I have called the problem of intuition in Husserl. The direction of Derrida's reading[2] has already been anticipated by the discussion in chapter 2 and I have alluded to it on a number of occasions in the previous chapter. It has a direct impact on my understanding both of the status of Husserl's theory of time, and on Derrida's later insistence on time's inherently metaphysical constitution, which I dispute. I shall spell out in greater detail Derrida's argument in this book and its implications for our inquiry as a whole.

There is a simple, schematic way to understand what is going on in the book. Derrida is claiming that the entire project of Husserlian phenomenology rests on the possibility of a realm of meaning that precedes public language and hence can provide it with a ground. It rests on the possibility of an evidential purity from which all exteriority, all inductive, indicative relations have been excluded. For Derrida such moves are illustrative of a much more general philosophical privileging of the value of "presence," which at times he suggests is a value inherent in philosophy itself. He shows, in detail, how Husserl attempts, and fails, to bring about this exclusion in two crucial ways. He is unable to complete the separation of expressive from indicative signs; if the former are always "contaminated" with relations of indication, then it is the very possibility of a foundation of meaning that is excluded. Second, Derrida argues that Husserl's own theory of time-consciousness undermines the possibility of a "present" as an intentional unity free from all temporal alterity. And without such a temporal present, the evidential sense of "presence" collapses too. In each case, Derrida's

strategy is to treat Husserl's texts as struggles (to exclude, to purify) that can
be only temporarily suspended, and that the texts themselves allow one to
resuscitate. Every such text contains the seeds of its own deconstruction.

The guiding question of this whole book is whether Derrida's claim that
the concept of time is inherently metaphysical can be sustained. Toward that
end I am considering in detail the most obviously relevant writings of
Husserl and Heidegger, and Derrida's readings of these texts, to trace the
course of his thinking. I have already suggested (at the end of our reading of
Husserl) that Husserl's own text betrays something of a recognition that,
when pressed far enough, the concept of Time (and even time-consciousness)
dissolves. Derrida does not follow Husserl on this route, for it takes us to
something yet more fundamental (Absolute Flux), which, whether or not it
could actually perform any useful philosophical function, would perhaps
have succeeded in detaching itself from the value of presence. Instead, in *La
Voix et le phénomène*, Derrida substitutes one temporal complicity for another.
The fractured bonds between temporal and epistemological presence, between the intentional unity of the temporal present and the subject's own
immediacy to itself, are replaced by another intimacy, one in which it is the
values of absence, difference, and delay that inform. It is entirely appropriate
that the English translation—*Speech and Phenomena*—should include in the
same volume the essay "Différance," for Derrida's explication of that term
shows clearly how, in its dual aspect (of deferring and differing), it unites a
temporal and a diacritical sense. To those who find this fusing confusing, the
reply must surely be that it has to be seen as a response to the condensation
effect to be found in presence itself. Derrida's subsequent denial of the
possibility of a nonmetaphysical concept of time can then be seen to be a
direct consequence of the *strategic* significance of the term *"différance."*[3] It is
not introduced as a concept, but as more of a device. To try to deduce from it
a new concept of time would be to attempt to reappropriate a transgressive
discourse, which is to misunderstand it.

However, not all of Derrida is transgression, and he cannot himself
finally rule out the possibility of a new discourse of temporality, as will
become clear. And even here, in his discussion of Husserl, there is some sense
in which he is making straightforwardly different *claims about language*, and
about the pervasiveness of certain of its salient features (repetition, play, the
trace, differentiation, delayed effect, and so forth), which, at one level at least,
supplant Husserl's. Derrida goes further than Heidegger, for example, when
the latter claims that he is not offering a new theory of language, only a new
way of relating to it, even if Derrida clearly inherits something of that
evangelical aim.[4] It is open to question whether it is possible successfully to
displace any incumbent philosophical position without at least implying
certain positive propositions of one's own.[5]

A final attempt to formulate the problem posed by this reading for the understanding of temporality presented here might be this: On the one hand it could be said that Derrida accepts the dependence of signification on temporality but disagrees with Husserl on the nature of that temporality, substituting a temporality of difference, of nonpresence, and so forth. Or it could be said that he sees the very idea of a distinct account of time or temporality as in itself a symptom of a metaphysical valuation of "presence" that can be formulated atemporally—for example, the positing of what he calls a "transcendental signified"—a source of meaning that lies outside of and escapes the "play" of language. There is no doubt that there is a strong topological dimension to what one might playfully call Derrida's guiding intuitions and interpretive schemas. But the belief that a postmetaphysical account of temporality can rise again after Derrida must surely be encouraged by his inability to keep temporally loaded terms out of his analysis.

I shall now discuss in more detail the argument of *La voix et le phénoméne*, where the topology of inside/outside, of separation and exclusion, or more revealingly, perhaps, that of purity, immediately plays such an important part.[6]

In section 1 of his *Logische Untersuchungen* (*Logical Investigations*) Husserl distinguishes two different sign relations or functions—*Ausdrück* (expression) and *Anzeichen* (indication). And by virtue of its relation to an ideal meaning, Husserl sought to privilege expression as the essence of language. However, in all *actual* communicative use of signs expression and indication are entangled together, so Husserl turns to the case of pure expression to be found in "solitary mental life," from which both the world and the other person are in principle excluded.

For Derrida, the direction of this argument reveals Husserl's silent presuppositions and commitments in certain crucial ways. The privilege of expression is tied up both with the idea that *what* is expressed is an ideal meaning, and with the assumption that the natural medium for such expression is the voice. Husserl's acceptance that, while expression and indication are always entangled, they are still *essentially* distinct presupposes, says Derrida, a distinction between *de facto* and *de jure* that is itself ultimately dependent on a certain understanding of language. In fact, however, Derrida immediately shows precisely how Husserl escapes such a charge of circularity by the reference to solitary mental life. For if "there is" such a thing as solitary mental life, then the "ideal" separation of indication and expression will actually be achieved. The precise status of this dimension must therefore be investigated.

Knowing Derrida's overall verdict on Husserl—that his phenomenology is the most vigilant example of metaphysics—it is something of a proof both of his candor and of his subtlety that he offers us not simply a "critical" reading of the moves just outlined, but also a sympathetic one. Critically one might say that for all his insistence on presuppositionlessness, Husserl never raises the question of the sign to an *ontological* level. And it is by neglecting this that he can privilege that indicative function (*Hinzeigen*) that supports his primarily logical conception of language (of language as the bearer of ideal meanings). But equally, and against this, it could be said that by beginning not with the sign, but with an original fracturing of the unity of the sign (expression/indication) Husserl is showing exemplary vigilance. And are there not very good reasons for avoiding the question *"What is* a sign?"—for the premature attempt to subject the "sign" to the regime of truth might well occlude the possibility that signification, without itself "being true," might "condition the movement and concept of truth." (p. 25) Derrida rightly sees that the interweaving of these two motifs makes the final judgment of phenomenology very difficult. Are these accounts of the production, the constitution of truth, on the one hand, and of phenomenology as a philosophy of presence reconcilable?

Derrida's comments here are enormously important. The very same ambiguity will be discerned again in his discussions of Heidegger when he identifies both a renewal of the value of presence in Heidegger's concept of Being[7] and yet, insofar as Heidegger poses the *question* of Being (and, one might add, to bring out the parallel with Husserl more sharply, replaces "presence" with "presencing" in, for example, "The Anaximander Fragment"[8]), it also points in the other direction. And the formulation that "the activity of signification . . . although it has no truth in itself [might] condition the movement and concept of truth" will have an important bearing on our understanding of terms like "*différance*," "trace" and so forth, of which it can be said that they make meaning possible without themselves *having* meaning. And most important, Derrida's denial of a *primordial* time will be based on the same kind of consideration—that when one unearths what makes truth (meaning, time) possible one must leave behind the language applicable to the constitut*ed*. Later I will discuss whether Derrida can avoid the accusation that he renews transcendental thinking (see pp. 293–319).

Strangely, the generosity of Derrida's double reading is then suspended as he shows how Husserl attempts the *reduction* of "indication." This is achieved by linking the indication/expression relation to that between facts and essences, empirical/logical and so on. Perhaps the most interesting suggestion Derrida makes is that in the distinction between indication and expression is rooted the whole problematic of the reduction.

In his chapter "Meaning and Soliloquy," Derrida takes us through Husserl's argument for the claim that the possibility of expression without indication is demonstrated by solitary mental life. But if expression is to be thought of as an exteriorization (ex-pression), how can it survive the excision of the external world, of the world of communication? Derrida argues that Husserl's account of solitary mental life anticipates his later account of consciousness as a noetic/noematic structure, and that the inside/outside relation still preserved within solitary mental life is that between the act of expression and the ideal object it intends. The significance of the term "expression" for Husserl lies in its association with the voice (which has to be a silent "phenomenological" voice in solitary mental life) and with the idea of voluntary intention—a connection made much clearer in the French (in which *Bedeutung* becomes *vouloir dire*, literally, "wanting to say").

But the importance of this account of solitary mental life lies in the way it embodies the value of *self-presence*, an achievement the ideality of which is the product of all the exclusions (of the other, of the body, of the "sensible," and of the context of communication). In ordinary communication, the other has no immediate grasp of the meaning lying behind my words, and can at best perform acts of "analogical appresentation." My expressive acts must all be understood as "indications" of intentional states not directly available to the other. In solitary mental life, however, these problems are overcome. "The meaning is therefore *present to the self* in the life of a present that has not yet gone forth from itself into the world." In solitary mental life, mediation is replaced by immediacy.

Husserl quickly comes to see that what this requires is in fact the elimination of signs as such, and he comes to identify signs with indications as the nature of expression in solitary mental life becomes clearer. What role is there for words, then, in inner experience? The inner monologue that takes place in a living self-presence makes no use of *real* words, to which indicative traces would attach, but instead functions with imagined words, which are free from such implications.

The claimed necessary connection between communication and indication has the consequence that expression and meaning do not function in a communicative way at all in "solitary discourse." It is in his account of how Husserl explains the fact that we certainly seem to communicate to ourselves that Derrida inaugurates his first really subversive strategy.[9]

The argument in this short chapter ("Meaning and Representation") has momentous consequences, but it is surely defective, even if something of the same conclusion can be reached by other arguments (provided here). I shall first offer a summary of what is already a very dense section, and then offer some extended critical comments.

Derrida's response to Husserl's distinction between actually communicating to ourselves (which, in pure expression, we do not do, in his view) and representing to ourselves that we do so, is to argue that the structure of repetition is constitutive of signs in general, and hence for communication, and as it can be shown that "presence" is inwardly and already constituted by representation, the distinction between actually and apparently communicating to ourselves breaks down. The corruption of presence by representation will have "a whole chain of formidable consequences for phenomenology." (pp. 56–57) Moreover, he claims that the basis for these moves is to be found in Husserl's own writing.

For Husserl recognizes that signs in general operate within a structure of repetition (a unique sign is no sign). And so this must hold of *expressive* signs just as well. (At this point, Derrida seems to ignore his earlier recognition that Husserl in a sense sees the trap and begins to suggest that expressive signs might not be signs at all, in his phrase "signs, i.e. indications.") And yet repetition surely involves or implies or has as its basic element representation. The argument, in essence, is contained in the following lines:

> [Something] . . . can function as a sign, and in general as language, only if a formal identity enables it to be issued again and to be recognised. This identity is necessarily ideal. It thus necessarily implies representation: as *Vorstellung*, the locus of identity in general, as *Vergegenwartigung*, the possibility of reproductive repetition in general, and as *Representation*, insofar as each signifying event is a substitute (for the signified as well as for the ideal form of the signifier). (p. 50)

Surely, then, representation is *essential* to language, and no accidental addition or intrusion.

> Since this representative structure is signification itself, I cannot enter into an "effective discourse" without being involved in unlimited representation. (ibid.)

And yet when Husserl says that we only *take ourselves* to be or represent ourselves to be communicating to ourselves in solitary mental life, he is supposing that *actually* representation is excluded.

He *interprets* Husserl here as, in a very traditional way, trying to save presence and to reduce or derive (that is make derivative) the sign. (He is trying to ward off the consequence that the general repetitive structure of signification should apply to "expression.") For Derrida the position of the sign is paradoxical here. While the classical way of eliminating the sign (or reducing it, taming its dispersive power) is to make it derivative from a presence, if the sign then *has* that derived sense, the sign, is eliminated if it is made primary. This latter move, I take it, is his own, and its consequences can be seen in such a term as "trace" (which, he claims, is not a trace *of* anything).

Derrida is engaged here in a very general strategy of deconstruction, the first stage of which—that of reversal (making representation nonderivative)—has already been mentioned. The conclusion of this will be that "The presence-of-the-present is derived from repetition and not vice-versa." If Husserl never says this, Derrida claims it is consistent with both *The Phenomenology of Time-Consciousness* and the fifth of his *Cartesian Meditations*. (Of course part of the point of such a claim by Derrida is that while it may be consistent, it would take Husserl [or our understanding of Husserl] in an entirely different direction from the one intended.)

Husserl does link ideality with the possibility of infinite repetition, and this is one of the points on which Derrida fastens. It is critical because of the threefold ideality of the sign on Husserl's account (that of the sensible word, the meaning, and [sometimes] the object referred to). The question to be discussed shortly is whether that "linkage" can plausibly, let alone compellingly, be regarded as one in which repetition (or the possibility of repetition) constitutes ideality. But the version of the reversal just quoted mentioned the presence-of-the-present. How does the relation between ideality and repetition bear on the question of presence?

Ideality appears in presence in two forms: the ideal object that stands before us, and the ideal possibility of repetition afforded by a pure presence. Presence has the *value* and *privilege* it does because of the way it supplies the form for all experience, and because it too, as an ideal form, is infinitely repeatable. Not only does this apply to my present, it applies to *the* present as such. In valuing presence I value an ideal ground of evidence, of transcendental life, in which not only is the existence of the world accidental, but in which my own death is no longer an issue. If, with Derrida, the value of presence is seen to be tied up with the reduction of the sign, then it is clear that the reduction of the sign involves a tacit recognition of the possibility of my death.

Derrida expands this, in a Heideggerean vein, to a general claim about the relationship between a subject and his/her own death. While for Heidegger, Dasein is essentially a being-toward-death, Derrida's version is that if my being as a *subject* is understood to rest on this general relation to presence, and if this presence has a general sense, one that rises above the possibility of my personal disappearance or nonexistence, then each of us can say "I am = I am mortal."

I cannot here pursue the subtle way in which Derrida makes much the same moves with respect to both imagination and fiction as he has for representation, moves that again make use of Husserl's own phenomenological subtlety in dealing with these matters (see chapter 2) (" . . . the general distinction between the fictitious and effective uses of the sign is threatened.

The sign is originally weighted by fiction"). The consequence of such further moves is to allow Derrida to extend, or perhaps just to put in other words, his undermining of the opposition Husserl needs to draw between mediated and unmediated self-knowledge, one's true experience of oneself from ways of imagining or fictionalizing oneself and so on.

However, the immediately relevant conclusion of the chapter is that if representation lies at the heart of signs, then the idea of distinguishing true from merely represented (or imagined, or supposed) communication is suspect.

There are two obvious objections to this conclusion: that the foundational role he gives to repetition is *arbitrary*, and that the relationship this has to Husserl's question—about communication—is misconceived.

The *arbitrariness* objection would go something like this: Derrida correctly gleans from Husserl a relation between the ideality of a sign and its repetition. But his attempt to go further than Husserl involves a certain casualness in his understanding of the modality of repetition. And it involves an unjustified weighting of the relationship of interdependence of ideality and repetition. To expand on these points: Derrida's position on the modality of repetition seems implausible when compared to the argument Wittgenstein used (in his *Philosophical Investigations*) against the possibility of a logically private language. For Wittgenstein, to use a sign intelligibly, it must be possible for that sign to be repeated at other times and by other people.[10] That is a test for whether there is a rule governing its use, and with no rule, there is no meaning. For Derrida, the question of whether what is at issue is actual repetition or possible repetition is often unclear. He certainly usually refers to "repetition" without further qualification. But a relationship between ideality and this plain repetition is much harder to see. The reason for Derrida's vacillation here is surely obvious. The relationship between ideality and the *possibility of repetition* is hard to construe as constitutionally unidirectional. It is as easy to suppose that the *possibility of repetition* depends on ideality as vice versa. And yet clearly the interrelationship is very close. In the case of plain (*actual*) *repetition*, the relationship to ideality is harder to grasp, while it is easier to see how that reversal would make a difference to whatever the relationship is.

There is a historical parallel to Derrida's position here, that of Saussure's principle of (semiological) difference (to which, inter alia, Derrida refers in his essay "Différance"). Saussure's revolutionary gesture was to have insisted that the identity of signs be understood as derivative from the differences between signs (phonic, graphic, and so on); identity comes through differences. Language is a system of relations "with no positive terms."[11] Clearly there are difficulties in conceptualizing this, because no point can be thought of as fixed independently of the others (of which, in each case, the same is true). And one of the critical features of this notion is, of course, that although thought of as synchronic, these differences need never be copresent. This in itself would facilitate the transition to spatio-temporality of Derrida's "différance," but more modestly it gives one some insight into the virtuality underlying the idea of repetition. Repetition supplies the numerical difference on the basis of which ideal identity (that is, ideality) can be made to appear. Derrida's reversal of the relationship between ideality and repetition might be said to have completed in a temporal dimension what Saussure had begun from a synchronic perspective.

Suppose, for the sake of argument, that these worries about the modality of "repetition" can be overcome. The question remains as to whether we do actually communicate to ourselves, and what bearing Derrida's argument, were it successful, would have on that question. Husserl says we only think we communicate to ourselves. For Derrida, the distinction is inapplicable to language because it is confused *in* language. The distinction between actually speaking to ourselves ("effective" communication) and imagining or representing to ourselves that we are so doing could only apply if actual speaking were free from the structure of representation. But does the argument work? Can there not still be a difference between two things even if each presupposes the other? Would not Derrida's argument imply that a husband and wife could not in fact act independently because the concept of husband already presupposes a constitutive relationship with that of "wife"? Is not Derrida saying (wrongly) that we cannot distinguish between things that are constitutionally related? Suppose we agree that repeatability is essential to all signs, and that signs are the basis of communication. What that means is that other examples (tokens) of the same signs must equally be able to make sense. What it does not mean is that there is no difference between doing x and my supposing that I am doing x. The latter is compatible with my not doing x, the former is not. Surely Derrida confuses repetition and representation.

Is there perhaps a better explanation of Husserl's claim that we only think of ourselves as communicating? How about this: Husserl claims that there is *no point* in communicating to myself because it would involve telling myself what I was thinking, and to be able to do that I would already have to know what I was thinking, and so would not need telling. So I can only *suppose* that this is what I do. As I understand it, Husserl can mean only that we misdescribe inner speech if we think of it as communication. But he cannot mean that we perform an act of (false-) representation, because his very own argument would tell against such an act. It would be immediately obvious that we were doing no such thing if, as he believes, the evidence of noncommunication were right there in front of us. I take it that Husserl means that we unthinkingly describe inner speech as communication just because we think it must be communicative as it involves words. If this is so, Derrida has got Section 8 largely wrong. Does he not perhaps, for his own reasons, put too much weight on the literal sense of "stellt . . . vor"?

I have said that Derrida would on this account be confusing repetition and representation. The fact that we think of ourselves/suppose ourselves/ "represent ourselves" as communicating is not inconsistent with an inherent repetition involved in *any* sign (assuming this is conceded in some sense). Representation is an intentional act, repetition need not be. But there is an obvious Derridean response to this. Such a reference to an intentional act takes the very point in dispute for granted. The position being argued for here would be one of those Derrida describes as

> living in the effect—the assured consolidated, constituted effect of repetition and representation, of the difference which removes presence. (p. 51)

Derrida's point is that the idea of intention rests on what *in itself* lacks meaning, namely (the possibility of), repetition.

But could one still respond: yes, *repeatability* is indeed a condition of my intention being a *discrete* one (being this intention rather than *that*) and so it is a condition of my being able to suppose that I am communicating to myself that I am able to suppose so at other times too. But surely this does not invalidate the difference between communicating and merely *thinking that* one is communicating?

For the second relationship of representation or supposing is a "that" relation, which is not the case for repetition. First, the relationship involved in repetition is neutral with respect to whether the repetition actually occurs (the possibility will do), and second, there is consequently no constitutive, internal relationship between *actual* signs so repeated. Other *actual* signs would just be *proof* of repeatability and not in themselves required. Third, even if one acknowledges the dependence of representation on repetition, that is not a relationship of equivalence.

The proper thing to do with intentional acts is to apply to them the principle of identity through difference, to demonstrate their diacritical interdifferentiation. This surely does not mean that identity is undermined, but rather explicated.

What then can be concluded, about Derrida's position in this chapter? The relationship between repetition and the sign's identity or between representation and presence is not convincingly a *constitutive* one.[12] The interposing of "possibility of . . ." makes that clear, allowing the relationship to be understood in quite the reverse way. It may be of course that there are other arguments for Derrida's conclusion about the relation between presence and representation that would dispell the thought that the meaning of any sign rests on its relation to self-presence. Derrida again claims that Husserl himself provides them when one considers the question of self-presence from a temporal perspective, as shall shortly be seen.

In the language of sameness and otherness, I have discussed Derrida's attempt to show that representation lies at the heart of the sign's presence. The argument was found to be suggestive but not yet convincing. But Derrida offers another argument, already alluded to in my initial summary, for an intrinsic relation to otherness in the sameness of the present. Derrida argues, as I understand him, that even if Husserl does not make much of the connection between ideality and repetition, his constant invocation of the value of *presence* itself rests on its infinite ideal repetition—ideal because the present as such "concerns no determined being," is content-neutral. The present is, if you like, the ideal form of repetition. But again, one can ask, is such ideal repeatability *built in* to the meaning of the present, or is it just a consequence of its content-independence?

Derrida seems clearly enough to treat this ideal repeatability of the present as an idealization betraying a particular function it serves. He suggests that this ideal possibility of repetition is ultimately a "dissimulation of one's relationship with death," a dissimulation in that it hides the fact that this ideality is bought at the price of denying my mortality, which will bring the series of such "presences" to a close.

If so, then the ideality of "presence" itself would betray a concern that transcended it, one's awareness of one's mortality, of a possibility *other* than the actuality of the living present. The remark is surely suggestive, but hard to judge the truth of. One might rather consider *abandoning* this ideal content-independent present, and moving to one that takes its structure and dimensions from its content. Think of the difference in the "present" when listening to a complex chord sequence and a single tapping rhythm. If it were admitted that *some* "content," with contributions of its own to make to the shape and scope of the "present," were always to be taken account of, then one would have an argument for the *radical impurity* of the present—whether one called this radical impurity "representation" might depend on one's critical motives, but clearly for some "content" to lend shape to the present, that content has to be grasped as such. And that grasping might be called "representation." For a tune to shape time it must be experienced *as a tuneful sequence*. This "as" is surely the key to its being a case of representation. Would the same be true of the simplest beat, such as a heartbeat? Husserl, as was seen in the last chapter, is not sure about this, but the answer is pretty clearly yes. While one can clearly retain one beat while listening to another, the protending of a third, while itself not an act of induction, surely *rests on* some recognition that one is confronted with a series of beats such that this protention is motivated. Representation then, in Derrida's language, would not precede presence, but be part of it. The claim made in chapter 3 is being repeated and extended.

To sum up at this point: (a) I am not convinced that Derrida can treat repetition and the possibility of repetition as interchangeable. (b) I claim an "opacity of implication." Just because representation or repetition infects presence in itself, it does not mean that at another level one cannot distinguish between acting in a certain way and thinking one is acting in that way. The putative constitutional indebtedness of all intentional acts to preintentional conditions of repetition does not preclude a *further* relationship between such acts, or between acts and states of affairs, and so forth. Derrida wrongly treats repetitive constitution as a *leveling* procedure. (c) But in the critical question of the possibility of the evidential (and temporal) present, one can accept, via a somewhat different argument, Derrida's central conclusion, that the present is permeated by representation.

I shall argue (pp. 267–278; 335–360) that the consequence of this permeation by representation is not, as Derrida would have it, the collapse of any concept of time other than a metaphysical one, but rather the collapse of time (as presence) as a pure intuitional foundation. What is opened up is *a concept of time that theoretically embraces the necessary intrusion of representation.*

In the chapter "Signs and the Blink of an Eye," Derrida's more specific discussion of the temporal presuppositions of Husserl's theory of signification can be found. And it is vital to understand the setting of the theory of signs in evaluating Derrida's discussion of Husserl on time.

At the end of the previous chapter, Derrida had written:

> The self-presence of experience must be produced in the present taken as a now. And this is just what Husserl says: if "mental acts" are not announced to themselves through the intermediacy of a "Kundgabe," if they do not have to be informed about themselves through the intermediacy of indications, it is because they are "lived by us in the same instant" (im selben Augenblick). The present of self-presence would be as indivisible as the blink of an eye. (p. 59)

Derrida will argue that Husserl's argument is fundamentally threatened by an attack on the "punctuality of the instant, " and it is this that he engages in, apparently with Husserl's own help!

Derrida himself divides this chapter into three parts, preceded by this warning of danger:

> If the punctuality of the instant is a myth, a spatial or mechanical metaphor, an inherited metaphysical metaphor, an inherited meta-

physical concept, or all that at once, and if the present of self-presence is not *simple*, if it is constituted in a primordial and irreducible synthesis, then the whole of Husserl's argumentation is threatened in its very principle. (p. 61)

The three parts concern themselves respectively with (1) the importance of the now point to Husserl (and to philosophical thought in general); (2) (contrary to [1]) the important way in which Husserl's *Phenomenology of Internal Time-Consciousness* offers a more complex account of the present in which retention and protention are included; and (3) the claim that however much weight Husserl gives to the distinction between primary retention and secondary recollection, that is, representation, both of these, and indeed the ideality of presence itself, have as their common root "the possibility of repetition in general," or what he calls the structure of the trace. This would undermine the pure self-identity of presence.

Let us now pursue the argument in more detail.

First, the fundamental point, at the risk of repetition, is of course that "self-presence must be produced in the undivided unity of a temporal present so as to have nothing to reveal to itself by the agency of signs." (p. 60) Temporality constitutes a threat to this unity if its movement cannot be excluded from presence. Once temporal differences arise, the condition of immediacy and unity that would make signs unnecessary is lost, and the possibility of a prelinguistic immediacy seems to vanish.

But is Husserl committed to the idea of the present as a now? Surely the issues that Derrida is raising are precisely the ones that, as has already been shown at length, Husserl himself raises, and in the most sophisticated way. For Heidegger, too, hardly an easy thinker to please, Husserl's *Phenomenology of Internal Time-Consciousness* was a radical break with the traditional Aristotelian view of time, determined in accordance with such ideas as "now," "point," "limit," "circle." And does not Husserl enter into our study precisely as one who has attempted to distance himself from any such metaphysical conception of time?

Second, Derrida's blend of appreciation and critique amounts to what can best be called a double reading, one that untangles two conflicting motifs and tendencies in Husserl's work. The focus of Derrida's interest is what he insists is the ambiguity of Husserl's expansion of the now by reference to the concepts of retention and protention, which has already been discussed. The point instant may seem to have been breached, and yet each case of retention and protention is seen to share the evidential value of the original now point, and moreover is fundamentally attached to it. Derrida can be seen to be arguing (a) that the point instant is still the reference point from which retention and protention are subtended, and (b) that the evidential value associated with it is retained, if one can use that word. Proof that Husserl is still committed to a view that lies comfortably within the tradition can be seen in the impossibility of Husserl's accommodating, say, Freud's concept of the delayed effect, becoming conscious of a content that was never conscious.

Not for the first time, I have a certain sympathy with Derrida's attitude here, and indeed his conclusions, but his argument is surely wanting. Let us turn up the magnification and examine it more closely.

Derrida makes the critical point about Husserl's use of protention and retention in this way:

> . . . . the presence of the perceived present can appear as such only inasmuch as it is *continuously compounded* with a non-presence and non-perception, with . . . (retention and protention). (p. 64)

He admits, however, that Husserl says that actually retention is perception, "if we call perception the act in which all 'origination' lies." In retention "we see what is past." Here "perception" gives us a nonpresent. Derrida surmises that Husserl is at this point allowing himself to call retention "perception" because for Husserl the important distinction is between retention and reproduction. Derrida *seems* to try to discredit this assimilation by quoting Husserl's claim that there is "no mention here of a continuous accommodation of perception to its opposite." Now it is quite true that in Sections 16 and 17 of his *Phenomenology of Time-Consciousness* Husserl does talk earlier of the continual passing over from perception to nonperception, but the claim just quoted is taken by Derrida completely out of context. Husserl there is talking about *recollection*. It is in recollection that "there is no mention of a continuous accommodation of perception to its opposite." Husserl has not made some sort of slip in calling retention "perception."

The point is that Husserl is working with two different senses of perception, first as *impression*, and second as an act of "origination" (covering both retention and protention). Husserl usually uses "perception" in the first sense—as in the phrase about "passing over." As it happens, Derrida's conclusion still goes through, for however we characterize retention, it will be continuous with the impressional now, and that, for Derrida, implies that no rigid distinction between "now" and "not-now" can be made. This takes us to the point.

Third, assuming that Husserl is committed both to retention being continuous with the now, and also accessible only in a nonperceptual sense, Derrida then concludes that whatever their relative difference ("without reducing the abyss which may indeed separate retention from representation . . ."), retention and representation have more in common than might at first seem apparent. The distinction, he says, "is rather a difference between two modifications of non-perception." There may be phenomenological differences between the two, but this "only serves to separate two ways of relating to the irreducible nonpresence of another now." If the relation between the impressional now and the retentional consciousness into which it passes over is treated as a difference within presence, a relation to the Other within self-presence, one cannot simply say that the *sign* has entered into the structure of presence, for the classical understanding of the sign involves a relation to a pure presence, which has now been discredited. It is here that he suggests the term "trace" to capture that which cannot be called a sign, which lies at the heart of presence. If one can also see this relation to what is Other in the very ideal repeatability of the present, the present can be seen to have a primordial non-self-identity.

Derrida clearly thinks, in one sense, that he is only redescribing what Husserl himself could admit. But where Husserl sees the impressional now and retention as primitively unified, Derrida sees in this relation a primitive difference. And this, it is claimed, equally affects how we think of time. For one cannot continue to give the name "time" to this primitive movement to otherness, for that concept has always "designated a movement conceived in terms of the present and can mean nothing else."

Here, then, is one of the places in which Derrida is arguing for the conclusion taken as the starting point for this entire work. How plausible is it in this version? Surely not at all, for the argument is tendentious in the extreme. It supposes that the same/other opposition has some sort of automatic ascendancy over any other it happens to come across. Clearly Husserl notes a difference between the now and retention, but Derrida simply refuses to allow him his own account of the meaning of that difference, and, while acknowledging the radical difference between retention and representation, treats it as secondary to the same/other opposition. What Derrida should have discussed are the ways in which Husserl too describes the positive contribution of recollection (that is, representation) to consciousness. Furthermore, Derrida's assertion that "time" "has always designated a movement conceived in terms of the present" is surely importantly imprecise. There is a considerable difference between, say, thinking of time as essentially recuperative of the present, or of some value associated with the present, and time seen as the inevitable loss of such a present or such a value. And yet both are covered by the phrase "in terms of the present" ("à partir du présent"). And the latter is surely represented not only in popular thinking (think of the picture of Father Time with his scythe) but also in the history of philosophy (think of Hume, for whom one can at least say the future offers no guarantees of continuing identity, be it for things or selves). Finally, it should be repeated that one of the most puzzling things about Derrida's position here is knowing whether he is actually disagreeing with Husserl or just redescribing that on which they would both agree. Husserl is

clear that it is only because of the running-off of the now to retention that
there is any consciousness at all. Presence is then for Husserl a primitive flux.
Is not Derrida being obtuse in attributing to Husserl allegiance to the present
as now-point? It is even more puzzling that Derrida should want to reuse the
same sort of arguments (giving "primordiality" a value, while applying it to
"difference") as Husserl had used. There may be a rescuable strategic value
for (quasi-?) transcendental arguments, but there is a real danger that in
supposing one has achieved a reversal at some (fictional) transcendental
level, one neglects to notice the consequence one's position has at other
levels. As as been suggested before, the real question may well not be
whether we understand presence in terms of difference, but whether we can
adequately articulate the actual overlaying of structures of representation
onto any such primitive description of conscious life. Derrida still leaves
himself the problem of coming down the ladder, negotiating the passage from
the quasi-transcendental to the everyday. It may be that these remarks will
one day find their proper place in a positive revaluation of Derrida's relation
to phenomenology. Further material for such a study is to be found in his
account of the privilege of the voice for Husserl, where it seems Derrida is
offering us a phenomenology of the voice.

For Derrida, of course, the voice has a privilege in phenomenology, a
privilege that, he claims, leads Husserl to value the spoken word over the
written word, a privilege that is said to have its roots in the special relationship between speech, the voice, and *presence*.

In the chapter "The Voice That Keeps Silence" Derrida does not offer
us further evidence of Husserl's phonocentrism, but we are offered a description of speech that would make it plausible that a phenomenologist *would*
privilege speech. The immediacy with which we hear what we say and the
way speech fades into immateriality as soon as spoken could easily lead one
to consider speech as privileged. And if Husserl did not say it explicitly, there
are many classical references to the privileged connection of the voice and the
soul (for example, in Plato's *Phaedrus*). A lot more is at stake, however: the
relation between the *voice* and consciousness, the idea of autoaffection, and
the temporality of speech. I shall draw out only those relevant to the present
discussion.

Derrida has argued that for Husserl, absolute self-presence, phenomenological "silence" requires not only the exclusion of indication but even that
of expression insofar as that is outside that of "sense." It also involves the
exclusion of the relation to the other. What we are left with is an autoaffection, for which the key phenomenon is the "s'entendre parler," hearing
oneself speak. It is through this experience that we come to associate ideality
with sound.

> The signifier, animated by my breath and by the meaning-intention . . .
> is in absolute proximity to me. The living act, the life-giving act, the
> Lebendigkeit, which animates the body of the signifier and transforms it
> into a meaningful expression. The soul of language seems not to sepa-
> rate itself from itself, from its own self-presence. It does not risk death in
> the body of a signifier that is given over to the world. . . It can *show* the
> ideal object or ideal Bedeutung connected to it without venturing
> outside ideality, outside the interiority of self-present life. (p. 78)

The autoaffection of hearing oneself speak is credited by Derrida with being
the basis of a number of the metaphysical illusions of experience—especially
those associated with its apparent ideality. And he says of it much the same
as he said of the relation between retention and the now-point in the present,
that it constitutes a pure difference dividing self-presence. And the difficulty
noted before appears again with full vigor. This pure self-affection, with
which he associates différance (with an *a*), is credited with a whole range of
constitutive capacities, in particular it makes possible all those "things" we
thought it possible (and Husserl thought it necessary) to exclude from
autoaffection—including space, the outside, the world, the body and so forth.
The ontological significance of this notion is to be found in its primacy in
relation to all constituted identities, which has the double consequence that
it itself cannot be grasped "in its purity," and means that it does not belong
to the subject of which it is true; rather it produces the subject. Support for
the primitiveness of this temporal event, if it can be called that, is to be
found, as Derrida notes,[13] in Husserl's admission that properly speaking,
"names are lacking" for this primordial flux.[14] The justification of his
injection of the idea of "trace" into the living present, with its peculiar
"logic," is that we are dealing with a "movement" that precedes and
conditions presence, and hence identity, the possibility of identifying, and
hence naming, and even the intelligibility that would go with this.

> The living present springs forth out of its non-identity with itself and
> from the possibility of a retentional trace. It is always a trace. This trace
> cannot be thought out on the basis of a simple present whose life would
> be within itself; the self of the living present is primordially a trace. . . .
> Being-primordial must be thought on the basis of the trace, and not the
> reverse. (p. 85)

All this corresponds to a very traditional problematic. How can one
attribute to those conditions that are supposed to be conditions of possibility
for identity, sense, and so forth those properties themselves? And in the case
of time, the "movement" that would seem essential to any account of its

primordial form undermines the possibility of primordiality and either identity or chronological priority being compatible. One either speaks metaphorically, and thus risks misunderstanding, or one invents terms that are logically "indecidable." Derrida sees the very word "time" as what he calls "ontic metaphor."

> The word "time" itself, as it has always been understood in the history of metaphysics, is a metaphor which *at the same time* both indicates and dissimulates the "movement" of this auto-affection. (p. 85)

What is the difficulty with all this? Surely if the transcendental mode of thought has been discredited its language of constitution, production, primordiality, and so forth has to be abandoned. I shall argue later against the adequacy of the "sous rature" strategy.

This idea of autoaffection has important consequences for time, and it is these that have charted a way somewhat swiftly through the intricacies of this chapter. Now, however, Derrida will reassert his claims about the fundamentally metaphysical nature of time, and these will have to be considered again.

In the last few pages of chapter 6, Derrida brings to a focus the implications that his understanding of time as *différance* have for identity, ideality, presence, and the subject—in short, for the language of ontology. What Derrida does is to draw out of Husserl, with renewed force and determination, the recognition that temporality fundamentally undermines and does not sustain the idea of presence, and the very possibility of the interiority by which subjectivity has been traditionally thought possible. For Derrida, this will not only open up the present primordially to what is not present, but also open up the "inside" of subjectivity to the "outside" of the world.

I have previously suggested that it is in the primitive movement from impressional to retentional consciousness that the idea of difference inserts itself in the heart of self-presence. Derrida here argues that autoaffection (which he links to, without identifying with, difference) is to be found in impressional consciousness itself. And the evidence for this is to be found in Husserl's description of the primal impression as a "pure spontaneity," a pure production. The idea is that as such, it cannot be any sort of being, where "beings" are thought of as constitu*ted*. And it is in this light that the strange logic of "*différance*" and "trace" must be understood.

This move will be evaluated in more detail on pp. 293–318. For now it will suffice to put down a couple of markers. I shall be asking (a) Does not Derrida's problem arise from being initially seduced by the transcendental project? Might we not resist that move from the very beginning? Is there a way of avoiding it that equally avoids the naiveties of empiricism? Is there not *another way back*, should it prove inevitable? and (b) Is not Derrida converting a *tendency* (that may well exist in our understanding of language) into a limit of language itself? Is he not, in effect, working with a very restricted understanding of the capacity of language to convey ontological subtleties? To be specific, why could we not say "time" and recognize quite clearly that it was *no sort of thing?* Does not the word "temporality" already point in that direction?[15]

We said that the other consequence of time seen as *différance* is that it opens up the interiority of subjectivity to the "outside." This claim, if it can be plausibly maintained, would be an important response to those who accuse Derrida of a textual idealism (for example, for his claim that there is no "hors texte"). It is not clear whether Derrida offers a number of arguments for this, or just one in a number of different guises, but the suggestion that difference opens up subjectivity to the outside is not one that he convincingly sustains. He succeeds at most in showing that a certain spatiality can in a rather abstruse sense be located within subjectivity. And even the language has a disturbingly Hegelian ring to it. Derrida writes:

> Since the trace is the intimate relation of the living present with its outside, the openness upon exteriority in general, upon the sphere of what is not "one's own," etc. the temporalization of sense is, from the outset, a "spacing." (p. 86)

He continues:

> As soon as we admit spacing both as "interval" or as difference and as openness upon the outside, there can no longer be any absolute inside, for the "outside" has insinuated itself into the movement by which the inside of the nonspatial, which is called "time," appears, is constituted, is "presented." (ibid.)

At one level, Derrida is saying no more than he has already said about the structure or movement of difference lying at the heart of presence. At another, he seems to be trying to repeat the criticisms of the transcendental reduction previously offered by existentialists. But all his references to the exterior or to exteriority are in fact references to an exteriority *within* subjectivity. One need not conclude that there is an opening from subjectivity to the world, only within subjectivity itself.

It might perhaps be said that we are adopting a philosophically naive sense of "space." But Derrida's own notion of space, exteriority, externality seems itself to depend on *it* having undergone some sort of reduction. Suitably interiorized, it can then be wedded to temporality in a kind of primordial spatiotemporality. Derrida's indebtedness to phenomenological themes returns to haunt his solutions. It is hard to say whether he has undermined or simply refurbished transcendental thinking at this point. When he writes:

> As a relation between an inside and an outside in general, an existent and a nonexistent in general . . . temporalization is at once the very power and limit of phenomenological reduction. . . (p. 86)

one is tempted to respond that he is confusing structural analogy with some sort of constitutional dependence. Temporalization does indeed in some sense involve a relation between an existent and a nonexistent, an inside and an outside, but it can at best be said to *symbolize* the inside/outside relation involved in phenomenological reduction.

Derrida, it is true, refers, perhaps in elaboration of these remarks, to hearing oneself speak, in a way that might be thought to echo Merleau-Ponty:

> Hearing oneself speak is not the inwardness of an inside that is closed in upon itself; it is the irreducible openness in the inside; it is the eye and the world within speech. *Phenomenological reduction is a scene, a theatre stage.* (p. 86)

But the way out of interiority offered by such an experience is surely not most obviously via "the spacing in all temporalization," but rather via the embodiment of the speaking voice,[16] and the manifold ways in which the purity of that listening to oneself might get interrupted, distorted (from a sore throat, coughing, coping with eating at the same time, to various parapraxes of speech, in which the spoken word seems to have run forward, ahead of thought, or in which the monitoring has broken down, or in which alien words, or associations, have clearly crept in).

For the sake of formal completeness and because Derrida's focus here is still explicitly Husserl, and even though the issues will have to be discussed again, I offer in conclusion a discussion of his logic of supplementarity (in his chapter 7). However, despite, or perhaps because of, the wealth of its analyses, I shall select ruthlessly only those parts that directly contribute to Derrida's treatment of temporality. I shall put to one side Derrida's further discussion of Husserl's theory of language and concentrate on (1) the logic of supplementarity and its radically disruptive consequences for any under- standing of time as "succession"; and (2) the relation between time, pre- sence, and the end of metaphysics.

What, then, is "supplementarity"? The concept does not merely appear in this chapter. It plays an important analytical role in two chapters of *De la grammatologie*.[17] It is both a complex and a powerful concept. It is powerful in that it captures a structure of dependence that relates a number of Derrida's key concerns—speech and writing, expression and indication, to name but two. And it serves to summarize and stabilize the relationship in each case.

It is complex in that we have to jump the rails of our ordinary thinking to understand it. In particular, the kind of thinking we have to transcend is that which thinks of explanation as having the form of *tracing back to a first point*, which, unlike all the subsequent stages, will not be derivative, but will have some sort of fullness of presence. When Derrida talks of "original supplementarity" or "primordial supplementarity," what he is saying is that there is no such privileged first point. But there *is* something like a first operation of signification, a first movement—and that is supplementation.

As I understand it, Derrida derives the term from an innocent use made of it by Rousseau, "Languages are made to be spoken, writing serves only as a *supplement* to speech."[18] In this remark we have of course, on the Derridean thesis, the traditional appreciation of writing as an unproductive replication of speech.[19] For Derrida, of course, the problematic of speech and writing is subservient to the question of the sign. And indeed, at the beginning of this chapter he explains supplementation, again, with the same duality as before, as the primitive structure of signification, both completing and displacing presence.

Keeping for the moment with the traditional concept of the sign, what Derrida believes is that signs *do not* merely reflect pre-existing objectivities or meanings. Re-presentation is not a simple standing-for a preexisting presence. Signs give to whatever they signify whatever sense of presence they possess. Derrida even extends this idea of an addition that fulfills to the idea of "the for-itself or self-presence" (or self-consciousness) and concludes "by delayed reaction, a possibility produces that to which it is said to be added on."[20] In this respect there is an echo of earlier discussions of the "I" and its relation to death, and the rejection of "innocence." Death does not befall a preexisting "I," rather a relation to death constitutes the "I." But if this *completion* aspect of supplementation is important, one must not forget *substitution*.

The written word replaces, substitutes for, the spoken word in Rousseau. In Husserl, *indication* substitutes for *expression* in communicative discourse.

Derrida describes supplementation as an original structure of signification. What does that mean? I take it he means that it is *from* this structure that the metaphysical picture is derived, by a certain distorting transformation. One cannot, he claims, grasp the structure of supplementation "on the basis of consciousness," or within a traditional temporal framework. I assume he means that when he talks of "movements" he is not talking of conscious acts, and that this structure cannot be laid out as a succession of moves.

The link between supplementarity and *différance* is not hard to grasp. By *différance*, Derrida picks out the positive sense in which signs do not have their meaning by a relation to a presence, or, alternatively, that such a relation is an infinite *deferral*. And if supplementarity is seen to displace presence, it equally introduces into presence a play of differences.

How does this account of supplementarity threaten time? The claim must be both that it involves a contortion of the order of succession that no amount of fiddling round with will straighten out into something resembling a simple sequence, *and* that this matters. For the interrelations of a static *structure* might equally be said to resist a temporal interpretation, but unless it mattered to time one would think no more of it. The point here is surely that the logic of supplementation is such as to undermine any reference back to a primordial "presence" that would be seen as the origin of a meaning, or indeed of time itself. Supplementary logic says that the "first" is merely a shadow cast back by the second, so to speak, or that what "comes first" is the movement from first to second, that the fullness was not there from the start, but is created by what seems to point elsewhere. The idea of the supplement could be said to capture the thought that the value of origin may be an effect of desire, generated perhaps from the sense that temporal representations cannot be all there are, that time itself must have some underlying layer. Another name for this desire would be metaphysics.

At the end of Derrida's work, one can distinguish six motifs that have concerned him: (1) *presence* as the matrix of all the interdependent conceptions of phenomenology, (2) the interpretation of this system as a teleological structure, (3) the interpretation of the philosophy of presence as the attempt to eliminate *différance*, (4) the announcing of the closure of the history of being as presence, (5) the problem of how to carry on "beyond" absolute knowledge, and (6) the metaphor of the labyrinth, "which includes in itself its own exits." The scope of this set of topics makes our attempt to clarify the particular status of time and temporality somewhat difficult. The most obvious remark to make is that Derrida makes it unnecessary to demonstrate that the question of time is more than one question among the catalogue of philosophy's questions. Derrida's discussion of Husserl's theory of signification has traced that theory, and with it traced philosophy itself, to a primitive assertion of the temporal and evidential primacy of presence. Equally, if one is convinced that theories of time have all traditionally privileged presence, the undermining of that value undermines time. By "traditional theories of time," I refer to theories that, with the aid of "presence," dissimulate *différance*, repetition, the fact that there is no origin, there are no absolute point sources, and so forth. The very concept of "time," like all other concepts, is associated with moves that are essentially metaphysical. To try to change this involves changing all the rules of the game. The introduction of *"indécidables,"* the logic of supplementarity, perhaps also the strategy of *"sous rature,"* can all be seen as contributing to such a change in the rules.

But what repeatedly needs to be asked is whether Derrida is not overreacting to the specific claims of Husserl here. For Husserl, the bond between the temporal and the evidential present is indeed tight. Husserl's goals were foundational, and his method in part intuitive. But it would need more argument to show that philosophy in general had embraced these goals. Is there not a great danger in seeing Derrida as having invented a new positive vocabulary when in fact he has at best undermined another? Could one not leave behind all the talk of "fundamental *différance*" at the very same time as we leave behind the language of "presence"? Is not the way he explains his goal in *De la grammatologie*—to shake our complacency about presence—about right? Positive theory to replace it would be an error, at least at that level. Equally, it should not preclude, as it can seem to have done, the exploration of those overlaid temporal structures in which representation and the shadows of presencing it casts backwards cohabit, in which succession is one of many "orders" of temporality, one interwoven with delayed effects, anticipations, teleologies, and so on. One enormous danger of trying to transcend naivety is that one transcends its wealth too.

Apart from the questions of strategy and method to be picked up again later, a final word ought to be offered about Derrida's arguments, and the repeated claim in this chapter that they are inconclusive. In particular, his reversals have seemed at least twice to be arbitrary. Would it be sufficient to gloss their effectiveness as inconclusive in the following way: that in the face of a relationship taken *naturally* to operate in one direction, the plausible suggestion that it *even might* operate in the opposite direction has the effect of *actually* unsettling the naturalness of the original position? The suggestion that repetition might constitute presence, even if inconclusive, weakens the force of the originally assumed constitutive relation.

Part 3

Heidegger's Treatment of Time and Temporality: A Critical Analysis of Being and Time

1 "Intentionality": A Central Problem

First, a short bibliographical note. It was Heidegger who (somewhat per-functorily, Husserl is said to have thought) edited Husserl's lectures on time. His editor's foreword tells us very little. It is idle but fascinating to imagine what he would have written had he offered a substantial introduction to the lectures. There is, however, one remark that merits attention. After pointing out the central role of the concept of intentionality, and the intentionality of time-consciousness in the lectures, Heidegger goes on, in words that Husserl could formally assent to:

> Even today, this term "intentionality" is no all-explanatory word but one which designates a central *problem*. (*PITC*, p. 15)

Husserl could agree to this. He called himself a perpetual beginner in philosophy. Nothing was to cease to be problematic. And in many of his references to intentionality (for example, in section 84 of *Ideas*), Husserl identifies it with "a number of pervasive phenomenological structures" (to which a set of problems belong), or (in section 146) "the problem which in its scope covers phenomenology in its entirety." Its formula may be simple ("all consciousness is consciousness *of* something"), but nothing could be less simple than working out its meaning and implications.

But the historian of phenomenology cannot fail to read into Heidegger's words another suggestion—that "intentionality" is not just the name for a legitimate and central problem, but is itself, as they say, part of the problem, rather than part of the solution. Heidegger dates his editor's foreword April 1928, two years to the month after concluding the writing of *Being and Time*, if the date of its dedication to Husserl is anything to go by. But in that book there are only mentions of the word "intentional." The first appears when he is paraphrasing Scheler, the second is a little more interesting—another case of marginalizing in a footnote. This footnote is worth quoting:

> The thesis that all cognition has intuition as its goal has the temporal meaning than all cognising is making present. Whether every science, or

137

even philosophical cognition, aims at a making-present need not be decided here.

Husserl uses the expression "make present" (Gegenwartigen) in characterising sense-perception. . . . This temporal way of describing this phenomenon must have been suggested by the analysis of perception and intuition in general in terms of the idea of *intention*. That the intentionality of "consciousness" [sic] is *grounded* in the ecstatical temporality[1] of Dasein, and how this is the case, will be shown in the following Division [in fact unpublished]. (H363, n. 23)*

Here buried in a footnote Heidegger states *the problem* with intentionality. It is the pivot of epistemology, and the very project of epistemology is both grounded in and, with all this tunneling, subverted by Dasein's intrinsic ecstatic temporality, a temporality that cannot just be thought of as "making present." Heidegger's task in *Being and Time* is to spell out the nature and role of this ecstatic temporality by providing, via an analytic of Dasein, an ontological dimension that will undercut the concern shared by phenomenology and neo-Kantian philosophy with questions of knowledge.

In its simplest terms, Heidegger has to rescue time from merely being the servant of epistemology. In doing so, he takes a fateful step out of the framework he had already begun to trouble in his early essay "The Concept of Time in the Science of History."

The relation to Husserl is important. Heidegger adopts, and is sincere in modifying, the meaning of the terms "phenomenology," and his account of it occupies section 7, the longest section in his introduction.

Of course he has his sights not just on epistemology, but equally on that whole tradition of philosophy (including its so-called ontological versions) that suffers in different ways from the same deficiency. This failure is more than a problem of method, though it is reflected at that level. It is the failure to ask (and keep asking) the question of Being, that is, a failure to pose one's questions in the light of the thought of Being. It is in this context that Heidegger's interest in time arises. Time, he believes, will provide him with a horizon (*the* horizon, he claims) within which he can diagnose the various ways in which philosophers have fallen short by privileging the present, or indeed presence, without grasping its deeper grounding in Dasein's ecstatic temporality, and at the same time open up the possibility of thinking of Being itself. In *principle*, the question of Time seems subservient to the question of Being. If there were another way of reawakening the question of Being, one

---

* Throughout Part 3 all references to *Being and Time*, trans. Macquarrie and Robinson (Oxford: Blackwell, 1967), will use the German pagination that they kindly supply, designated by the prefix "H."

might avoid a consideration of time. But it soon becomes clear that the two questions are inseparable.

INTRODUCTORY SURVEY OF HEIDEGGER'S WRITING ON TIME

"Time" figures in the titles of many of Heidegger's writings and as one of the key foci of certain others. While a full treatment of the question of time in Heidegger would ultimately coincide with an exhaustive treatment of Heidegger himself, something less comprehensive will be attempted here. Nonetheless, it would be useful to take a brief look at some of the major places at which Heidegger has discussed time, or key aspects of time, following which a relationship may be discerned between the shifts in Heidegger's treatment of time, and his more general philosophical outlook.

THE EARLY ESSAY

In 1916 Heidegger published a short essay, one conforming in both its vocabulary and its focus of concern to a neo-Kantian framework, but one that points forward to the problematic of time in *Being and Time*. It was entitled "The Concept of Time in the Science of History."[2] The basic position Heidegger argues for in this essay is that the concept of time required for the study of history is quite different from the one required for natural science, in particular physics. Physics has theoretical goals ("to trace all phenomena back to the laws of motion of a general dynamics") for which a concept of time that will make measurement possible is required. That concept of time is of time as "a homogeneous ordered series of points, a scale, a parameter." History, he says in this essay, understood as a scientific, that is, methodical study, while not denying succession, has quite different goals:

> to represent the context of effect and development of objectifications of human life in its understandable peculiarity and uniqueness in so far as its relation to cultural values is concerned. (p. 8)

And consequently its concept of time is quite different. "Historical time periods differ qualitatively," he writes, and time in history does not have the homogeneous character found in science.

Now this essay could be said to confine itself to the epistemological level, concerned only with the conceptual requirements of different sciences, and different types of science. Is he doing more than repeating the distinction made by neo-Kantians, such as Rickert, between nomothetic and ideographic sciences? There are however one or two clues to a breakdown of any simply epistemological framework. He begins with a quotation from Eckhart ("Time is what changes and evolves; eternity remains simple")—hardly suggesting epistemological confinement, unless the confinement being considered should be that preceding birth. He goes on to discuss the limitations of the theory of knowledge in appreciating "the full meaning of the ultimaue question of philosophy" (without saying what that might be). Further, when discussing—in a very Nietzschean way—the importance of the interests of the present in interpreting the past, he makes a number of ontological moves, The past does not exist, he reminds us. It affects us in its "qualitative otherness." And it forces the historian into an *overcoming* of the gap between past and present. "Time must be overcome and one must live one's way through the temporal gap from the present to the past." The advocacy of a living through that overcomes time suggests a clearly existential dimension to what might have seemed simply a question of scientific conceptuality. And he makes two important foundationalist moves. He argues (though I am here actively construing a claim that is less clear in print) that the possibility of understanding the past rests on the fact that its "otherness" is such only in relation to the "objectifications of human life." As living, creating beings, we have access to a preobjectified realm from which alternative objectifications can be grasped. And finally, he makes a move that presages a move he will make in *Being and Time* in suggesting that there is a dependency of computational time on historical time, one to be found at the beginning of any time series. Such series-beginnings are always invested with meaning and value. Calendars are obvious examples. They begin with the founding of Rome, the birth of Christ, and so forth.

Fundamentally in this essay, Heidegger injects a little Nietzsche into a neo-Kantian approach to the logical structure of the different sciences. In doing so, the qualitative time of the study of history, and indeed the very business of reckoning time, are suggested, albeit in passing, to require an existential grounding and are certainly not reducible to any other sciences.

The general question of history will at this point be deferred. It would be much more fruitful to discuss the question in the light of Husserl's *Origin of Geometry* (1939), Heidegger's later discussion in *Being and Time*, Derrida's reading of Husserl, and Nietzsche's early essay on the *Uses and Disadvantages of History for Life* (1874). But it is worth pointing out, following a schema of bifurcation that could be allowed to recur endlessly, that the question of history is not merely one among many for Heidegger, it is not just a topic of special interest. Many of the claims he has made in this short essay will find a place in his general understanding of the *History of Being*, the setting for much of the thought of *Being and Time* which both discusses and enacts this overcoming of the past. It calls it "destruction" at one point, and "repetition" at another. But in this preliminary sketch we cannot extend these considerations further.

BEING AND TIME: THE ORIGINAL PLAN

Naturally, it is on *Being and Time* (1927) that our attention must focus not only for the analyses of time that it contains, but also for the perspective it provides for the various other of Heidegger's writings on time we shall consider. For these other writings can be seen, in various ways, to be the working out of the complete structure of the two-part treatise, never in fact officially completed. By "these other writings" I refer to *Kant and the Problem of Metaphysics* (1929), *The Basic Problems of Phenomenology* (1975, but given as a lecture course in 1927), and *Time and Being* (1962). In the first part of the whole treatise, the two published sections of which were entitled *Being and Time*, Heidegger offers us, in his own words:

The interpretation of Dasein in terms of temporality, and the explication of Time as the Transcendental Horizon for the Question of Being. (p. 7)

In fact in 1927 this explication of time as transcendental horizon for the question of Being was not offered. Indeed the subject is broached in that work only in the last paragraphs, and the last sentence of the book is a question, "Does time itself manifest itself as the horizon of Being?" In the schematic outline of the whole treatise (section 8, H39–40), this missing third section had a shorter title—simply, "3 ⊙. time and Being." One can suppose in a preliminary way some sort of fulfillment of this section's promise in the lecture *Time and Being* given in 1962.

Part 2 of the treatise was to have looked like this:

Part II:
1. Kant's problem of schematism and time, as a preliminary stage in a problematic of temporality;

2. the ontological foundation of Descartes' "cogito sum," and how the mediaeval ontology has been taken over into the problematic of the "res cogitans";

3. Aristotle's essay on time, as providing a way of discriminating the phenomenal basis and the limits of ancient ontology.

Kant and the Problem of Metaphysics, conceived[3] if not written in 1925–26, before *Being and Time*, if published only in 1929, could be thought of as the working out of the first section. One could treat part 1 of *The Basic Problems of Phenomenology*[4] as a way of working out the second section, substituting Kant for Descartes, and part 2 as the working out of his relation to Aristotle's view of time, a question to which Heidegger could devote only a footnote (his longest) in *Being and Time*.[5] (I discuss on pp. 269, 317 Derrida's long treatment of this footnote in "*Ousia* and *Gramme*."[6])

There are other works by Heidegger in which time is either discussed or plays an important role either as a subject in its own right or as presupposed by the questions being considered.[7] The only other writings considered in any detail here will be Heidegger's *Nietzsche* volumes[8] and *What Is Called Thinking?*[9] for his further thoughts about Nietzsche, and in particular Nietzsche's understanding of time.

This merely schematic assimilation of various other of Heidegger's writings to the plan laid down in *Being and Time* has value only to the extent that it serves to focus attention on the ways in which the project of *Being and Time* was not completed at the time of publication, and the various ways in which the problems encountered at that time are solved, or at least approached, in these other works. To say that they fulfill the promise of the original design is really mistaken, for in the end (for example, in *Time and Being*) that plan is put in question.

The discussion of Heidegger here will be guided by the following six questions. And I do not claim to resolve all these issues, only to have considered some of them seriously.

1. How successfully does *Being and Time* overcome the traditional metaphysical understanding of time?

2. In what ways do the various other writings on time successfully resolve deficiencies in that original account?

3. How does the shift in Heidegger's understanding of time correspond to his (and our) understanding of his more general "reversal," "change of emphasis," "turn," and so forth from *Being and Time* (1927) to *Time and Being* (1962)?

4. Is there any sense in which (a) there are *other ways* of proceeding beyond *Being and Time* (or even avoiding it) while endorsing the same goal (of rethinking metaphysics)? or (b) there remains what is *still* a metaphysical motif in the very movement that Heidegger makes? (The work of Derrida is obviously what we have in mind here.)

5. Suppose that the history of our understanding of time is the history of our inadequate ontological grasping of it (because ontology has always itself privileged *the present*), might it not yet be that time needs releasing from its rescuer, from its continuedly ontologico-transcendental framework of understanding? (Not only Derrida, but also Levinas and Nietzsche are relevant here.)

6. How unproblematic is the association of metaphysics with "philosophy of presence"?

TIME AND PRESENCE IN THE HISTORY OF PHILOSOPHY: HEIDEGGER'S GENERAL VIEW

The question of time as has been seen is rarely absent from Heidegger's writing in some form or another. The term itself figures in one of the handiest symmetrical frames by which the bulk of his work can be organized—the path from *Being and Time* (1927) to *Time and Being* (1962). Heidegger is one of a long line of philosophers for whom the question of time is not merely one of great philosophical interest as a topic for philosophical inquiry, but one that bears on the nature of that inquiry itself. This was in fact true for philosophers before Kant, but with Kant, and then Hegel, Nietzsche, and Husserl—to mention only the most obvious—this reflexive status of the question of time became thematic.

No philosopher who attempted to deal with the question of time without confronting the way in which temporal valuations, ways of thinking of and relating to time, are woven into the very practice of philosophy could any longer deserve to be taken seriously. If such a demand were posed as a test of philosophical seriousness, Heidegger would pass with flying colors. But an important question remains. Can the fate of our understanding of time be linked so closely with its place in our understanding of philosophy? Is it not possible that the power of this insight might swamp and occlude certain possibilities of understanding time itself, even as it opens them up? But what about this "time itself"? Does such a notion still make sense? Although it is somewhat difficult to recognize, Heidegger does not lose sight of the aim of restoring to time a sense free from metaphysical determination, and such accounts appear at each stage of his penetration into the question of Being. For with Heidegger, the engagement of time with philosophy itself appears in the form of the relation between time and Being, and in particular the question of Being.

For Heidegger, the importance of an investigation of time is derived from its role in our understanding of Being. The importance of the question of Being ("What is the meaning of Being?" is the shortest form of the question) is twofold. It plays a vital role in Heidegger's reading of the history of philosophy, and it also serves as the basis for awakening awareness of certain existential possibilities. For Heidegger, the importance of the question is that while philosophers have always philosophized in its shadow, they have posed this question to themselves inadequately, if at all. The reason for this is that a way of answering the question has become historically so canonized as to obscure the fact that the question ever existed. Heidegger attempts to reopen it in essentially three ways: first, by an existential analysis of that being for whom the question of Being, at least in a certain form, is still a live issue; second, by a reading (a "destruction," best read "de-struction") of the history of philosophy, drawing out the points at which Being is silently interpreted without that fact being acknowledged; and third, by meditation on the work of art, on the language of poetry, and on language itself. The consequence is not, in fact, a renewal of philosophy, not its revitalization, but the marking out of its limits, the recognition that philosophy is not so much limited by its interpretation of Being as constituted by that interpretion, and once one begins to think beyond or beneath that interpretion, one begins to leave philosophy (that is, metaphysics) behind to its own devices.

How is it that Being has always already been interpreted? In broad outline, Heidegger's answer never varies. The history of philosophy is the history of the interpretation of Being as presence (*Anwesenheit*), that is, in a particular limited and limiting temporal determination. Heidegger writes of

> the treatment of the meaning of Being as *parousia* or *ousia*, which signifies "presence" (Anwesenheit) in ontologico-temporal terms. A being is grasped in its being as "presence" (Anwesenheit)—this means that it is understood by reference to a determinate mode of time, the "present." (H25)

But what of time itself? What Heidegger says of St. Augustine applies to the interpretation of time in general.

> The essential nature of time is here conceived in the light of Being and, let us note it well, of a totally specific interpretation of "Being"—Being as being present. This interpretation of Being has been current so long that we regard it as self-evident.

And he goes on:

> Since in all metaphysics from the beginning of Western thought, Being means being present, Being, if it is to be thought in the highest instance, must be thought as pure presence, that is, as the presence that persists, the abiding present, the steadily standing "now." (*What Is Called Thinking?* p. 102)

These words, which date from the years 1951–52, complete the circle. Being is understood in terms of "presence" guided by a valuation of the temporal "present." When time itself comes to be reflected on explicitly, it is understood in relation to Being already determined as "presence." These words bring out sharply a number of important themes. Heidegger does not and cannot, as Husserl was wont to, treat self-evidence as evidence of truth. The reason he offers here is simply that self-evidence may just be the mask of habit. But there is another more important reason. The standard of self-evidence is itself the perfect embodiment of the value of presence. What is self-evident shows itself to me as it is now. To ask self-evidence to pass judgment on presence is to ask it to evaluate itself. The suggestion is not merely that this would be improper, but that this accounts for the self-sustaining nature of the interpretation of Being as presence. But there are other explanations for the self-evidence attached to the interpretation of Being as presence. In particular, when Heidegger writes of time "conceived" in the light of (a specific interpretation of) Being, and when he says that throughout metaphysics "Being means being present," he is in neither case describing a conscious choice among a set of possibilities. The concept of time was not made the object of a fundamental interrogation until the very end of Greek philosophy, and by then its association with the idea of "something present" (*ousia tis*) had already been established via Greek physics.

Clearly, if one wants to talk of the question of Being as "forgotten," no psychological sense of forgetting will do. Heidegger's project is a project of radical interpretation, uncovering not just what has been forgotten, but what has not been thought, or better, the unthought that lies in all thought. Collingwood's suggestion that we treat all philosophical propositions as answers to questions is helpful here. Heidegger's project is a revival of the question that, unknown to itself, metaphysics has consistently been trying to answer.

Presence and the present are each *determinations of* time. Heidegger's suggestion is that if we can gain access to the *temporal horizon* itself, within which such particular determinations of time, and thence of Being, occur, we will be able thereby to reopen the question of Being in a fresh way. In *Being and Time* Heidegger projects two different, though because Dasein is itself "historical," related ways of going out about this. First, as he puts it:

> The interpretation of Dasein in terms of temporality, and the explication of time as the transcendental horizon for the question of Being. (H39)

This, by and large, is what is offered in *Being and Time*. The second way is what we have called the radical interpretation or destruction of the history of metaphysics, with, as he puts it, "the problematic of temporality as our clue." With the exception of his brief treatments of Descartes and Hegel, this latter approach was not in fact carried out in the published version of *Being and Time* but, as has been seen, accomplished instead in later lectures and published works (*Kant and the Problem of Metaphysics, The Basic Problems of Phenomenology* [Kant, Aristotle], *Hegel's Concept of Experience*, "Plato's Doctrine of Truth," his two-volume work on *Nietzsche*, and so forth). Heidegger also shows that the question of Being can be raised in relation to the understanding of a work of art ("The Origin of the Work of Art"[10]), which, although unable to raise for itself the question of its own Being, as was true of Dasein, nonetheless works as a work of art, not by being a mere thing or being, but by "opening up a world." In his meditation on the language of poets and on poetic language, Heidegger tries yet a different way. In poetry, the referential confidence with which we ordinarily use language is broken. In poetry, language's openness to Being can be shown to be what is always at stake—certainly in the work of the great poets (Hölderlin, Trakl, Georg, Rilke). That poetry, which is always on the verge of silence, is a witness to what Heidegger will later call the "mystery." All along, while he has moved away from the project of describing the temporal horizon through an existential analytic, Heidegger has nonetheless tried to find a way of talking about the relation between man and Being that avoids the trap of supposing that either can be adequately identified separately. For such a move, even without the use of such labels, would lead back into the old opposition between subject and object, however subtly conceived. *Being and Time* had seemed to suggest that an authentic taking up of the temporality of human existence might be a task for each individual—a view inherited perhaps from Kierkegaard. The "later" Heidegger, however, influenced undoubtedly, as Vail suggests,[11] by his reading of Hölderlin and by a continuing and deepening confrontation with Nietzsche, now attempts to name this relation in a way that first reverses the direction of activity (with the "es gibt") and then moves away even from naming Being at all (with "ereignis"), naming with this word only the relationship between man and Being, seeing the terms as derivative from the relation itself.

In the course of Heidegger's thought, he moves away, as has been suggested, from the existential analytic of Dasein. It is often said that his attention moves from man to Being. And yet the advent of the "es gibt"[12] and "ereignis"[13] and the crossing out of the word "Being"[14] all point equally to his drifting away from Being.[15] Or rather, he drifts away from "man" and "Being" in the same way—that is, from the dangers of objectifying, conceptual, representational thought with which the very use of these words is fraught. Heidegger talks of man? He is offering us a new (existential) anthropology or humanism. Heidegger talks of Being? He has become a mystic, as we were warned he might. These responses are understandable, if mistaken. From the point of view of getting a grip on Heidegger's treatment or treatments of Time, these shifts are important. What Heidegger has at various time said publicly about the "Kehre" in his work will shortly be discussed. It remains true that if the later writings, and especially *Time and Being*, are to be seen as offering samples of what can properly be said about Time once one has thought through its previous metaphysical determination, then doubts about the value of Heidegger's path of thinking for understanding time do not need much encouragement. It becomes imperative to look carefully and closely at what is going on in *Being and Time* and to ask whether the interests of understanding Time are best served by the path Heidegger took after writing it, or whether that path was in fact followed in pursuit of a different problem.

BEYOND THE TRANSCENDENTAL: A SPECULATION

In discussing *Being and Time* a number of issues will be addressed. It will be asked what positive contributions it makes to the project of rescuing time from its metaphysical determination, whether it supports the idea that the concept of time is intrinsically metaphysical (which would mean that rescuing it from metaphysics would be like rescuing a fish from water), and whether or not Heidegger's treatment of time in this work exhausts the possibilities of understanding time available in the "everyday conception of time" or whether it in fact covers them over. And, as has been suggested, the possibility will be considered of questioning the way in which Heidegger develops his insights about time after *Being and Time*.

Consider again the words with which Heidegger actually describes what for the most part he is engaged in in this book:

> The Interpretation of Dasein in terms of temporality, and the explication of time as the transcendental horizon for the question of Being. (H39)

In fact, the second half of even this characterization of the published part 1 properly corresponds to the division he did not include (3. *time and Being*). The last sentences of the book put the completion of the task of the third division in some doubt:

> The existential-ontological constitution of Dasein's totality is grounded in temporality. Hence the ecstatical projection of Being must be made possible by some primordial way in which ecstatical temporality temporalizes. How is this mode of the temporalizing of temporality to be interpreted? Is there a way which leads from primordial time to the meaning of *Being*? Does *time* manifest itself as the horizon of *Being*? (H437)

With these closing sentences a life's work is foreshadowed. If the book, the torso, has been successful, it will have offered a rather special account of man—namely, of man as a being for whom his Being is in question—and will have shown that it is the temporality of his existence, in particular his way of taking up the fact of his finitude, that gives that Being its fundamental meaning. Before embarking on an analysis of the book, it is worth pausing to take note of the language Heidegger is using in those closing sentences. He talks of "grounding," of "constituting," of the "primordial," of "horizon," of "making possible," and of the "transcendental." Much of the interest of Heidegger's writing lies in the way in which he transforms the sense of these terms from their often Husserlian origin. And yet the question must remain—Do they not still represent a strongly neo-Kantian/Husserlian way of posing problems? Even if the method of analysis is different, is not Heidegger still wedded to a transcendental, foundational approach, one that ultimately derives its legitimacy from an idealism he cannot accept? This difficulty will not escape him, and much of the subsequent development of his thought after *Being and Time* will be an attempt to deal with it. This fundamental matter is mentioned now not in criticism, but simply to offer some focus to our reading. The question "Is there a more basic frame of reference or horizon or context within which something we think we already understand appears and derives its possibility of appearance and all its intelligibility" is an enormously productive one in the history of philosophy. But whether it *fully understands itself* is another matter. It often presents itself as a *vertical* principle. A certain surface makes more sense in relation to a deeper, hidden layer. And where the vertical dimension is missing—one might say that the concept of horizon explicitly tries to avoid it—the power of the analysis still seems to rest on an exclusive and unidirectional ordering of the levels of description involved. An account of the temporal horizon uniquely illuminates Dasein's existence. The relation is asymmetrical.

But what if, to use its own terminology, the possibility of such transcendental grounding itself rested on a certain doubling up, or superimposition of language, in which something is thought *in terms of* something else?[16] It is not simply a question of translation, but of transformation. The possibility of a transcendental grounding is discursively dependent on an operation of *mapping* of one discourse on to another, of the language of temporality on to that of existence. This may be true, but what is gained by such a formulation? The gain is this: there are no a priori restrictions on what mappings are possible, and no guarantees whatsoever that certain incestuous superimpositions and wild intermappings will not occur. This very intervention into the language of grounding, which treats relations of possibility, of ground, of constitution as special cases within a more general horizon of doubling up, superimposition, mapping, transformation, and so forth, demonstrates graphically the disturbing power of such mappings. By offering a *neutral way* of describing what is going on when one gives a transcendental grounding for a phenomenon, it allows us to focus attention on the ways in which claims about the privileged status of such a transformation are made. Here a return can be made to such terms as ground, constitution, and so forth. They can be treated as attempts to *order relations of mapping*.

Can such ordering be justified? Will it not end in theology? ("We have not abolished God if we still believe in grammar," wrote Nietzsche.) It may be suggested that the test of proper order is power to illuminate, the shedding of light. This candle must be grasped with both hands. For what is so interesting about illumination is that (a) it does not jealously proclaim its own uniqueness (it certainly *need* not), and (b) it does not require that one establish *vertical series* of deeper and deeper levels of profundity. Sidelighting is often far more effective. Shadows often reveal quite as much as they obscure. The only objection to all this would come from someone who held there was only *one* source of light, whether or not that was to be found in the *one* proper discourse.

PRELUDE TO THE LATER DISCUSSION OF HEIDEGGER

In the discussion of *Being and Time* that follows, the attempt will be made to show the possibility of (and need for) a double reading, in which tendencies toward unity and dispersion are interwoven. The concept of authenticity will provide an exemplary focus. In his later writing there is analogous tension. For on the one hand he reminds us frequently of the tentative nature of his writing, and yet what he consistently aims for is the "proper word," the final story, to gather up the light into a single source. But if accounts that offer transcendental groundings are treated essentially as exercises in the transformation of one discourse into another, of the mapping of one on to another, then the "vertical" flavor of a transcendental grounding would dissolve into a horizontal relation of transformability. What consequence would this have? There are two sorts of possible consequence. One could adopt a deliberately countertranscendental attitude to language, an openness (one that involves danger, risk, and so forth) to the phenomenon of superimposition, that is, of the undermining or illumination of one or more of one's present discourses by another one. Corresponding to Heidegger's talk of "where words break off,"[17] would be the experience of using words (or not using them) known to be vulnerable to engulfment by another discourse, and yet still. . . . Heidegger himself, after all, does not stop writing but continues to take the risk of losing his previous insights (previous discourses) by developing ones that he can superimpose on them, and that threaten to conceal them. There is a second more direct consequence for reading the later Heidegger. One could try to interpret the apparent orientation to the unique word as the very principle that generates constant revision. For it is unachievable! One of the names of this principle would be "desire." To that extent one could justify belief in truth as a regulative ideal, not because one thereby gets closer to it, but because it generates a multiplicity of texts, and in the succession of such texts, and the light generated by their constant displacement of one another, is the only truth.

All this, as must be clear, is somewhat speculative and programmatic, but it does give a sense of how the mapping principle might function critically. And this principle has a more direct application to this present treatment of time. For at a certain point it will be necessary to explain how it might be possible to transcend "ordinary temporality" without resorting to a more primordial level of primitive elements, and the mapping principle will play a significant role here. All this will of course have to be tied in with what Derrida calls deconstruction. And room will have to be found for the move from a valuation of "grand récits" to what Lyotard calls "petits récits" and his corollary principle of productivity.[18] Heidegger writes, for example:

> If we penetrate to the "source" ontologically, we do not come to things which are ontically obvious for the common understanding; but *the questionable character of everything obvious opens up*. (emphasis added) (H334)

What if one came to think of the true significance of fundamental ontology in a *purely consequential* way?

HEIDEGGER'S OWN INTRODUCTION TO *BEING AND TIME*

One of the most extraordinary features of *Being and Time* is its structure. After an introduction that justifies renewing the question of Being, provides a methodological perspective on the whole treatise (including the parts never published), and offers a radically new interpretation of phenomenology, the book is divided into two parts. The first half, which offers a "preparatory fundamental analysis of Dasein," postpones at every turn the question of the *temporal* dimension of the phenomena it deals with, and yet it interprets them in such a way that a temporal determination is what is most called for. This textual epoche of the time question is all the more fascinating because of the radical displacement of traditional (epistemological) accounts of our relation to the world that occurs in the first division. To accomplish such a displacement without reference to time or temporality, given the fundamental role that these will be seen to have in Heidegger's thought, is quite extraordinary. He is operating, as it were, with one hand tied behind his back.

What does this "preparatory fundamental analysis" achieve? At the beginning of Division 2, Heidegger sums it up, essentially, in four points: (1) Dasein is a Being-in-the-world whose "essential structure centers in disclosedness." (2) Taken structurally as a whole, Being in the world revealed itself as care. (3) *Dasein* "exists" in that it has "an understanding potentiality-for-Being." (4) Care has been seen to be concretely connected with Dasein's facticity and its falling. Heidegger's account of care (*Sorge*), on which we shall focus our attention, is the structural hinge of the book. One and the same structure of involvement is revealed by Angst (in Division 1) and by the anticipation of death (in Division 2); that structure is care.

In his preface to the seventh edition, Heidegger writes of *Being and Time* that ". . . the road it has taken remains even today a necessary one if our Dasein is to be stirred by the question of Being." Immediately after the page on which this sentence occurs is another, which begins by interrupting Plato's *Sophist* at the point at which the Stranger is talking.[19] Heidegger's point is that the meaning of Being is today *doubly* concealed, and that double concealment requires a philosophical strategy adapted to it. For not only do we not have an answer to the question of the meaning of Being, we do not even find the issue perplexing. We have to learn to ask the question before we can begin to answer it.

In the first part of the introduction ("The Necessity, Structure, and Priority of the Question of Being"), Heidegger begins by sketching out the historical scenario by which the question might be thought to be both forgotten and important. It is a version of the view that the history of philosophy is the history of footnotes in Plato. While Plato and Aristotle were in some sense aware of the questions, their successors are more interested in these philosophers' answers than in the space of questioning from which they arose. Heidegger dispels, or at least wards off, the obvious objections to the question. But how is one to begin to ask it? It is by reflection on this question that Heidegger formally at least justifies the whole endeavor of *Being and Time*. He begins by claiming that it befits what is claimed to be a *fundamental* question that it be made transparent in itself (and not merely asked casually). That means it must (minimally) be asked what a *question* is. His account is a structural one. A question has a topic (that which is asked about), an object (that which is interrogated), and a goal (that which is to be found out), and it is the behavior of a questioner, and thus characterizes the Being of that being, in that questioning is not itself a "thing" but the "how" of a thing. So if the question is to be asked at all, we must have some *grasp* of it. This he supposes we do have—albeit in the form of "a vague average understanding of Being"—evinced by our apparently unproblematic use of the parts of the verb "to be." ("The water *is* rough.") This "vague average understanding" is very far indeed from being determinate; for Heidegger its very unclarity is a sign that it is Being that we are dealing with. What he claims is clear is that the Being of things is not *itself* a thing but may nonetheless only be accessible through things.

But how do we then choose *which* particular entities to interrogate? When we pose this question of choice, however, and begin to consider the problem of different kinds of access to Being and the like, different modes of investigation, we are *already* discussing the Being of a particular entity— namely, we the questioners. So, Heidegger argues, if we are to make the question of Being structurally transparent, which *looked like* a mere methodo- logical preliminary, we have "first" to consider the Being of the questioner, of "Dasein" (". . . this entity which each of us is himself, and which includes inquiry as one of the possibilities of its Being . . ." H7). Is this not circular? His response is basically that phenomenological exhibition of fundamentals is not concerned with such formalistic objections.[20]

In the third section of his introduction ("The Ontological Priority of the Question of Being"), Heidegger asks what the point of this question might be. Initially his answer is very like Husserl's account of regional ontologies.[21] The discrete natural and human sciences in themselves are ontic disciplines that each require an *ontological* foundation—a fundamental clarification of their categories and concepts.[22]

But this ontological enterprise itself (and here Heidegger is clearly and directly marking his distance from Husserl) rests on a priori conditions, which it is the more general task of the question of Being to consider. So talk of the "ontological priority" of the question of Being is somewhat ambiguous, for at its *deepest* level, it has a priority *in respect* of being ontological.

It is in section 4, the final section on "the ontical priority of the question of Being," that the fullest and most convincing explanation of the place of the analytic of Dasein in the question of Being is to be found, one which, if it does not attribute "reflexivity" to Dasein in the old metaphysical sense, certainly does make all sorts of reflexive moves in the analysis itself. The key to Dasein's ontical priority, that is, its priority as an entity among other entities, is that it is *doubly inscribed by Being*. What is distinctive about Dasein is that *its very Being involves a relation to Being*, "its Being is an issue for it." The passage is critical:

> . . . its Being is an issue for it. But in that case, this is a constitutive state of Dasein's Being, and this implies that Dasein, in its Being, has a relationship toward that Being—a relationship which is itself one of Being. And this means further that there is some way in which Dasein understands itself in its Being, and that to some degree it does so explicitly. It is peculiar to this entity that with and through its Being, this Being is disclosed to it. *Understanding of Being is itself a definite characteristic of Dasein's Being.* Dasein is ontically distinctive in that it *is* ontological. (H12)

The word "disclosed" will turn out to be the important one. There is no question here of the kind of reflexive self-transparency in which a subject is aware of itself as a subject. Dasein is open to possibilities of self-understanding which are existential rather than conceptual. "*Existence*" is, as it was for Kierkegaard, the distinctive "determining character of Dasein," and in the brief remarks Heidegger makes here, the later theme of authenticity is adumbrated.

. . . in each case it has its Being to be, and has it as its own. . . . (ibid.)

Dasein always understands itself in terms of its existence—in terms of a possibility of itself: to be itself or not itself. Dasein has either chosen these possibilities itself, or got itself into them, or grown up in them already. (ibid.)

Only the particular Dasein decides its existence, whether it does so by taking hold or by neglecting. (ibid.)

What is clear from these remarks, which signal a significant intensification of language on Heidegger's part, is that when he says that for Dasein, its Being is an issue, he is talking not about a matter of mere curiosity or fascination for each of us, nor is he talking of choosing between a variety of possible life-styles. The level of possibility he indicates here concerns the stark alternatives between grasping and not grasping the distinctiveness of having the relation to Being that each Dasein has, the truth of (one's own) Being. The radical possibility he offers is to be (become) "what" we are, which means to gather ourselves up into our distinctiveness as beings that "exist."

Formally, the ontical priority of Dasein lies in its being ontological. But this "ontological" status affects not only its existence but also its relation to other entities. Heidegger's argument is that we are not metaphysical subjects but Beings-in-the-world, and as such have some sort of grasp of the Being of entities we encounter in that world. This gives the existential analytic of Dasein the role of providing (in itself) that "fundamental ontology from which all other ontologies can take their rise." This he calls Dasein's ontico-ontological priority.

All this means that if a choice has to be made about *which entity* to scrutinize in pursuing the question of Being, it will be Dasein. Heidegger, most interestingly, concludes by a further linking of existence with philosophical inquiry itself. If for Dasein (its) Being is an issue, philosophical inquiry is only a radicalization of existence. And for philosophy to be successfully pursued it must be "itself seized upon in an existentiell[23] manner as a possibility of the Being of each existing Dasein." Passive readers stop here!

The immediate importance of this introduction is threefold. It offers a compelling route back into thinking about the forgotten question. It makes clear the role that the analysis of Dasein is to play in *Being and Time*—that is, one subsidiary to the general question of the meaning of Being. And it announces many of the major themes to be taken up in the first division of *Being and Time*—Being-in-the-world, authenticity, and disclosure. But reflection on its movement reveals more. When John Sallis asks, "Where Does

Being and Time Begin?" he begins at the literal beginning—Heidegger's quote from Plato—but ends by reflecting on the real place at which Heidegger begins, the place he begins from, and begins with, and returns to: "the place of the disclosure of Being."[24]

In a later essay Heidegger wrote that his aim is not to "get anywhere, but for once to get to where we are already."[25] This could be called self-collection, or self-gathering, the thoughtful recovery of the ground of our Being. Such thinking has of course its own temporality: it is the temporality of the circle, the circling back of the movement of recovery. With brief references to Parmenides, Aristotle, and Aquinas, Heidegger gives this a historical dimension too. But temporality makes no thematic appearance in this introduction. And this is proof, if proof were needed, that Heidegger's central focus in this book is not time but Being. The horizontal functions of time and temporality are part of the solution, not the problem. Fundamentally they supply a way of describing the dimensionality of the "horizon" of what for Dasein is its "disclosure of Being."

The two halves of Heidegger's introduction anticipate, though not in any organized way, the division of the book itself. While the first half does not concern itself with temporal considerations, the second half bristles with them, at least in its first two sections, which can be provisionally[26] distinguished as existential and historical in their focus. In particular, in section 5, four different levels or regions can be distinguished: (1) temporality as the dimension of Dasein's Being by which the move from the everyday to the "authentically ontological" level of analysis is accomplished; (2) time as "the horizon for all understanding of Being," and in terms of the temporality of (1); (3) the ordinary understanding of time, which *also* stems from the temporality of (1); and (4) the *naive* ontological function time traditionally plays—chiefly in distinguishing temporal and nontemporal or supertemporal entities.

The very title of section 6, "The Task of Destroying the History of Ontology," embodies at the level of particular project what Heidegger argues is the general structure of historicality. Initially one could say that Dasein *is* its past rather than that it merely *has* a past. But this "is" needs articulating. For Heidegger, as for Nietzsche,[27] the past supplies the ways in which we understand ourselves, and it is in the light of tiese "possibilities of Being" that we project the future. It is this *necessary* historicality that makes possible the thematic study of history.[28]

These remarks, moreover, have direct reflexive application to the very project being engaged in here. For the possibility of thinking about the meaning of existence, or of Being in general, and indeed of the historicality of our Being, is itself characterized by historicality. And, in order that "by positively making the past our own, we may bring ourselves into full possession of the ownmost possibilities of such inquiry," we must concern ourselves explicitly (historiologically) with the history of the question of Being.

But "tradition," certainly when merely part of the taken-for-granted world, and even when recovered explicitly, all too easily conceals both the originating power of the past it delivers, and an understanding of the "most elementary conditions which would alone enable it to go back to the past in a positive manner and make it productively its own."

> If the question of Being is to have its own history made transparent, then this hardened tradition must be loosened up, and the concealments which it has brought about must be dissolved. We understand this task as one in which by taking *the question of Being as our clue*, we are to *destroy* the traditional content of ancient ontology until we arrive at those primordial experiences in which we achieved our first ways of determining the nature of Being—the ways which have guided us ever since. (H22)

But destruction is not negative:

> . . . we must on the contrary stake out the positive possibilities of that tradition, and this always means keeping it within its limits . . . to bury the past in nullity is not the purpose of this destruction; its aim is positive; its negative function remains unexpressed and indirect. (H22–23)

This introduction was designed as an introduction to *Being and Time* as projected in its entirely in section 8. (H39–40) Part 2 was to consist of "basic features of a phenomenological destruction of the history of ontology with the problematic of temporality as our clue," and it was to deal with Kant's doctrine of schematism and time,[29] with the unthought ontological foundations of Descartes's "cogito ergo sum" and with Aristotle's seminal essay on time.[30] In this section 6, Heidegger sketches out something of the line he will be taking (would have taken). Descartes, in sections 19 to 20, and Aristotle, in his longest footnote,[31] are returned to later in the foreshortened published form of the book, but further discussion of Kant's schematism has to wait for Heidegger's "thoughtful dialogue" with Kant in *Kant and the Problem of Metaphysics*.[32] And Hegel, who is not discussed in this section, does figure prominently later on in section 82.

Clearly his brief discussion of these three figures is meant as an example of the practice of destruction. But they are not randomly chosen. Heidegger maintains on many occasions—even where it verges on the implausible— that all of Western thinking on time depends on Aristotle. And he makes this claim here too. (H26) Moreover, the choice of Descartes is quite critical, and instructive. For despite the fact that Descartes clearly did reflect on the nature of the human subject, he did not, Heidegger claims, determine "the meaning of the Being of the 'sum'." Heidegger's diagnosis is essentially that Descartes unquestioningly inherits a scholastic/theological notion of substance, which, when interpreted as "needing no other being for its existence," obscures our Being-in-the-world. Descartes is the inheritor of the most ancient way of interpreting Being—as "presence." His valuation of certainty is the most obvious sign. Kant's importance lies in the fact that he more than glimpsed the significance of temporality. But he finally failed to raise the question of the *Being* of the subject in such a way as to successfully link time and the cogito. His chapter on the schematism in which he (Kant) writes of the "obscurities deep in the human soul" is treated as a monument to this (brilliant) failure.

These brief sketches of Heidegger's own (mere) indications are clearly of very limited value. They do, however, illustrate Heidegger's understanding of the history of philosophy. For Heidegger there are original and powerful formulations that as the congealed *products* of thought influence every successive generation and *prevent* certain fundamental questions from being even considered, let alone answered—particularly Aristotle's treatment of time. There are, correspondingly, blind receptions, failures to fully examine what one borrows from the tradition, and unconscious repetitions—Nietzsche writes of "an invisible spell"[33]—Descartes being a case in point, and there are revolutions that finally do not fully exploit their own insights, as was the case with Kant. Elsewhere[34] Heidegger insists that every new disclosure is accompanied by a further concealment, and *necessarily*. The midday sun that casts no shadow will always elude us. This suggests that philosophizing (and perhaps "theoretical" writing in general) has a general structure of disclosure/concealment, so that the "de-struction" of the history of ontology should indeed be thought of as exploring particular instances of this structure (or play) of disclosure/concealment in the light of the question of Being. This raises the question, to be returned to later, of whether Heidegger's own enterprise is contained within, or escapes from, this structure. At the end of section 6 Heidegger makes some tantalizingly ambiguous remarks:

> In any investigation in this field, where "the thing itself is deeply veiled" one must take pains not to overestimate the results. For in such an inquiry one is constantly compelled to face the possibility of disclosing *an* even more primordial and more universal horizon from which we may draw the answer to the question, "What is 'Being'?" We can discuss such possibilities seriously and with positive results only if the question of Being has been reawakened and we have arrived at a field where we can come to terms with it in a way that can be *controlled*. (emphasis added) (H26–27)

The ambiguity lies in the words underlined. Is it suggested that there is *one* or more primordial horizon that we are constantly on the verge of glimpsing, or that different, ever more primordial horizons can keep opening up? On the first interpretation, the name of that *one* horizon is temporality (locally, for Dasein) or time (more generally, for Being). On the second interpretation, the possibility arises that horizon-seeking thinking could even go beyond temporality and find new horizons. The reference in the last sentence to "coming to terms with the question of Being . . . in a way that can be controlled" may perhaps serve as a clue, all the better for being unintended. The ideal—"coming to terms with a question in a controlled way"—may *itself* constitute a *limit* of Heidegger's thought, beyond which . . . Heidegger's later writing certainly suggests that the ideal of *control* might well have faded. But where Derrida can explicitly celebrate this loss of control without handing over control to another responsible agent,[35] Heidegger seems to be prepared to release control only because of a faith that the ends it sought will be more surely realized by letting language speak. A *responsible* language. These questions will all be taken up later, and with specific reference to the problems of speaking (and writing) of time.

As has been said, these introductory sections (6 and 7) bristle with temporal considerations in a way absent from the earlier sections. Having discussed the historicality of Dasein and the way it makes possible Heidegger's destruction of the history of ontology, it is clear that no single separation can be made of the concerns of sections 5 and 6 into "existential" and "historical," and this provisional distinction must now be discarded. Historicity is just the name for existential temporality in so far as that precedes and grounds historiology. The structure of a future-projective past that historicity consists in is a specification of general features of Dasein's temporal existence.

The term "reflexivity" was used earlier in connection with the possibility that Heidegger's own writing might (and might necessarily) be characterized by its own "unthought." And this section brilliantly demonstrates the way considerations of method are self-applicable. Heidegger is perhaps right to think of this as a circle. And in connection with historicality, something like the same structure emerges. (The inquiry into historicality is itself historical.) If historicality is just (a dimension of) existential temporality, and if it is possible (by Heidegger's "destruction" or by some other means) to recover a sense of the horizonal significance of temporality, then what one must contemplate (and indeed do more than contemplate) with amazement is a kind of layering of superimposed temporalities. What Heidegger will call (anticipatory) resoluteness (*Entschlossenheit*) will exhibit just this intensity of overlaying. But what in such a set of layers is basic?

What is the position of everyday time in all this? Are representations of time, reflections on the course of time, to be excluded, integrated, or what? To answer these and other questions, it is now necessary to turn to the body of the text. The strategy will be to comment initially on those features of the first half that will figure most importantly in the temporalizing rerun of the second, developing, in particular, the hinge of the book, the concept of care.

2 The Existential Grounding

At the beginning of Division 2 in section 45, Heidegger reviews the story so far and issues in broad outline a prospectus for the future course of the book. He has already provided a "preparatory" account, but "our existential analysis of Dasein up till now cannot lay claim to primordiality." (H233) In particular, he questions the primordiality of his earlier account of care. He had at one point called it "the formally existential totality of Dasein's ontological structural whole," (H192) and more commonly Dasein's "primordial structural totality." But at the end of Division 1 he had asked, ". . . *has* the structural manifoldness which lies in this phenomenon presented us with the most primordial totality of factical Dasein's Being?" (H230) Division 2 is based on the recognition that this "structural totality" of care itself rests on another horizon.

> . . . The primordial ontological basis for Dasein's existentiality is *temporality*. In terms of temporality the articulated structural totality of Dasein's Being as care first becomes existentially intelligible. The Interpretation of the meaning of Dasein's Being cannot stop with this demonstration. The existential-temporal analysis of this entity needs to be confirmed concretely. We must go back and lay bare in their temporal meaning the ontological structures of Dasein which we have previously obtained. Everydayness reveals itself as a mode of temporality. (H234)

And where *Angst* had disclosed the "structural totality" of care, it will now be the anticipation of *Death* that provides care with a deeper, temporal dimension.

His discussion in Division 2 is clearly central, but the enormous *wealth* of analysis in Division 1 (cf. H334), and the role that plays in the structure of the book, means that some of this material must be selectively brought forward, even where it may already be very familiar to the reader, and at the risk of being accused of omitting certain vital considerations. Such selection will be guided by the basic question of the possibility of a postmetaphysical understanding of time.

THE DEFERMENT OF THE QUESTION OF TIME

The role of the existential analytic (Division 1) can be explained like this: There "is" no Being independently of man's understanding of Being (even if "understanding" is never the whole story). But man's understanding of Being is shrouded both by the practical concerns of everyday life, and by traditional philosophical interpretations of both man and the world. The pursuit of Being has to proceed via an account of man's "existence" that discerns both his *essential* relatedness to Being, and the various levels at which that relatedness itself can be lived and understood. In talking about "Dasein" rather than "man," Heidegger already marks out the essentially *ontological* orientation of his analysis. He *is* concerned with man, but only *as* Dasein, a particular kind of Being—Being-there—with a (problematic) relation to its own Being, with an openness to other Beings, and to the question of Being. Heidegger's method is hermeneutical, and phenomenological. It is phenomenological in the sense that it begins with what is most common and everyday—namely, Dasein's everydayness. It tries first to show how this always has an existential dimension, and then how this existential dimension is to be interpreted as a relation to Being.

A number of tasks are being concurrently carried out in this first half. Heidegger is describing Dasein's everydayness via his various "existentialia," he is transforming the framework of ordinary philosophical inquiry, and he is setting the scene for a temporal reinterpretation of this same material in the second half. The bracketing out of time from consideration in the first half has a number of explanations. Heidegger wants to be able to single *out* the temporal as a distinct necessary further *horizon* for the interpretation of Being. This is most effectively done by going as far as he can without it. (One is reminded of the way Paul Klee refused to use *color* until he had exhausted the possibilities of black and white.) But it could also be said that Heidegger's first half already had quite enough to do without introducing the complications of temporality. Heidegger's first half is engaged in a transposition, a radical reworking of a whole tradition independent of the "subject," "the self," the world, knowledge, our relation with others, the possibilities of self-realization, the nature of language and of truth. That this can be achieved, even in a preparatory way, without discussing temporality, when that is a perfectly obvious surface characteristic of human existence, quite apart from being the ultimate horizon of our Being, is quite extraordinary.

The preliminary explanation of how this was possible is that in the first part Heidegger is essentially manipulating the question of Dasein's "limit," "boundaries," "spaces," inner and outer relations, and so on, ways in which Dasein's identity *opens on to what might be thought to be other than it.* "Opening onto" is a radical undetermining of self-subsistence. It is quite extraordinary how Heidegger manages to take positive account of the relational aspects of human being while at the same time enhancing, and not diminishing, the sense of individual human possibility. And this "opening onto" is not just *a fact about* Dasein's Being, but the most luminous dimension of it (prior to any discussion of temporality). Another way of putting this would be to say that Dasein is a being for whom there are horizons—of significance, of potentiality for Being, of understanding, of disclosedness. When Heidegger talks of Dasein's "existence," it is surely an ex-sistence, a standing outside of itself, that "opens onto" such horizons. The reference to horizons, a term adopted from Husserl, converts the negative sense of opening on to "what . . . [is] other than it" into a positive feature. From the point of view of a Cartesian subject, for instance, the "world" is *oiher* than it—and not even "external" in any sense shared by the two relata—but ontologically alien. For Heidegger, man opens onto the world, Dasein is a Being-in-the-world. And a measure of the primordiality of this relatedness—that it does not leave its terms unmoved— can be found in his account of knowledge as *a* mode of Being-in-the-world, one *founded on* a more primitive involvement.

An Anglo-Saxon approach to such an account would be to talk of knowing-how as more primitive than knowing-that. Heidegger is talking about knowing-*that.* Knowing-how would be part of a more general orientation of equipment, things ready–to–hand. At the same time, Heidegger's account of the derivativeness of "knowing" is aimed very generally at the epistemological centeredness of most modern and much traditional philosophy, more specifically at Descartes, and indirectly at Husserl.

FROM HUSSERL TO HEIDEGGER

Whenever Heidegger discusses various possible ways of understanding "phenomenology" he makes the same move: phenomenology is fundamentally *ontology.* And it is no accident that in section 10 Heidegger has first reworked in his own terms the antipsychologistic arguments Husserl offered both in his *Logical Investigations* (1900), in *Philosophy as a Rigorous Science* (1911), and elsewhere. And while discussing Heidegger's implicit references to Husserl, it should be noted that the whole of his discussion of Being-in-the-world must be seen as a response to Husserl's insistence on the need for phenomenology to begin with a reduction that bracketed out the world. Heidegger accepts that we stand back from our practical involvement in the world, but only so as to articulate and interpret its significance, not to put it "in brackets" or "out of play."

One could also see Heidegger's whole discussion of Being-in-the-world (and of Being-in, Being-with, Being-there) as a radicalization of the principle of intentionality (that all consciousness is consciousness of something). The move from Husserl to Heidegger could be said to be the move from *of* to *in* (and *as*). With intentionality, *consciousness* is opened onto its objects; with the analytic of existence, consciousness itself is drawn back into its Being. This step back into a concern with Being, with the clearest but silent reference to Husserl, can be seen in section 11:

> we shall not get a genuine knowledge of essences simply by the syncretistic activity of the universal comparison and classification. Subjecting the manifold to tabulation does not ensure any actual understanding of what lies there before us as thus set in order. *If an ordering principle is genuine, it has its own content as a thing* (Sachgehalt), *which is never to be found by means of such an ordering, but is already presupposed in it.* So if one is to put various pictures of the world in order, one must have an explicit idea of the world as such. And if "world" itself is something constitutive for Dasein one must have an insight into Dasein's basic structures in order to treat the world-phenomenon conceptually. (H52)

This is surely an ontologizing expansion of the frame of reference in chapter 1 of *Ideas*.

The question of time is of course displaced. It could hardly be otherwise. And yet there is an interesting parallel in the place given to time in both Husserl's and Heidegger's respective thought. For Husserl, no less than for Heidegger, temporality is brought to bear on a scene that has already been described *without* reference to time (the realm of transcendental subjectivity, of epoche, of structures of intentionality). Husserl offers first a structural account of the space of intentionality, and then brings time in. And in *each* case it *then* becomes clear how inadequate the pretemporal account really was. One might also try to show that the way the scene was first set in each case determined the way time entered into it. For Husserl, time-consciousness is (initially at least) a form of intentionality, and for Heidegger, temporality is a horizon, an "opening-onto." And if that could be supported one might be thought to be paying a higher price than was realized for the methodological exclusion of the temporal from the first stage analysis, if it establishes the basic framework within which time can make its appearance.

OPENING ONTO THE WORLD

Heidegger makes a number of other important moves in Division 1. It has already been remarked that Heidegger reworks Husserl's antipsychologism. The form it takes in Heidegger is an emphasis on the error of taking the Being of things in the world (which he calls *Vorhandenheit*, present at hand) as the model for understanding Dasein. Man is not a thing. And yet the usual ways of showing this end up treating man as nonetheless possessing the same sort of being. Heidegger brings out the status of Being-in-the-world with reference to two obvious truths about Dasein—that Dasein *exists*—that is, bears itself in relation to its Being, and that for Dasein, it is in each case (for itself) "mine," and so grasped adequately or not (authentically or inauthentically). These are both, he claims, ways in which Dasein's being takes on a definite character, and as such, they must presuppose the domain in which such definite characters are possible—namely, Being-in-the-world. And the radical *difference* between this *in-relation* and that spatial inclusion relation found amidst ordinary things, "present at hand," is brought out by a list of ways in which we are in the world:

> having to do with something, producing something, attending to something and looking after it, making use of something, giving something up and letting it go, undertaking, accomplishing, evincing, interrogating, considering, discussing, determining . . . (H56)

Heidegger says all these are characterized by concern (*Besorgen*). Perhaps the word "*in-volvement*" (literally, being rolled up in) would help to preserve the "in." Concern here is an *existentiale*—that is a fundamental dimension of Dasein's Being—one that does not come and go, but rather has positive and deficient modes. The chief point is that this *involved relatedness* to something other is an *essential* intrinsic (*in*-trinsic) feature of Dasein's Being. And it serves as an example of the way in which the metaphysical conceptualization of subjectivity and selfhood will be undermined. *Being-in-the-world* is not something that an independently definable being does or has as an appendage, or an additional quality. Rather, Dasein is *exhausted* by such *existentials*, such ways in which it "opens onto . . ."

But what, then, is "the world?" Heidegger, like Wittgenstein,[1] draws a distinction between the world and any mere collection of things, including that collection by the name of nature. Again, it is Descartes's notion of "res extensa" that is the most obvious point of differentiation. For the *worldhood* of the world is a *phenomenon*, albeit one it is easy enough to "miss." For we tend to focus on things *in* the world, and the world itself is not a thing in the world.

But this *formal* point does not itself reveal the phenomenon of worldhood, even if it encourages its further pursuit. Heidegger's key move is to distinguish between two modes of Being of things encountered in the world—presence at hand (*Vorhandenheit*) and readiness–to–hand (*Zuhandenheit*). It is through a consideration of the latter kind of Being that the worldhood of the world will emerge. The best examples of Zuhandenheit are *tools, equipment, devices*. And yet we do not grasp them in the fullness of their Being simply by an observational encounter. Our most direct contact with things ready–to–hand is to *use* them. And when we reflect on the use to which we put them, we realize that such uses, functions, instrumentalities are significant only within a whole network of such functions. This nexus of "references and assignments," as he comes to call it, constitutes a background against which any particular function is set. When our tools break down, that referential context stands out. If I lock the keys inside the car, I cannot get into it, I cannot drive it, I cannot get to work on time, I cannot tell students what to read for next week, and so forth.

> The context of equipment is lit up, not as something never seen before, but as a totality constantly sighted beforehand in circumspection. With this totality, however, the world announces itself. (H75)

The "world"—and what is always at issue here is the world in its primary *everyday* sense—is the world of our significant involvement, a world structured by such relations as "toward-which," "for the sake of which," "in order to," and so on. It is in this sense of the world that Dasein is a Being-*in-the-world*. And it is such an understanding of the world that proceeds via the *Being* of things ready at hand, that allows our understanding of Dasein, at this everyday level, to escape reduction to the mere thinghood of the present-at-hand.

In the sections in which Heidegger compares his own account of the world with that of Descartes, he brings out the way Descartes inherits a scholastic and theological notion of Being as *substance*, one to be understood as linked to the idea of God the "ens perfectissimum," which "needs no other entity in order to be." Such a notion of Being as autonomous substance is extended to characterize things in the world that, if not relative to God, at least relative to one another, have this independence. It is just such an ontological autonomy and discreteness that Descartes relies on—but does not himself inquire into—in formulating the idea of a *res extensa*. And it is just such a notion that Heidegger characterizes as present-at-hand. Similarly, compared to Descartes's "thinking substance," Heidegger's Dasein radically displaces the idea of subject as substance, as autonomous source that, leaving aside the vertical relation to God, Descartes held man to be. The worldliness of Dasein is the opening out of the Being of "substance," a radical articulation of Being as ec-sistence.

THE TOPOLOGY OF SELFHOOD

It has been suggested that the key to Division 1 is Heidegger's unfolding of the "limits" of Dasein, the bestowal on man of horizons of disclosedness (and self-disclosedness) that explode the traditional space of its topological representation, that transfer the location and, indeed, locatability of the limits of Dasein's Being. Confirmation of this insight is found in chapter 4 ("Being in the world as Being-with and Being-one's-self. The 'they.'") A few short quotations will bring out something of the pattern of Heidegger's thought here:

> . . .others are encountered *environmentally*. This elemental worldly kind of encountering, *which belongs to Dasein and is closest to it*, goes so far that even one's *own* Dasein becomes something that it can itself proximally "come across" only *when it looks away* from "experience" and the "centre of its action." (H119)

Dasein in its oneness is declared decentered:

> Dasein finds "itself" proximally in *what* it does, uses, expects, avoids—in those things environmentally ready-to-hand with which it is proximally concerned. (ibid.)

> It could be that the "who" of everyday Dasein just is *not* the "I myself." (H115)

> Perhaps when Dasein addresses itself in the way which is closest to itself, it always says "I am this entity," and in the long run says this loudest when it is "not" this entity. (ibid.)

> By "others" we do not mean everyone else but me—those over against whom the "I" stands out. They are rather those from whom for the most part one does *not* distinguish oneself—those among whom one is too. (H118)

One of Heidegger's moves here is to undermine the privilege of the "near," and to argue, as he has often done, against the truth of the obvious, the "self-evident." Self-appropriation must make the detour of the worldly *articulation* of Dasein.[2]

The parallel with Hegel here should again not be missed. For Hegel, spirit has to make the "detour" of historical articulation. And to the extent that the parallel holds, one can ask of Heidegger the question that has often been put to Hegel: Does not the need for the detour presuppose the validity of the destination? It is quite *true* that self-understanding, self-knowledge, and self-appropriation are not available to immediacy. But is there not a danger that the articulation of modes of mediation will only reinforce the value of the ideal, rather than put it in question? The scheme of Heidegger's thought is surely such as to call for questioning. It is because everydayness is only the necessary *mediation* of our authenticity that the danger of losing oneself in it arises. But what if the very *value* of selfhood, and hence authenticity itself, were ultimately bound up with a philosophy of identity that after Heidegger we have come to call metaphysical?

The other move is in a sense in the opposite direction. What I take to be other ("others") is in an important way *not* the other at all, but part of, or a level of, my own existence. Again, the boundaries of Dasein are being transfigured. It is obvious that others are out there. What is less obvious is that at some level "I" am "out there" too. . .

These remarks are subject to a quasi-spatial articulation. It is, however, interesting that the same point can be, and is made in temporal language in this same chapter.

"The others" *already* are there with us in Being-in-the-world. (H116)

Not only is Being towards others an autonomous, irreducible relationship of Being: this relationship, as Being-with, is one which, with Dasein's Being, *already* is. (H125)

What has been called Dasein's articulation is not something that happens at some point to a worldless Dasein: it is always already . . .

SEEDS OF DOUBT ABOUT AUTHENTICITY

Heidegger's chapter on Being-with is the source of many important formulations. In particular, he develops the idea of the inauthentic "they"-self which identifies itself with Dasein as a "Being-with" and treats this not as the basis for developing either an authentic selfhood, or an authentic relationship to others (see authentic solicitude), but rather treats it as a limit. Inauthentically, Dasein does what "one" does. And yet Being-with in principle sets no such limits. To speak a public language is merely a necessary condition for poetry, not an obstacle to it. But it *is* a necessary condition. And it is vital to remind ourselves that authenticity cannot be simply *opposed* to everydayness, but only to that mode of Being that takes everydayness to be the standard. Close to the end of the chapter, Heidegger puts it like this:

Authentic Being-one's-self does not rest upon an exceptional condition of the subject, a condition that has been detached from the "they"; it is rather an existential modification of the "they"—of the "they" as an essential existentiale. (H130)

This general principle—that authenticity *modifies* everydayness rather than breaking with it altogether—will be returned to later when discussing what "authentic temporality" could consist in. Clearly it will be a *modification* of everyday time, not a *denial* of it. The idea of authenticity was already found in Heidegger's introduction—where he says that Dasein can either choose itself or lose itself. It is fashionable among some sophisticated readers of Heidegger to pass over the concept of authenticity. Doesn't Heidegger himself largely drop the term after *Being and Time*? But this passing over is a mistake. If there are difficulties with the term, they are difficulties that penetrate Heidegger's philosophy more generally. If Heidegger drops "eigentlichheit" one should not miss the return of the "eigen" in "er*eign*is." To leave aside the concept of authenticity would be to fail to confront what can be seen as a central issue for a modern reading of Heidegger—whether his thought aims at a phenomenological *restoration* of such values as identity, totality, and "presence." or, on the contrary, radically puts them in question. And there is a third position— which is that he makes this question in some way undecidable. These questions must not be lost sight of, but the question of authenticity proper must await the temporal dimension of Division 2. (See esp. pp. 183f; 221f) However, the question of how we judge Heidegger's treatment of time and temporality may well depend on just such a question about the "fundamental tendency" of Heidegger's thought. Is existential time just a way of taming "the time that destroys" by finding secular meaning in death? (See pp. 203–4.)

CARE AND DISCLOSEDNESS: INTIMATIONS OF TEMPORALITY

"Care"—what has been called the hinge of the book—is the subject of chapter 6. The question of Dasein's "wholeness"—a question that guides his discussion from the outset—is one which will be subjected to close scrutiny. But an essential final step in articulating Dasein's Being-in is to be found in Heidegger's discussion of *moods* and of *understanding*—two distinct modes of the existential constitution of Dasein's "there," that is, of Dasein's manner of Being as disclosure. Only the particularly salient features of these two analyses will be dwelt on: (1) the way "moods" can reveal Dasein "as a whole" in a special way, and (2) the proto-temporality of understanding.

"Moods," understood as ways of being attuned, and thus not just good or bad moods as we usually think of them, are revealing both for Dasein and for the philosopher. They are said to disclose ourselves as-a-whole to each of us in a way that no attempt at a comprehensive self-knowledge could do. ("The mood has already disclosed Being-in-the-world as a whole . . ." [H137].) Through moods we do not learn facts about ourselves, rather, Dasein learns "that it is and has to be." This is an *existential* disclosure of our thrownness. It achieves a kind of self-confrontation that knowledge cannot. Moreover, the philosopher reflects on the fact that Dasein turns away from its moods so that time is an essential concealment in this disclosure. Heidegger's analysis of Befindlichkeit is denser and at points seems overdone, even acknowledging the genuine revolutionary insight on which it is based. If they can be held apart, Heidegger adds to the disclosing of thrownness that states of mind also provide a "current disclosing" of "Being-in-the-world-as-a-whole" and opens up the possibility of circumspective concern, the awareness that things *matter* to us. As he will say about "being in the truth," the possibility of going wrong here is not an objection to the analysis. What is at stake is, if you like, the possibility of a certain *shape, modality,* or *character* of disclosedness, not the accuracy or otherwise of its content.

If Befindlichkeit is one, understanding is the other existential structure in which the Being of the "there" maintains itself. It could be said—although further discussion must be postponed until we consider the disclosive relation between Angst and Sorge—that Heidegger's account of Befindlichkeit offers a perfect example of the *problem* of judging whether the elaboration of an *existential* sense of being-a-whole allows him to escape the charge of taking the integrity of Dasein for granted, as a telos. The concept of de-severance (*Entfernung*) (discussed, for example, in his section 23) is another place at which this question can be raised. A decision on that question—should it prove possible—could give a firm direction to this interpretation of his general account of time. Although that question is strictly speaking excluded from this division, in Heidegger's discussion of Understanding (and interpretation and discourse) temporal structures are already pressing forward. What is so interesting about this eiscussion, of course, is that the subject of understanding has direct reflexive application to the method of phenomenology itself, and hence to the method being pursued in this book (including this very section on understanding). Something like the model of understanding he will offer the reader was simply laid out for acceptance right at the outset in his analysis of the structure of the question (of Being). The ideas of disclosure, phenomenology as "letting show itself," and the hermeneutic circle benignly conspire to allow this.

It has often been said that moral responsibility rests on one's ability to have acted otherwise than one did. Without freedom no praise or blame makes sense. We do not censure the hungry mosquito. Heidegger's account of authenticity, while not a moral concept, shares a structurally parallel relationship with the concepts of understanding and possibility, or being able to be (*Seinkonnen*). It is only because we are beings for whom both ordinary (ontic) possibilities and possibilities for Being are intrinsic features, that understanding and authenticity are possible.

The notion of "possibility" (being-able-to . . .) used here is quite central to our grasp of the way Heidegger's account of Dasein transcends that of any being merely present at hand, and it begins to make it clear how any idea of Dasein's other "presence," or living in the present, would have to be radically distinct from the presentness of, say, a stone, or a machine, or an ant. To be sure, a difference in man's way of being present has been clear to all philosophers who took memory, expectation, imagination, and self-consciousness seriously. The thought that one's Being could be adequately tied down to a single-dimensional bodily/perceptual being present at an instant is not one that withstands a moment's scrutiny. But the difficulty remains that the attempts to expound the way these "faculties" expand our "presence" have typically taken for granted the categories devised for our understanding of things. Heidegger's sustained effort in *Being and Time* is devoted to the attempt to avoid this. His discussion of "possibility" is a case in point. Certain *positive* features of Heidegger's concept—that possibility is *intrinsic* to Dasein, and that it *precedes* reflection on what its possibilities are—demand its clear separation from (a) empty *logical* possibilities, (b) mere *factual* possibility, and (c) Aristotelian *potentiality*. Understanding, in Heidegger's sense, has as its *object* "Being-possible"—or a being for whom *possibility* is an *intrinsic* feature. But of course for Heidegger there are at least two different ways of Being: authentic and inauthentic, as well as (ontically) various degrees of lucidity about our involvements with the world and relations with others. Understanding and possibility are really to be thought of here *not* as "achievement" or "success"—words—but as dimensions—existentials—that point to a manner in which Dasein is to be elucidated. Heidegger speaks later of their "transcendental 'generality.'" (H199) This point can be made over and over again in respect of Heidegger's existentials. The same must be said, for example, of both *disclosure* and *truth*—each is a dimension in which Dasein's Being is worked out, not the name for an achievement in itself. Heidegger's account is ontological in that he determines the dimensions within which subsequent ontic differentiations can be made.

He explains the relation between understanding and possibility in terms of "projection" (*Entwurf*). It is not clear whether anything essentially new is added by this concept. It has a range of "meanings" that deepen our grasp of the inherent *towardness* of Dasein's Being. As "thrown toward," it links up with "thrownness" in emphasizing that we are always *already* in this state, that it precedes any *explicit* forward planning. (In this combination of an "already" and a "forward" the [ultimately temporal] structure of care begins to be articulated.) Macquarrie and Robinson suggest the geometrical sense[3] (of which perhaps Mercator's projection of the surface of the globe is the most celebrated). "Projection" in this sense means the way Dasein's Being is *mapped onto* its possibilities.

Understanding in general is understood in terms of projection. And the notion of "projective understanding" is used by Heidegger as a way of rethinking the idea of "sight," which he links, at the same time, to "disclosure," and "clearedness." The force of this is to provide a way of deepening our understanding of the conditions on which *any* kind of seeing must rest. And this allows him, almost in passing, to mark out, as sharply as he possibly could, the basic difference between existential and Husserlian phenomenology. As was seen in the last chapter, Husserl's understanding of time is limited in principle (and so *for him* not limited at all) to what pure intuition reveals. For Heidegger both pure intuition (*Anschauen*) and the "intuition of essences" (*Wesensschau*) are *deprived* of their methodological priority by being "grounded in existential understanding." (H147)

Understanding is not, however, *itself* devoid of possibilities, which get developed in and by interpretation. What is tacit in understanding is made explicit in interpretation. But this is perhaps misleading. While *Being and Time* is itself a work of interpretation, it is not actually necessary for interpretation to take a linguistic form. For Heidegger the move from understanding to interpretation is one in which the "as" structure appears. And while this is clearly accomplished in *linguistically* explicit forms of interpretation, it can equally be accomplished by the move from an understanding engaged in the world to the level of "seeing-as." "The 'as' makes up the structure of the explicitness of something that is understood." (H149)

It must be admitted that it is in the end difficult precisely to locate the point at which the "as" structure, for Heidegger, begins. He offers an account of the kind of presuppositionless seeing on which science is often supposed to depend, which argues (as he has argued earlier when discussing the "present-at-hand") that such seeing is a very considerable achievement— being a privative *derivation* from seeing-as. And as such, of course, it does not have the primitive status it claims it needs. Many philosophers of science (for example, Popper, Hanson, Feyerabend) now agree that the foundationalism that inspires this appeal to a presuppositionless seeing is not only impossible but undesirable and (for Feyerabend) positively dangerous. The fact that they do *not* however, typically, call for an existential analytic to replace it does not mean that this is not called for.

The basis for this claim about the primacy of the "as-structure" is of course Heidegger's claim about the primacy of our involvement in a world of things ready at hand. Even if not every item in the world is something we *use* as a tool, everything is encountered in the light of its *significance* to us—even in "deficient modes." Things that *baffle* us may be very important. Things we find very significant may be repressed, and so forth. "An interpreration," Heidegger sums up, "is never a presuppositionless apprehending of something present to us." (H150) But Heidegger is prepared to go further and articulate the forestructure of the "as" of interpretation. His three levels have been translated as forehaving (*Vorhabe*), foresight (*Vorsicht*), and foreconception (*Vorgriff*), which seems to indicate that the various different levels of explicitness all cumulatively contribute to the forestructure of interpretation. It is not simply a matter here of "pre-*supposition*," but of an articulated array of levels of "alreadyness."

Heidegger goes on to defend the "circularity" of interpretation, which seems to deliver only what it has already taken for granted. Understanding *requires* the circle. There can be no question of trying to avoid the circle, rather what matters is "to come into it in the right way." (H153) For, "in the circle is hidden a positive possibility of the most primordial kind of knowing." (H153)

It is in this same pursuit of primordial existential grounding that Heidegger offers readings of "assertion" (as derived from interpretation and understanding) and of discourse—the latter, it turns out, being existentially on a par with state of mind and understanding. Discourse is understood as the significant articulation of the intelligibility of Being in the world. In the course of these accounts of assertion and discourse, the possibility of thinking of language as having a fundamental *logical* structure is ruled out by understanding *logos* itself as rooted in Dasein's existential analytic. What it suggests, and is meant to suggest, of course, is that *all* structures and all *representations* are to be so understood, including the structures of time.

Chapter 5 concludes with an account of our everyday being there—in such forms as idle talk (*Gerede*), curiosity, and ambiguity. These together point to an important aspect of Dasein's everyday Being—that Dasein is for the most part alongside and absorbed in its world. He calls this, quite neutrally, "falling" (*Verfallen*)—and distinguishes, very interestingly, a whole range of ways in which we hide from ourselves and flirt with the possibility of authenticity that fallen everydayness conceals (temptation, tranquilizing, alienation, and entanglement).

WHOLENESS AND METHOD

So far in this division, Heidegger has been spelling out in a reasonably ordered way the most general "existentials" of Dasein's Being-in-the-world, filling these in with more specific characteristics. But the moment of analysis is counterbalanced by one of synthesis, and Heidegger begins his chapter on *care* with section 35 headed "The Question of the Primordial Totality of Dasein's Structural Whole." The first sentence says it all: "Being-in-the-world is a structure which is primordially and constantly *whole*." Why does Heidegger put so much weight on this wholeness? It is, after all, an issue that repeatedly recurs in the book. A preliminary answer would be this: that Heidegger has ventured a radicalizing articulation of man, of the self (as

Dasein) that does away with any basis for unity there might previously have been, such as the permanence of a substance, or some cluster of features (grasped as present-at-hand). In opening man onto the world, Heidegger has risked *loss* of identity because he has dissolved the traditional limits (for example, simple self-presence). Moreover, in the account he has given, he has continually stressed that Dasein can exist in two different ways— inauthentically and authentically—and he has argued for "possibility" as an intrinsic feature of Dasein's being. At every turn, one might say, identity in any traditional sense is harder and harder to credit.

And yet, as the wealth of his analyses piles up, it is easy to see how the demand for an understanding of Dasein *as a whole* can arise. Through his analysis of care, and then Being-toward-death, Heidegger will supply an answer to the question of "the primordial totality of Dasein's structural whole." What has to be asked, however, is *on what ground* this projection of wholeness is based.

Heidegger is right to suppose that if Dasein is to exhibit a "structural wholeness," it must be of a somewhat original constitution. But that Dasein should have such wholeness, and what the source or point of that assumption might be, is still unclear. Heidegger assumes that as a work of interpretation he is merely bringing out structures latent in Dasein's (understanding) existence. But might these not be exigencies of interpretation quite *alien* to the ground from which they arise? To put the question another way: Might it not be that the demand for an account of Dasein in its wholeness is the demand that a certain sort of *account* makes—a certain rather traditional account perhaps—rather than a demand that actually articulates the existence it interprets? Were *that* so, one would be at least as much interested in investigating the existential basis of this demand (inauthentic turning away from finitude?) as in trying to satisfy it. Much of course will rest on an assessment of his account of death, but before that a series of questions have been outlined through which we can read his chapter on Care, which, with his famous disquisition on Truth in section 44, brings Division 1 to its end.

A clue to the nature of this demand comes early on in section 39: "If the existential analytic of Dasein is to retain clarity in principle as to its function in fundamental ontology, then in order to master its provisional task of exhibiting Dasein's Being, it must seek for one of the *most far-reaching* and *most primordial* possibilities of disclosure—one that lies in Dasein itself. The way of disclosure in which Dasein brings itself before itself must be such that in it Dasein becomes accessible as *simplified* in a certain manner. With what is thus disclosed, the structural totality of the Being we seek must then come to light in an elemental way."

So the *requirement* that Dasein's Being be able to be "simplified" is one that stems from the role Heidegger has given to the existential analytic in approaching "the question of the meaning of Being in general." There is no *a priori* guarantee that this can be accomplished. ("Can we succeed in grasping this structural whole of Dasein's everydayness in its totality?" [H181]) But given the premise that "an understanding of Being belongs to Dasein's ontological structure," it seems plausible to look for a privileged moment of such understanding. Heidegger's difficulty is that he has had to employ the scaffolding of the language of *structure* to organize the complex relationships among the various primordial features of Dasein's existence he has distinguished, and to prevent them from being a mere heap of insights. And yet he denies that this structural approach actually yields a totality in its own right. Why? " . . . the totality of the structural whole is not to be reached by building it up out of elements. For this we would need an architect's plan." (H181)

This is ambiguous. It could mean that as Dasein is not the product of design, its "totality" will not be able to be found by drawing up such a ground plan—this would be to confuse a *representation* of totality with the totality itself. Or it could mean that without such a plan we would never be able to make sense of the complex relations among the parts. It is pretty clear as he continues that Heidegger is opting for a version of the first alternative. A representation of Dasein as a structural whole is not an exhibition of his Being. What is needed is "a single primordially unitary phenomenon which is already in this whole in such a way that it provides the ontological foundation for each structural item in its structural possibility."

THE HINGE OF CARE

The simplification Heidegger seeks is found in care, which is revealed through Angst. Care, as will soon be apparent, has a triadic structure that waits only for the kiss of Division 2 for its temporal significance to be made explicit. The importance of this *structural condensation* of the existential analytic into the structure of Care via a fundamentally disclosive experience (*Angst*) is that it makes it possible for Heidegger already *within* the covers of *Being and Time* explicitly to move away from "the special task of an existentially a priori anthropology." (H183) The *structural condensation* of Care, achieved via the primordial disclosure of Angst, prepares the way for the absorption of that structure into a dimension that *precedes* all structure and representation, namely, temporality. Care itself is the hinge on which the thoughtful progress of the book turns.

Put very simply, the experience of Angst does not (as does fear) have an object in any ordinary sense; rather it "discloses, primordially and directly, the world as world." (H187) The importance of Angst for Heidegger lies in his ontological interpretation of it. It puts *in question* our Being in the world, renders us "unheimlich," forces us to face our potentiality-for-Being. If Angst reveals "Nothing," it is as "no-thing"—nothing in particular—that is, our Being in the world. And most important, Angst distinctively reveals to us our *individual* Being-in-the-world.

How does Angst play the role of revealing the structure of Care? Angst, we suppose, is a "single mood." But in his account of it, Heidegger, not implausibly, had displayed its nature on three axes:

> Angst as a state of mind is a way of Being-in-the-world; that in the face of which we have anxiety is thrown Being-in-the-world, that which we have anxiety about is our potentiality for Being-in-the-world. (H191)

This can be abbreviated further:

> The fundamental ontological characteristics of this entity are existentiality, facticity, and Being-fallen. (ibid.)

Shortly afterward, this gets *reexpanded* in a different formula:

> The being of Dasein means ahead-of-itself-Being-already-in (the world) as Being-alongside (entities encountered within the world). This Being fills in the signification of the term "Care" (Sorge). (H192)

Clearly any term that is to bear the weight carried by "Care" will have difficulty in shaking off all the "ontic" senses it brings to mind. In particular, despite the value he places on anxiety (*Angst*) he does not intend that "Care" be understood in such a way. Care is understood, as he says, "in a purely ontologico-existential manner." This does allow him to distinguish between our "care" for the different kinds of entities in the world—our "concern" for the ready at hand, and our "solicitude" for others. But it is hard to find any *ontic* paraphrase of the concept of care in Heidegger's discussion. As a way of emphasizing its ontological sense, this is clearly understandable, but it raises the question of why the word "care" should be used in the first place. Is "Care" just a convenient label for the structure (ahead/alongside/already . . .) already mentioned, or is that structure an *analysis* of the ontological *concept* of care?

It could be replied that the whole point of the discussion of Angst was to give us access to *care* itself as what Angst disclosively brings into focus. But Heidegger's *interpretation* of Angst is surely not conclusive. That it is possible to draw out the features he does draw out is clear. But whether they are the *only* features, or the ones that some *other* interest would have focused upon— that remains *far* from clear. The move from the interpretation of Angst to the structure of Care seems somewhat preordained. And when it turns out that the structure of Care can be given an all too easy *temporal* reading, this need not be thought of as "confirmation" of anything except Heidegger's own prescience.

The second difficulty that arises out of trying to understand the meaning of Care while maintaining its *ontological* significance, and radically eschewing its ordinary ontic significance, is that this is surely an *impossible* task. The word is *chosen* because of its everyday meaning (with the etymological connection between Sorge/Besorge helping). Does not this suggest a radical inverse dependence of the ontological on the ontical, one that Heidegger would find disturbing?

In *Being and Time* it seems that Heidegger is content to keep the senses apart rather than to pursue the difficulties involved in articulating their interrelationship. Elsewhere, however (for example, in *Letter on Humanism* and his essay *Language*), he explicitly refuses to describe his use of certain *apparently* ontic terms as metaphorical (language is the *house* of Being, the *flower* of the mouth). The suggestion is that the relationship of one-way *transference* of meaning (from ontic to ontological) belies the fact that the ordinary *ontic* meaning of a word may well be open to a *renewal*, or revitaliz- ing, one that an ontological transpositon can provide. Heidegger suggests that understanding language as "the house of Being" might deepen our sense of "house," "dwelling," and the like. Another way of putting this would be to say that language has an ontological dimension even when being used ontically, even if that may often be *unapparent*, concealed. The difficulty with this position is that the possibility of a retroactive illumination of a literal sense by a metaphorical one can be accommodated without giving the relationship an ontic/ontological status. And this "ordinary" retroactive illumination might well be enough to explain the effect Heidegger describes. "House" could *still* be a metaphor.

What about the case of "care"? Why does Heidegger choose such a word, rather than opt for some emptier, abstract term? It is no accident, I think, that the terms with which he *compares* "care"— unfavorably for that term—are "willing" and "living," which can respectively be traced back to Nietzsche/Schopenhauer and to Dilthey. "Care" is a competitor in a field of other candidates for the title of "legitimate successor to the epistemological paradigm," that is, to "knowing." He has already argued for knowing as merely one mode of Being in the world, among others. Here he argues that *willing*, when considered ontologically, exhibits the structure of care:

> In willing, an entity which is understood—that is, one which has been projected upon its possibility—gets seized upon, either as something with which one may concern oneself, or as something which is to be brought into its Being through "solicitude." (H194f.)

He then articulates the structure of this a priori possibility in terms that allow him to conclude that "in the phenomenon of willing, the underlying totality of care shows through."[4] And this same form of dependence is claimed for ordinary "ontic" care too. (The "existential condition for the possibility of 'the cares of life' and 'devotedness,' must be conceived as care, in a sense which is primordial, that is ontological." [H199])

Allusion has already been made to Heidegger's justifying references to the "transcendental 'generality'" of care, and to the (apparent) "emptiness" and "generality" of such descriptions. But even if one accepts his structural analysis here as persuasive, illuminating, and so forth, there is a sense in which the *abstractness* is not just the price one has to pay for a priori virtue, but is a sign of *incompleteness*. There is a sense in which a *further* story is needed about *why* there are just these three factors (existentiality, facticity, falling), a story yet to be told.

In the face of this urgency, the discussions of *reality* and *truth* seem almost distractions. Heidegger's discussion of truth, by its very brilliance, interrupts the course of the book, delays the moment of time. If it has a *logical* place here in the book, it can be found in the way the word "disclosedness" keeps "appearing" in the analysis of care. If the idea of trying to "preserve the force of the most elemental words in which Dasein expresses itself" is applied to the Greek word λόγος from which the traditional philosophical understanding of truth derives, one can trace back a series of steps (through uncoveredness and uncovering) to Dasein's *disclosedness* as "the most primordial phenomenon of truth." (With this existential grounding of truth one can say that "Dasein is in the truth.") And Heidegger can then show (via disclosedness) the intimate relation between truth and care.

It has been claimed that there is a certain *delay* here (in the last four sections of Division 1). At the end of section 41, Heidegger is already pointing forward, playing with the moment at which *structure* becomes *time*, a moment caught in the word "articulated" (*gegliedert*) which deserves a special place in philosophy.

> In defining "care" as "Being-ahead-of-oneself-in-Being-already-in . . . —as Being-alongside . . ." we have made it plain that even this phenomenon is, in itself, still structurally *articulated*. But is this not a phenomenal symptom that we must pursue the ontological question even further until we can exhibit a *still more primordial* phenomenon which provides the ontological support for the unity and the totality of the structural manifoldness of care? (H196)

For all the artifice involved in withholding the temporal, Heidegger has succeeded in giving an account of the fundamental structure of Dasein's

Being that is both plausible and yet deficient, hugely so. Clearly the *art* has been to anticipate in structural form a triad susceptible of temporal transformation. But before proceeding to the gratification and fulfillment of Division 2, there is one question that ought at least be adumbrated here, even if it must return when discussing Heidegger's own assessment of *Being and Time*—a single question even if it occupies a number of sentences. Is it really clear what is meant by an existential grounding? By treating Dasein's existence as capable of serving transcendental functions? Is it (and if not, does it matter?) really possible in principle to pin down the logical interrelationship between the multiple claims to primordiality (even being "more primordial")? How satisfactory is the logic of deficient modes—which has the consequence that phenomena apparently detached from their proper ground are actually (though negatively) attached? (Might this not be an instance of an ill-advised *conversion* of a powerful interpretive schema into an a priori assumption?) Is there not a *bad* circularity involved here, in that by assuming the universal possibility of grounding one convinces oneself that apparent exceptions are deficient modes, and thus *misses* the possibility that *groundedness itself* might be a limited phenomenon? It is worth recalling the suggestions made above, when introducing Heidegger, that it is the very impossibility of a primordial ground that incites a multiplicity of cross-mappings, translations, and so forth. Vertical impossibility generates horizontal diversity.

In the light of the problematic of time, the question to be put to Heidegger is this: Is it not possible for the existential grounding of our ordinary understanding of time to *overestimate* its own significance? To suppose that there *is* one ordinary concept of time already seems contentious. And does it not run the risk that all such transformations take, that of being silently determined by the shape of what it overcomes? Finally, and this is meant as a hint of the source of these doubts about transcendental grounding: Is it really possible to exclude *representation* from our fundamental account of existence? Heidegger is much more sensitive to the positive status of the everyday than some of his critics suggest, but it remains true that if an authentic *modification* of everydayness is possible, it would, as I understand it, involve an exclusion of representation. In particular, it would exclude representations of time from any role in our understanding of authentic temporality. Is this even in principle possible?

3   Death, Resoluteness, and Care

THE TEMPORAL REWORKING

Heidegger never misses the opportunity to make methodological remarks. The ease with which his contemporary readers drew merely "anthropological" conclusions from his work fully justifies this practice. No less does the revolutionary nature of his project. At critical points he must remind us—the point of all this existential analytic is to answer the question of the meaning of Being. It is necessary to go through this existential stage, because

> . . . to lay bare the horizon within which something like Being in general becomes intelligible, is tantamount to clarifying the possibility of having any understanding of Being at all—an understanding which itself belongs to the constitution of the entity called Dasein. (H231)

In each of these methodological discourses, Heidegger channels our questioning in the right direction—toward the ontological, toward the primordial, and toward a conception of the question that progressively incorporates the insights already obtained. This last feature is one of the most impressive of the book, one that gives it unusual unity and penetration. It happens that section 46 provides a prime example of this textual roll-on.

The overabstractness of the structure of Care, the felt need for another "story" that would make sense of this structure (why *three* parts?) have already been addressed. Heidegger begins his second division by a clarification of the ultimate question (the question of Being) and by a demonstration that our account of Dasein is so far *inadequate*. And he does so by (1) reminding us that the *ontological* interpretation in which he is engaged is a *species* of interpretation and as such subject to its *forestructure* (forehaving, foresight, and foreconception) and hence to the demand that this be clarified as far as possible, and (2) insisting that more needs to be done on this score.

Consider "foresight." In treating existence as potentiality-for-Being, attention was directed to its necessary but incomplete *inauthentic* form. And how can one claim "forehaving" without an assurance that Dasein has been grasped in its wholeness? "Care" seemed to provide us with an account of Dasein's Being as a whole, but surely when Dasein is considered from the everyday temporal point of view, this must be questioned. Indeed, as Dasein is never complete until it *ceases to be* Dasein, as it almost seems to be defined by an incompleteness (potentiality for Being), one might wonder "whether 'having' the whole entity is attainable at all," and hence whether a primordial interpretation of Dasein is not rendered impossible by the very nature of its Being. Heidegger is using the standard of "wholeness" as a vantage point from which to point to deficiencies in the story so far.

> If the interpretation of Dasein's Being is to become primordial as a foundation for working out the basic question of ontology, then it must first have brought to light existentially the Being of Dasein in its possibility of *authenticity* and *totality*. (H233)

It is at this point that Heidegger's concern with death is brought into focus. The argument goes like this: *if* Dasein is to be grasped as a whole, then we have to come to some understanding of death such that the temporal incompleteness of a living Dasein does not prevent us from grasping its Being-as-a-whole. The basic move is to argue that death is significant for Dasein only in his Being-toward-death, and far from sabotaging the possibility of wholeness, this serves, albeit peculiarly, to constitute it.

Heidegger's discussion of death, most particularly as it appears in the first chapter of Division 2, is offered as an answer to the question of how we can understand Dasein "as a whole," and *that* question is critical because it is, for Heidegger, a precondition of being able to move from the existential analytic (Division 1) to the ontological account proper (Division 2 onward) that we have such an understanding.

The explanation for this concern for Dasein's wholeness may well be that Heidegger's approach to understanding man's Being has always been in some sense dispersive, so that what metaphysics could solve at the beginning (with for example, the idea of a self-identical substance) is continually postponed, while its solution becomes more and more difficult to envisage. All our everyday models of wholeness have been taken away from us. (And yet Dasein cannot *surely* be just a fragmental dispersion, scudding clouds in the sky?) It might also be said that without something like a *complete* and *full* account of Dasein there can be no adequate account of Being, given that Being only *is* through Dasein's understanding and our understanding of that rests on a complete understanding of Dasein. This argument seems less persuasive; perhaps it doesn't carry the same *urgency* as Heidegger's continued insistence would require. And perhaps, too, it is possible to think of a complete account that did not rest on Dasein's wholeness. One can have a complete archaeological inventory of the remains of an old building without that giving us an understanding of the building in its "wholeness."

There is perhaps another explanation for Heidegger's repeated insistence on our getting a grip on Dasein-as-a-whole, which is that such a question does indeed *force* us to confront the very *peculiarity* of Dasein's Being, in comparison to other modes of Being. And it allows him to integrate a considerable number of themes (possibility, care, finitude, nothingness, anxiety, and so forth) into the question of death. It supplies the central *thrust* for Heidegger's discussion of death.

In order genuinely to invert the order of explanation, one would have to show that Heidegger for quite independent reasons wanted to give death (and our authentic response to it) a central role, and then prepared the ground for it by inventing or at least emphasizing the question of Dasein's "wholeness." It is hard to see how to do this, for it is clear that the capacity of "death" to focus many of the issues he needs to focus in Division 2 is quite genuine, and these issues are a logical progression from those of Division 1. Nonetheless it can and must be borne in mind that the question of wholeness *might* just be a *residual* question, one that has been left over from metaphysics, and that it is not the business of a fundamental ontology (let alone anything "after" that) to consider. Coupled with that is the key question that will guide this analysis of his account of death—Is it a subtle form of the "appropriation of death"? Or is it, in fact, the site of the most far-reaching renunciation of the value of *presence*? In fact, as will become apparent, it is almost impossible to subject Heidegger to this either/or. There will instead be offered what might be called, after Derrida, a "double reading."[1]

Heidegger's first chapter in Division 2 is entitled "Dasein's Possibility of Being-as-a-whole, and Being-towards-Death." It has been the source of much misunderstanding of Heidegger's philosophy in general and of serious disagreements. Even now, with the benefit of these various discussions, it is possible to misunderstand what Heidegger is saying, and yet it is not, in essence, very difficult. Whether or not it is right is another matter.

The key to Heidegger's discussion of death lies in the initial move he makes to resolve the problem Dasein's Being in time and toward death seems to pose for the project of understanding Dasein's "Being-as-a-whole." The *problem* is simple. From the point of view of how much of our allotted time we have eaten up, Dasein is always either (a) incomplete (still living, not yet dead) or (b) dead (and no longer Dasein). How can Dasein *be* a whole when (s)he is always, in fact, "becoming"? The move Heidegger makes is to remind us that what this means is that for Dasein death affects us existentially—that is, it is in our *Being-toward-death* that differences will be found. The question then becomes—How can we find a basis for Dasein's wholeness in Being-toward-death?

Heidegger's central move here is to convert a *temporal limit* into an *existential/ontological* one. My death is at one level an *event*, and from an external point of view it can be dated, perhaps commemorated and so forth. When I say it "is" an event, I mean that I can envisage it as happening, as about to happen, as having happened, and so on. This way of thinking about death is very important for Heidegger in that it (potentially) characterizes our everyday way of thinking of death. Opposed to it is the idea of death as "possibility." This may seem odd. Surely the ideas of "possibility" and "event" can be conjoined. Do not insurance companies deal precisely with "possible events"? Is that not what lies behind many safety precautions—life jackets on boats, seat belts in cars, helmets with motorcycles—just in case a certain possibility (an accident) should materialize, and to ward off another possibility (accidental death)? Such a way of thinking of death as a possibility is, however, (severely) limited. It presents death as an event in a series of events of which it is the conclusion. Such a "possibility" is always understood as a future actuality. It always and essentially gets (or could get) *cashed out* as something present, something actual. However, for Dasein, death as possibility has another meaning, and it is a meaning that is only *in part* oriented to the future. Death is primarily understood by Heidegger as the possibility of my non-Being—where "possibility" is not contrasted with necessity or probability. To live with a grasp of one's mortality, to be a Being-toward-death, is to grasp in as full a way as possible that one's existentiality, one's potentiality-for-Being, is fully disclosed only in the light of the possibility of one having no (more) possibilities. To say that death is possible "at any time" is not simply to say that one never knows when one will die, but that human life itself in its every instance derives its fullest significance from its *contingency*; I *need not be* at all.

What is the status of the future here? For Heidegger, the future is the *privileged* ecstasis, and it is so because of the role that "possibility" and "potentiality-for-Being" play in the distinctive constitution of Dasein's Being. And if the inauthentic attitude to death (which he calls "Erwarten" [expecting]) treats it as a *future event*, the *authentic* attitude cannot deny its futurity. But what is the future? *Existentially*, and that is the perspective from which Heidegger takes his orientation, the future is significant as the dimension in which our potentiality-for-Being is exercised. And it looks very much as though Heidegger understands the future via the notion of possibility. The *authentic* orientation to death he calls *anticipation*, which does *not* treat it as an event, but rather is receptive (in anxiety) to what death means for my potentiality for Being.

It has been suggested in the quite proper cause of stressing the continuity of Heidegger's thought,[2] that the central idea of his treatment of death in the early work *Being and Time* is already a kind of *Gelassenheit*[3]—a letting be. Certainly many of Heidegger's sections here are devoted to *releasing* death from the grip of representative, "thingly" thinking. But surely one needs a word that captures the *intensification* that an authentic Being-toward-death is clearly meant to bring about. Moreover, death does not *have* any Being in its own right; it is irrevocably attached to Dasein (like Being itself, and *Time*).

It is worth noting, before passing on to an assessment of these claims, how seriously Heidegger takes the need to distinguish negatively an existential understanding of death from any other understanding. Heidegger explicitly distances himself from any other disciplinary approach that would treat death as an event. One cannot help wondering whether matters are quite that simple. Psychoanalysis certainly would not treat death primarily as an event. Thanatos is no *event*. And a biology that ignored the death implicit in life (for example, in disease, in aging) would be a poor biology. Heidegger also labors the point that the "not-yet" of Dasein's death is quite unlike the various other kinds of incompleteness we can find in the world. Here Heidegger would almost qualify as an analytic philosopher.[4] The many examples Heidegger uses are genuinely interesting in their own right, but cannot be pursued here.

DEATH AND REPRESENTATION

An evaluation of Heidegger's discussion of death can usefully begin by focusing on a fascinating theme he brings up early on in the chapter—that of Representation (*Vertretbarheit*), for it offers both a way of further clarifying his position and also a way of beginning to ask how satisfactory it is.

Heidegger associates the claim that Being-toward-death is (what he will come to call) my own-most possibility, with the idea that no one else can die *for* me (see H240). There are, he rightly suggests, many ways in which one man can stand in for another. Andy Warhol is said to have responded to the numerous requests for him to make a personal appearance by sending a look-alike, often without detection. But the relationship more commonly does not involve deception, but simply some form of taking the other's place—carrying out someone's instructions, as does a secretary (who may even sign another's name). Or, further, one can represent one's fellow countrymen or one's school or firm at a conference or in some public affair such as the Olympics. The varieties of this kind of representation are very wide—even steering clear of the sense of representation as *idea* (*Vorstellen*), as the translators point out. Heidegger's claim, positively speaking, is that *this* form of representation is not only common but in some sense *constitutive* of everyday Dasein. Here he draws on the idea that Dasein is, in particular "das Man," a

Being-with-others. Modern role theory would confirm this dramatically. And yet, however many contexts there are in which someone else can represent me, there is *one* in which no one can represent me.

> "This possibility of representing breaks down completely if the issue is one of representing that possibility-of-Being which makes up Dasein's coming to an end, and which, as such, gives to it its wholeness. No one can take the Other's dying away from him.

> By its essence, death is in every case mine, in so far as it "is" at all . . . death signifies a peculiar possibility-of-Being in which the very Being of one's own Dasein is an issue. In dying, it is shown that mineness and existence are ontologically constitutive for death. (H240)

Strictly speaking, the words "own-most possibility" do not occur here. One has to wait until a little later (H250), at which point the word "non-relational" is added, which suggests something even stronger than "incapable of being represented by another."

This all sounds thoroughly plausible. There are certainly things for which it makes no difference who does them. There are things that someone else can do as well as me. There are things that someone else with my permission and at my instruction can do as well as (or better than) me, and so on. And there seem to be things that even if someone else could in some sense *do* them, it would matter that it was not *I* who was doing them. But Heidegger seems to move *very quickly* over this question. One can only speculate as to why.

He claims or admits that someone else can "go to his death for another," but he adds, "that always means to sacrifice oneself for the other in some definite affair." This remark sounds much more like the kind of position that might be taken by the early Sartre concerned with the dilemmas facing members of the French resistance, or someone reflecting on war. But one most obvious reference of such words is surely to Christ's Crucifixion. And is it not a quite extraordinarily mean interpretation of the idea that Christ died "to save us" that this relates to "some definite affair"? Indeed, one may explain the *circumstances* of many a martyrdom without reducing what is *at issue* in such deaths to "some definite affair." (How "definite" was what was at stake in Socrates' death?) And when, as has been documented, valiant souls substituted themselves for those in line to be liquidated in concentration camps, is it really possible to define the "affair" in question? What if the affair was the question of life itself (its value, or one's individuality)?

There are other directions in which Heidegger's move must be queried, and Sartre points the way.[5] If one *accepts* that in some sense no one can die for me, is it not equally true that no one can urinate, eat, feel joy, understand a philosophical argument, or fall in love for me? Surely anyone who replies that this objection misses the point must at least be pressed to explain how the privileged status of dying arises. At some point here it may be said that what is at issue here is an *interpretation* (with the form of a hermeneutic circle), not a *proof*, so that to make (such) an "objection" is a misunderstanding. However, Heidegger does reach a very strange conclusion—that representation is *excluded* from Being-toward-death, and this *is* the conclusion of an *argument* that can, more or less, be reconstructed.

The argument is circumscribed by the general question of totality—the problem of "getting a whole Dasein into our grasp" (see section 47). The suggestion is initially made that we might be able to get around the problem of how to grasp our own totality (we always seem to be *either* alive [incomplete] *or* dead) by some analogical transfer from the death of others, which we do indeed experience. Heidegger's answer is that we only at best experience the death of others as a loss—to *us* who remain alive.

> In suffering this loss, however, we have no way of access to the loss-of-Being as such which the dying man "suffers." (H239)

And no psychological intimacy with the other would suffice:

> we are asking about the ontological meaning of the dying of the person who dies, as a possibility-of-Being which belongs to *his* Being. (ibid.)

It is in the light of his demarcation of the *failure* of this enterprise that Heidegger focuses on the seemingly *more* plausible case that we dealt with before: that in which someone else *dies for us* where the other is, if you like, no mere "object" in the world, but in which the "for me" is doubled up.

But is Heidegger really careful enough in his analysis of what the death of others can provide? One thing that does not seem to be adequately accommodated is that it is only from the fact that others *have* died that we "know" of our own mortality, just as, should it come to pass, the technologically assisted superlongevity of others will convince us of our own (possible) immortality.[6]

Heidegger might reply that if this is a "condition" for our being able to have attitudes toward our own death it neither limits the scope nor determines the direction of our understanding of our *own* mortality. This is not implausible. More serious, for his case, is the way in which he seems to be *restricting* the possible ways in which the dying of others can guide us, by focusing on the meaning, to us, of the *event* of the death of the other. This may be of limited importance. We have indeed *become accustomed* to and tranquilized about the death of others, especially as the range of our vulnerability to such information increases (newspapers, radio, famine relief appeals, and so on). But Heidegger wants to make *his* point, without explicitly saying so, about the fact or event of the death of those near to us. (The "deceased" has been *torn away* from those who have "remained behind." (emphasis added) (H238) And yet he seems to do justice *neither* to the death of those near, *nor* to the death of those distant. Consider briefly, two cases: the death of a loved one (my spouse, my child, my father, my mother), on the one hand, and the death of a stranger in a biography (including autobiography), on the other. For what Heidegger says to be convincing he has to show that neither of these can offer (or constitute for) Dasein any kind of ontological illumination. This assumption appears as a rigidity in the distinction between ontic and ontological, to which the distinction between *representation* and *existence* is subsumed and must surely be questioned.

Consider first the death of a loved one. It has already been pointed out that Heidegger *seems* to be concerned with the deaths of those close to us. Yet it is strange and perhaps symptomatic that he ends his discussion of the possibility of existential illumination through the death of others by supposing that what is at stake is whether just any randomly selected other will serve as a substitute for Dasein's own Being-toward-death (see H239). Why this impersonalization? Might he perhaps be trying to avoid the question of love?

If love were an ontological event affecting my very experience of myself, might not the potential significance of the death of a loved one rest on the adequacy of one's analysis of "mourning," "loss," or "commemoration"? If such loss is merely an ontic event, perhaps susceptible to "psychological" investigation, what kind of story would Heidegger give of the withering away of the bereaved? (We do not even have to consider those cases in which the loved one dies at his or her own hand.) Surely the lofty response that these are of merely *ontic* concern to Dasein in its everydayness is quite insufficient. But why?

A number of possible responses suggest themselves. Is there not something quite as fundamental as one could ever want about being born from, creating with, and giving birth to, something that implicates those actual beings of whom it is true (or deemed true) in a way that quite transcends "the body"? To be of one flesh: this "flesh" is already "outside itself." Its exuberance[7] already exceeds the way the body itself transcends simple self-presence (intermeshing rhythms, its scars and its passions, its sensory organs and its eroticism). It might be replied: This is all very fine and moving, but what more does it do than remind us of the primitiveness of being-with, and the temporal structure of relations between the generations? How does it *approach* the ontological concern with the meaning of *one's own* Being? The answer, surely, is that it does not so much *approach* that question as perform a certain "molework" on it.[8]

When a father shields his child against the sword, and in so doing dies, it is perhaps neither courage nor "instinct" but a sign of a fracture in the heart of "own-ness." It is not so much selflessness, as a sign of the original rupture of the self, and one that, surely, is *not* reducible to the ultimately limiting "being-with." The reader could be forgiven for reading into these remarks a certain evocation of the spirit of Levinas.[9]

It is in that respect interesting that Levinas does not seriously (if at all) take account of Heidegger's analysis of authentic solicitude, which surely goes out of its way to *liberate* the Other from objectification. However, it does *not* liberate the Other from him/herself, but rather, charged with the weight of a renewed responsibility, *returns* it to him.

Is this all something of a sidetrack? *Not* if Heidegger would otherwise be *allowed* to return to the totally individuating nature of one's relation to one's death. It has been suggested that he cannot succeed in establishing this without a satisfactory account of love. This has not really even begun to be provided here. The writings of Marcel, for example, would have to be mentioned. But it should perhaps be clearer where Heidegger's ice is thin.

It is equally thin at the point at which he dismisses the value of the death of strangers (and hence, one would suppose, of biography and autobiography). As has been said, he accomplishes this by an implicit narrowing down of the meaning of "the dying of others" to the fact or event itself. Clearly it is then all too easy to contrast the Being-toward-death of *my* Dasein. But what then of (auto-)biography? Surely the ideal is precisely to reveal the existential choices and circumstances that shape a life, and many of the most interesting (auto-)biographies are of those whose lives are conducted with an eye on their own temporal finitude, their own mortality. Biographies of martyrs, saints, zealots, great soldiers and adventurers provide only the most graphic examples. Now it is of course true that there need be nothing adventurous about reading the biography of an adventurer, and one gets no closer to the flames by reading about Joan of Arc. To read about the comportment of others in the face of their own mortality does not *eo ipso* make one's own attitude authentic. Indeed it *can*, quite the contrary, be a form of escapism, a deferment of self-confrontation. Heidegger's claim, however, must be that it can have *no* ontological value, and that seems highly unlikely.

Perhaps we are being unfair to Heidegger. He might *simply* be saying that *understanding* what it means to be-toward-death and *acting and experiencing* at that level are two different things. There is a *radical* difference between my reading about St. Augustine having his doubts about becoming Christian removed by reading Paul, and *my* doubts being so removed. But even if this were true, the general distinction between *understanding* and *acting/experiencing* will not do, because Heidegger is at this stage concerned precisely with *understanding what it is* for Dasein to be "toward death." And when one considers, for example, the extraordinary power of the Gospels—of the accounts of the teachings and sufferings of Christ—to affect men's lives, one begins at least to consider how far possibilities-for-Being can be thought of entirely in isolation from *prior example*, from the choices others have made in different circumstances, from the historical stock of significant ways of Being. The reference to biography is only the most obvious way in which the horizon of such "ways of Being" is radically expanded beyond what our local community may offer even as possibilities. The suggestion is *not* that we choose our own Being-toward-death as we would a commodity in a super-market. The claim is, rather, that to think of our Being-toward-death as something that *inherently individuates us* need not and perhaps cannot exclude some sort of grasp of the horizon of possible choices—which derives from tradition, in the widest sense. It is not obvious that it is possible to *separate*, in the way Heidegger would require, the level at which the understanding takes place, and the various thematized and unthematized *possibilities* that have perhaps always offered themselves to us.

What is beginning to be questioned is the way in which Heidegger's concept of authenticity, his references to authentic solicitude notwithstand-ing, *essentially excludes the other*, and not only the personal other but also the other as Sign, as Representation. Might it not be that Heidegger's reference to death as Dasein's "own-most non-relational possibility" has *retained* the goal of Husserl's reduction to a "sphere of ownness,"[10] transforming it from an epistemological to an existential/ontological setting?

FINITUDE AND MORTALITY

Clearly all these points must be redeemed (or otherwise) in a further discussion of the concept of authenticity, but before moving on to the topic, it is worth perhaps considering another of Sartre's major points of disagree-ment with Heidegger on the subject of death. Put bluntly, it is that Heidegger confuses, or at least conflates, finitude and mortality. Surely we would still be finite beings even if we were, or believed ourselves with reason to be, immortal.

What is it to be finite? To be finite surely has its most distinctive manifestation *not* in our inescapable death, but in the "death" involved in every instant as one *has to* choose (even by not choosing) to realize some possibilities and not others. Even immortal beings have to live *somewhere* at *some time*, have a particular body, sex, culture (and a particular order of their acquisition), and so on. Does this not show that finitude and being mortal are different because we can imagine one without the other? The situation is a little more complex than this. (Even "immortal" beings [as we have defined the term] may be *vulnerable* and are *capable of* dying so the possibility of their death, even if never brought about, could still be a source of anxiety.) Would it not follow that even "immortal beings," beings for whom death is not an inevitability would still be able to think of themselves as wholes only in relation to the possibility of their nonexistence? Is that not what Heidegger would/could respond? And that would seem to suggest a difficulty in conceiving of a finite being that was not at the same time capable of an authentic Being-toward-death. It is even more difficult to think of a mortal being that would not be finite. So are the two concepts not, contra Sartre, after all intimately connected, and perhaps not even to be distinguished at all?

This certainly does not follow. Many of Heidegger's "existentials" are equiprimordial, hard to think of as occurring separately, and yet "conceptually" quite distinct. Surely just this is true of finitude and mortality. But if it *were* so, what would hang on it? These critical remarks run the risk of being all too finite, but might it not be that Being-toward-death can function for Heidegger as the answer to how Dasein can "be a whole" only if finitude can be *absorbed into* being mortal. And if it cannot, might it not be that Dasein's Being was *inherently* incomplete, or perhaps "undecidable," for Being-toward-death would be no resolution of our everyday finitude.

One last dangling thread still needs to be taken up. Heidegger does not want to treat death as an event, but he does say that it *cannot be circumvented*. (Let us suppose that the sense of "cannot" here is stronger than a philosopher's confident assertion in the 1930s that man could never see the other side of the moon.) *How important is it that death is inevitable?* The whole tenor of Heidegger's discussion of the *existential* dimension of death is to play down death as an actual future event, and to accentuate death *as possibility* in a different sense. The force of this idea is surely that *existence* is most radically to be understood as a multileveled relationship with *the possibility of not existing*. It is multileveled because the mere passage of time involves the closing off (the death) of "openings-onto" (the future), while the fact that we *need not* exist (both in the sense that one's *birth* was not necessary and that nothing *guarantees* my persistence) gives further intensity to every such opening. It *happens* that for none of us can this possibility of nonexistence be circumvented. But surely that is to go further than Heidegger really needs. Does not Angst for example actually reveal not our Being toward *certain death*, but rather that we can *never rule out the possibility of death*. Indeed, so tempting is it to rethink Angst as disclosive of contingency or finitude that one should not rule out the possibility of Beings certain of this immortality suffering from it, assuming that they had been *born* and had once *not* existed, and that their existence had never been "necessary." They could still feel Angst at their own finitude and "groundlessness."

What has been claimed in this section? Notice has been given that Heidegger's account of Being-toward-death as Dasein's ownmost nonrelational and uncircumventable possibility is open to doubt in each respect, and that those doubts provide the opening for a reading that treats Heidegger's discussion of death as an *exclusion* of the other—the other person and the other as "sign." However, though these doubts will be developed further, they will in no way amount to a wholesale rejection of Heidegger's remarks on death and authenticity. What is brought to light in this double reading is an unresolved tension in Heidegger's thought between the necessity of grasping Dasein as a whole and the recognition of time as the possibility of the other, beyond appropriation. If Heidegger is right in thinking that the move from an existential analytic to the ontological sphere, to the question of Being, rests on grasping Dasein as a whole, then the very structure of the book is here being shaken.

AUTHENTICITY

There are a number of key features of authenticity, as Heidegger presents it. As they are often misunderstood, it would be well here to state them clearly, whether to correct these misunderstandings or to reveal those being assumed here. Authenticity (*Eigentlichkeit*) is centrally linked to the idea of intense individuation, ownmostness, even if it can be plausibly said to redefine rather than presuppose any sense of "individuality." Authenticity is utterly dependent on everydayness, of which it is nonetheless a distinctive modification. The "choice" is not between everydayness and authenticity, but between *inauthenticity* and *authenticity*. Inauthenticity is perhaps best seen as the self-satisfaction of everydayness. One might indeed prefer authenticity to inauthenticity, but it is not for all that an ethical concept, but an ontological one. As such, of course, it might well be essential for any ethics to take account of it. Authenticity makes no concrete prescriptions, nor is it (as Hegel said of Kant's ethics) merely a *formal* notion. Rather it is an *existential* notion, one to be understood as a certain *way* of Being, a certain ontological self-relatedness, in which one relates to oneself as the kind of Being one is. Although one rarely gets a hint of it (Heidegger's discussion of poetic creativity at the end of *The Origin of the Work of Art* might be a case in point), a certain universality and anonymity is built into the whole discussion. To be oneself (rather than losing oneself) is to have achieved a rather difficult form of self-understanding and to direct oneself in that light.

It would be easy to conclude, especially before having read him, that authenticity was for Heidegger what Absolute self-knowledge was for Hegel (and for Spirit). Negatively they agree that what is at issue is a nonobjectlike kind of "wholeness," and that the resulting whole presupposes rather than "destroys" the level(s) through which it must pass for its realization. (A discussion of the intermediation of Kierkegaard, though a risky move to control, might be thought to seal the matter.)

Does not Hegel insist that man must accept "determinate Being" or have his light fade away without making any mark? Does not Hegel's account of the struggle to the death in the section "Lordship and Bondage," in the *Phenomenology of Spirit*, give the same or a similar status to death as Heidegger? (He who lives in the possibility of his own death [non-existence] wins?) Do they not both share an affection for Hölderlin's tender mortality? These are all interesting and important issues. And certainly Heidegger's words did not fall from heaven (any more than did Hegel's). But any prior understanding of Hegel's problem and solution can at best be used as a handrail confirming that what is at issue is not *entirely* new. Moreover, it *is* some help to consider the importance for Hegel of the appropriation of the Other, the reduction of the Other to the same. It is the point of the dialectic to offer a method and a justification for this. And it is for that reason that both Nietzsche and Heidegger have little good to say about dialectical thinking. What *can* be asked, however, is whether this motif—of the appropriation of the Other (what is *different*) into a final *whole*—is not still at work in Heidegger. And it is with this question in mind that, in a more detailed textual manner, Heidegger's discussion of authenticity shall be opened up.

Before Heidegger can address the particularly temporal nature of Dasein's authentic Being-a-whole, there is a preliminary step in the exposition that must first be gone through. The need for this step rests on a sense of the *limited* nature of what Heidegger has, up to now, managed to show. It is Heidegger's methodological scrupulousness, one might say, that reveals itself here, even if there will be occasion to cast certain doubts on the precise moves he makes.

For what he claims he has shown so far, in discussing "the ontological possibility of an existentiell Being-toward-death which is authentic," is "the possibility of Dasein's having an authentic potentiality for Being-as-a-whole." But, he adds, in a way that stresses this as a limitation, "*only* as an ontological possibility." (H266) In other words, we lack any assurance that it is a real "existentiell" possibility for Dasein rather than "a fantastical exaction." Heidegger believes we need to uncover an experiential *attestation* of the possibilities of authenticity, and, interestingly, and without apparent justification, sees this as evidenced by Dasein's *demanding* authenticity for itself. What Heidegger has in mind is a treatment of *conscience*.

His treatment of conscience might be thought somewhat marginal to the investigation of his understanding of time. But if that understanding belongs at least in part (for example, in his discussion of Being-toward-death) to the project of trying to answer the question of how we can have a grasp of Dasein *as a whole*, then one should take a serious if brief look at sections 54 to 60, in which he takes one further step in attempting to attain this. As has been said, Heidegger's interest here is in a *phenomenological* attestation of the possibility of authenticity—in conscience. A mark of the *limitation* of his treatment in this chapter is apparent in the final paragraph (section 60) thirty-seven pages later, where he writes that

> . . . as an authentic potentiality-for-Being-a-whole, the authentic Being-toward-death which we have deduced existentially *still remain a purely existential project for which Dasein's attestation is missing.* Only when such attestation has been found will our investigation suffice to exhibit (as its problematic requires) an authentic "potentiality" for Being-a-whole, existentielly confirmed and clarified—a potentiality which belongs to Dasein. (H301)

The discussion of conscience and resoluteness contributes to the question of authenticity, but not insofar as it relates to death. So in a sense, these sections represent a pause in the temporalization of the problematic of Dasein that is the distinctive feature of Division 2. Its appearance here confirms the suggestion that Heidegger's treatment of time must be understood as subservient to the wider problematic of grasping Dasein's wholeness (which itself is directed toward the "question of Being"). All this is both to *place* this discussion of these sections, as well as to justify a certain brevity in our treatment of it.

Our central contention will be that Heidegger does not here resolve our question about whether he is ultimately committed to an appropriation of the Other. It makes it clear that he cannot be said to be committed to this in any *simple* way, but it leaves open the question of whether he can, at some deeper level. It is now important to focus on the question of "what" conscience can be said to disclose, and in particular, whether Heidegger successfully disposes of the suggestion that its function is (merely) *ontical.*

THE DISCLOSURE OF CONSCIENCE

The importance of conscience for Heidegger can be explained very simply. Lost in the "they," Dasein already possesses, but is not explicitly aware of, an authentic potentiality-for-Being-itself. Indeed, as this "lostness" prescribes a self (the "they" self), criteria of significance, and a level of discourse, it is easy to suppose that there is "no way out" of it. What is needed is an experience by which *inauthentic* Dasein can be drawn out of its complacency. And this is provided by the "call of conscience." This calls Dasein to "its ownmost potentiality-for-Being-its-Self."

Heidegger insists the greatest *care* be taken over this phenomenon. In particular he wants to stress the difference between a *disclosive* experience and something merely present-at-hand. *Conscience,* as befits a fundamental experience of Dasein, is such a disclosive experience, and any attempt to understand it in the ordinary way distorts this. At this point Heidegger's language gets rather strong:

> The demand that an "inductive empirical proof" should be given for the "factuality" of conscience and for the legitimacy of its "voice," rests upon an ontological perversion of the phenomenon. This perversion, however, is one that is shared by every "superior" criticism in which conscience is taken as something just occurring from time to time rather than as a "universally established and ascertainable fact." Among such proofs and counter-proofs, the fact of conscience cannot present itself at all. This is no lack in it, but merely a sign by which we can recognize it as ontologically of a different kind from what is environmentally present at hand. (H269)

The language is strong, but it betrays an important weakness in Heidegger's position. And the discussion of this weakness will be one of three critical moves made in preparation for a more detailed probing of Heidegger's analysis of conscience.

Heidegger's argument here begins by making its case on the easiest ground (against the demand for an "inductive empirical proof") and then generalizes the argument to "every 'superior' criticism." There is not much of an argument offered here, but surely it *is* an argument, and a thoroughly weak one at that. It rests, at its most elementary, on the claim that no inductive procedures can demonstrate the "legitimacy" of its "voice." Mere facts cannot determine "essential" truths, and certainly not evaluative ones. But even if this is accepted, nothing shows that all "superior" criticism (which may attempt to understand "conscience," from within, say, some theoretical framework) is equally inadequate. Heidegger's contrast between what is present-at-hand and what is "ontologically of a different kind" is surely question begging. And so too is his reference to the "Fact" of conscience. (emphasis added) These remarks aim at seducing us into supposing that there is *one correct* ontological understanding of conscience, by castigating in a broadbrush way attempts that fail to meet that standard. What it suggests is that there is *no room for argument* about the meaning of conscience, and that, surely, is unacceptable.

This same imperiousness had emerged, unfortunately, only on the previous page:

> That the very "fact" of conscience has been disputed, that its function as a higher court for Dasein's existence has been variously assessed, and that "what conscience says" has been interpreted in manifold ways, all this might only mislead us into dismissing this phenomenon if the very "doubtfulness" of this Fact—or of the way in which it has been

> interpreted—did not *prove* that here a *primordial* phenomenon of Dasein
> lies before us. (H268)

This argument is a stronger version (in force, but not in validity) of the one
he used to justify continuing interest in *the question of Being*, despite the
wrangling over whether it is a serious question. But does it really work? It is
quite true, and *important*, that disagreements about the status of conscience
are *quite compatible with* it being a primordial phenomenon (but equally one
that seems important only from time to time for certain extrinsic reasons).
And it is true that if one had established on independent grounds the
primordiality of conscience, then one might well come to treat alternative
accounts symptomatically (for example, as attempting to cover up what it
was unacceptable to believe). But it is, surely, quite unjustified to suppose
that disagreement about the status of something proves it is "a primordial
phenomenon." It shows only that its status is contentious. And if, indeed, the
issue at stake was whether it was or was not primordial, it would surely be a
perversion of reason to conclude that it must therefore be primordial.

The third difficulty with Heidegger's account becomes apparent in
section 54, the very first of this chapter. Conscience is on the agenda because
it offers a way out of the self-enclosure of Dasein's "lostness in the 'they'"; it
is supposed to attest in an "existentiell" way to the possibility of Dasein's
"authenticity." But in Heidegger's analysis

> conscience will be taken as something which we have in advance
> theoretically, and it will be investigated in a purely existential manner,
> with fundamental ontology as our aim. (H268)

This suggests that the understanding of conscience will *not* be confined to the
individual Dasein's experience of it, but will engage in an analysis of the
meaning of that experience. But this would not show how it was that
conscience *actually* does succeed in letting lost Dasein rediscover itself. For
everyday Dasein is familiar with the "voice of conscience" and has its own
everyday interpretation of it. Heidegger's account would be quite unsatisfactory if he had to say that whatever the everyday account of conscience was,
the authentic story was different, because that would give conscience no
radically disclosive power at the existentiell level, where it is needed.

From these difficulties two questions can be extracted for the critical
guidance of our discussion of Heidegger's treatment of conscience: (1) Might
we not allow the importance of a phenomenological treatment of conscience
while arriving at conclusions different from Heidegger's? and (2) Does
Heidegger adequately clarify the way in which the voice of conscience can
actually appeal to everyday Dasein? It is worth following through, selectively, some of Heidegger's presentation here.

Everyday Dasein is disclosed to itself "understandingly" in terms of the world in which it is involved. This is a public world, dominated by the public interpretations of the "they." In being attuned to these, Dasein loses touch with itself. In *listening* to "they," Dasein no longer hears itself. Into this situation, the *call* of conscience serves to *break* or *interrupt* this listening that listens-away (*hinhören*). Heidegger describes this call as one that not only redirects our hearing, restoring a lost possibility, but also one that does so "unambiguously," leaving no foothold for "curiosity," and with these qualities it is radically distinct from ordinary world-directed "listening." This claim is soon repeated, and elaborated.

How, it might be asked, can the call of conscience make an impact on the "they-self," which is essentially and by definition blind to such possibilities? Heidegger's answer is twofold. First, that the disclosedness of Dasein's world is always accompanied by *some* sort of self-disclosedness. Dasein *always* has some self-understanding, even if it is "the they-self of concernful Being-with-others." Second, Heidegger claims, the call of conscience *appeals* not to this they-self, as a whole, but only to the *Self* itself.

[The they-self] . . . gets *passed over* in this appeal; this is something of which the call to the Self takes not the slightest cognizance. And because only the *Self* of the they-self gets appealed to and brought to bear, the "they" collapses. (H273)

He further suggests that in being passed over, the "they"(-self) is also affected (—almost as if snubbed?), and "the Self . . . gets brought to itself by the call."

Now it has to be said that this account is almost magical in its absence of any kind of explanatory model. The ease with which the self is detached, when appealed to in this way, from its public involvement, is startling. Heidegger is quick to insist that there is no question here of any kind of worldless ego, but rather of a Self that is still essentially (though in a different way) Being-in-the-world. What the call in each case discloses is Dasein's "ownmost possibilities." As has already been mentioned, Heidegger claims this happens without there being any ambiguity. This claim is expanded and underpinned by the distinction he draws between the *content* and the *direction* of the call. Thought of as having a *content*, the call might arouse a kind of inner dialogue, but the call essentially tells us nothing, it just points to (summons us to) a possibility of Dasein's Being. (Understanding the call is an ontological, not an ontic, affair.) Heidegger strongly dismisses the possibility that the call of conscience might properly induce a reflexive self-scrutiny. (Such possibilities are discussed in fact in section 37 on *Ambiguity*, and they have a place only in everyday Dasein.) Can nothing go wrong with the call?

Heidegger does make two important claims at this point, though without giving them much emphasis.

1. ... what the call discloses is unequivocal, even though it may undergo a different interpretation in the individual Dasein in accordance with its own possibilities of understanding.
2. When "delusions" arise in the conscience they do so . . . only because the call gets heard in such a way that instead of becoming authentic understanding, it gets drawn by the they-self into a soliloquy in which causes get pleaded, and it becomes perverted in its tendency to disclose. (H274)

The first claim is simple. The call discloses in each case an ownmost-potentiality-for-Being, but we each give content to it in a different way. Allowing that disclosure and interpretation can be kept distinct, this claim could be allowed. But the second claim is surely not so easily dealt with. *The call of conscience does not always succeed.* The problem is that the case in which it does not succeed is precisely the one that one would suppose was *standard* and predictable. The "they-self" interprets the call in its own terms—as it does everything else. This, for Heidegger, is a *perversion* of conscience's disclosive possibilities. But if we think back to what was described as his magical account of a successful call of conscience, there is no explanation of why and how it succeeds on some occasions and fails on others. It is not that the *call* is inadequate, he claims. The explanation must lie in *the way it is heard.*

THE IDEALIZATION OF CONSCIENCE

Suppose one asks *why* the call succeeds here and fails there. Heidegger can say no more than he has said. But an explanation can be offered: that there is no question of a disclosive event that may or may not succeed, but rather, on the one hand, an "ideal" possibility (which may nonetheless, or perhaps thereby, never be actualized), and, on the other, the ordinary messier types of self-understanding to which the call of conscience can be assimilated. How then could one understand the *authentic* self-disclosure supposedly wrought by conscience? It would be the *elimination*, the *expulsion* of all that would sully the purity of a silent and wordless grasp of one's authentic Being-as-a-whole. When Heidegger was describing the interruption of Dasein's "listening-away," he says that "[this] possibility . . . lies in its being appealed to *without mediation*." (emphasis added) If this is right, Heidegger would be presenting as a real possibility what is in fact an ideal accomplishment of reflexive thought, one, to be sure, that responds to a genuine *demand*, as will be seen.

A brief parallel to Husserl is again justified. The innocent reader of, say, Husserl's *Ideas* comes away with the impression that bracketting-out, *epoché*, putting to one side one's everyday and theoretical views of the world, is an operation it is possible to carry out at a certain time, by taking extreme care. Three problems (at least) arise. First, it is far from clear what "attitude" one is in (the natural attitude? the "reduced" or "transcendental" attitude?) when one performs this. Second, it is unclear how *within* the complacency of the natural attitude the motivation to transcend it can arise. And third, the time and thought needed to spell out what the reduction involves (not just its consequences, which might indeed be complex, even supposing it were a simple act) makes it hard to see how the reduction could be any sort of simple operation. There is surely an instructive parallel to be drawn with Heidegger's account of the power of the call of conscience to effect an unambiguous, unmediated, *separation* of the "they" from the "Self." Clearly, the fact that the call of conscience is said to come from "beyond" (though it turns out that this "beyond" is none other than the authentic self as understood by the they-self) is meant to explain how the cosy complacency of everyday Dasein can be interrupted. And, as has been indicated (though it falls short of an *explanation*), Heidegger will soon talk of the *demand* for a conscience, which supplies the motivation. But the parallel with the problematic status of the *epoché* for Husserl is the most instructive. Husserl came to see what Merleau-Ponty made quite explicit (see his preface to *The Phenomenology of Perception*)—that the reduction was not a one-off operation, but something to be repeated indefinitely. One is misled, perhaps, by a spatialized topology (for example, entering "the *sphere* of the transcendental reduction") to suppose that one can, in these matters, actually *be* in one place rather than another. Is this not just as plausible in Heidegger's case? For it *is* clear that the various forms of self-understanding that Heidegger carefully *distinguishes* from what is disclosed by the call of conscience are quite as common, if not far more so. They include, in their wealth of diversity, various modes of self-objectification, self-interpretation, self-scrutiny, self-criticism; they are all, indeed, "mediated," by models, by theories, by values, by *language*. If one supposes that what is so unsatisfactory about these modes of self-interpretation is precisely the same as what is unsatisfactory about "signs"—that we are led from one sign to another, without end—then one might come to think that what Heidegger has done for "the call of conscience" is to have cleared a site for its appearance as a mode of self-interpretation that *precedes all signs*, and radically excludes them. On this reading, the name of this desire would be *metaphysics*, the metaphysics of presence. Clearly the possibility of such a reading was ruled out by the methodological remarks with which Heidegger begins this chapter, but as has been shown, they did not in fact justify any particular determination of the experience of conscience. In any pursuit of this critical reading one would clearly have to offer some account of the experience other than the one Heidegger provides.

It may be that a critical discourse has been launched into prematurely. On the one hand, the scope of these remarks is such as would disturb the whole of Heidegger's discussion of authenticity, and indeed the very "question of Being" that illuminates this book. On the other hand, there are so many responses that Heidegger could make, or that could be made on his behalf, that it will seem perhaps to be doing him an injustice not to pursue them at this very instant. There are, however, further important things to be said about his particular discussion of conscience.

FURTHER REMARKS ON CONSCIENCE

It is worth noting, if only in passing, that Heidegger's discussion of conscience as a *call* has in common with his later discussion of "es gibt"[11] that he operates a kind of *transcendental abstraction from a transactional term*. In particular, while the call seems to come from beyond, there is in fact no caller distinct from Dasein itself; nor is there any real voice, nor are words actually used, nor is it a question of "anything like a comment." The parallel here to what Husserl calls "solitary mental life" in the *Logical Investigations* is most striking.[12] It turns out that the caller, like the called, is none other than Dasein itself—not any external power.[13] (H278) Again, the "caller" is not to be understood or identified in everyday terms. Rather it is Dasein "in its uncanniness."

The "call," as has already been indicated, is stripped of its literal everyday linguistic significance. Now it is stripped of any suggestion that it requires a relation to another being. Heidegger is offering an interpretation of the call of "conscience" that, at first glance, might be called "internalized," but of course Dasein is not, for him, "internal" at all. The question of whether Heidegger effects some sort of metaphysical *closure* of Dasein's Being will ultimately hinge on how one assesses his alternative to a conversation in inwardness. His secret, of course, is that he can work a dialogue—or at least a relation between two aspects of Dasein's Being—as care, Dasein's existence (*projective of possibilities*), and Dasein's facticity (thrownness). This is what is going on in his otherwise somewhat odd discussion of Guilt and being Guilty, the significance of which is brought out in his discussion of what might be called Dasein's existential self-grounding. The call of conscience leads to the recognition of our "Guilt," but this guilt is no ordinary matter that could have been avoided, or that one could make amends for. It is a *primordial* condition of Dasein that results from Dasein's facticity, from his "thrownness." And the basic move that Heidegger makes here is that Dasein's Being has its *basis* in the fact that, as thrown, it has *no* ordinary basis but rather exists in its projection of possibilities. Heidegger's position here can be seen, dialectically, as a denial *both* of anything like an ego cogito, *and* of a theological location of the ground of my Being "outside" me, in another power. Kierkegaard's discussion of Despair and the Self is on precisely these same lines, but concludes that man cannot be his own ground.[14]

To talk about Dasein's primordial *guilt*, to say that it is this that the call of conscience finally reveals, is to talk of the *constitutive* negativity or nullity of Dasein. It is as difficult as it is important to pin down a precise sense to this Being guilty, but the following senses, at least, should be included, that: (1) Dasein is characteristically "lost" in the "they." In this sense it exists as a falling short of its true possibilities. This is one sense of negativity. (2) Dasein always has itself to be—it is as projected toward the future. This is nullity in a positive sense (for example, Dasein *is not* an already established substance). (3) Dasein does not *choose to be*, nor that it one day will not be. (4) If Dasein always understands itself in terms of its possibilities, then every action— every actualization of itself—will rule out other possibilities, and that particular cases of this may result in remorse is based on this more primordial necessity to choose, and so to exclude.

It has to be said that Heidegger's discussion here presents the greatest difficulties of assessment, not just of comprehension. In particular, Heidegger clearly cannot be accused of offering an ontological *fundament* in the sense of some full and self-subsistent ground. *Guilt* is a primordial ineliminable negativity in Dasein and, as with so many of the existential phenomena Heidegger deals with, would seem both to prevent any simplistic (for example, ontic) sense of Dasein's Being a whole, *and* to make even more radically difficult the construction of another sense.

It was suggested earlier that Heidegger's analyses of conscience might perhaps be understood as accounts of an ideal possibility, one that had eliminated all those features (such as ambiguity, sign and so forth) that would threaten its purity. Such a clarity of self-understanding would be thoroughly *desirable*, and it is surely no accident that Heidegger's account of what it is to understand the appeal should lead to the idea of "*wanting* to have a conscience."[15] Not unexpectedly, Heidegger immediately tells us not to think of this *wanting* in any voluntaristic sense, but rather as a sort of "preparedness for" or being "ready for" the appeal. This is, of course, a very common theological notion. But it is not easy, in Heidegger's own terms, to pin it down. What would it be for the "they-self" to be ready for the appeal? And what would it be for Dasein's authentic self to be so ready? Indeed, if the authentic self were dissatisfied with its lostness in the "they," why would it need the call of conscience to effect a break? Is it not quite as plausible that the desire, the wanting, might be the metaphysical desire on the part of the philosopher for an idealization that would satisfy the requirements of not being subject to the doubts, ambiguities, and uncertainties of the sign? It is worth noting that this desire would, beyond all doubt, be distinct from anything psychological, and would have no less "primordiality."

Heidegger's response would surely be that, apart from anything else, this interpretation would surely have conveniently to ignore that the features of conscience he has stressed are disclosed in experience. They are not an invention of the philosopher. There are two responses to this. (1) Disclosures are always subject to *interpretation*, which is not without its risks. Certainly Heidegger is no friend of self-evidence. And his question at the end of section 58 might be taken as more than a rhetorical one. ("Is, then, the phenomenon of conscience, as it 'actually' is, still recognizable at all in the Interpretation we have given?") (2) The question of the status of our experience of conscience (and/or of Heidegger's interpretation of it) *cannot* be settled without an account of its temporality. It is strange that this question has been deferred, but it is clearly important. A single example will suffice. Conscience reveals Dasein's authentic Being-for-self "unambiguously." But if doubts subsequently arise, this unambiguousness becomes subject to scrutiny. For Heidegger, it might be said, the very thoroughness of the penetration of the concept of "authenticity" with Angst, uncanniness, wish, and so forth allows him to relegate doubts, uncertainty, and even criticism to the inauthentic, as a kind of second-rate anxiety. Critical questions about "authenticity" itself are then symptomatic of a basic misunderstanding. But this would be sheer intellectual hubris.

The question of the temporality of conscience, of course, is aligned to Heidegger's own line of questioning. The concept of resoluteness (*Entschlossenheit*) is an obvious advance here, and will be turned to shortly. But first, a brief discussion of the *alternative* to Heidegger's account of conscience mentioned at the beginning—that of conscience as *critique*, and Heidegger's dismissal of this.

CONSCIENCE AS CRITIQUE

In section 59 Heidegger's discussion of conscience as critique, arguably the most plausible account, is set within a broader discussion of a distinction between the existential and the vulgar (everyday) understanding. The suggestion that conscience has a critical function is *one* of four alternative possibilities he deals with, the one he mentions first but deals with last, and this itself may suggest he takes it most seriously. His general argument in this section is that these vulgar interpretations all have their place, but they are all ultimately derivative from the more fundamental account he has given. This will be questioned, focusing on "conscience as critique." Heidegger argues that (1) understanding conscience as critique *presupposes* that the call is related in each case "to some guilt-charged deed which has been factically willed," (H293) and (2) this (perfectly genuine) experience (that conscience relates to particular deeds and so on) does not permit it (conscience) to "proclaim" itself fully. (ibid.) However, neither of these two claims is satisfactory. The first fails by overstatement, and the second simply begs the question of what it is for conscience to proclaim itself "fully" and seems disinclined to assist such a process.

These points deserve expansion. Heidegger does not explain what he means by conscience as critique, so here is a brief characterization. To understand conscience as critique is to see it as a withdrawal, a standing back, a detachment from some particular involvement, or some mode of involvement in the world, one in which one asks whether one wants to continue with that involvement or act otherwise, and one in which what is at stake is whether that involvement can be fully avowed. Let us now look at Heidegger's argument. (1) The "critical" element is found in the combination of detachment with questioning. Now clearly one need not be concerned with a "guilt-charged" deed, or even a "deed" at all, or one that has been "willed." These are all possible, but one may equally be concerned with "where one is at" ("am I not wasting my life as a writer . . .?") without any prejudgment of *guilt*, without reference to a *particular* deed, and without a particular act of willing. (2) In the light of this one *could* argue that conceived of as narrowly as Heidegger does, this relationship of conscience to particular deeds may not fully bring out the meaning of conscience. But it is surely possible to conceive of it *less* narrowly.

The essential difference between this way of thinking of conscience as critique and Heidegger's understanding of conscience as disclosing our authentic potentiality-for-Being centers, as Heidegger seems to realize, on whether and in what sense conscience reveals anything "positive," and/or whether its disclosive power must be thought of negatively. Heidegger admits the value (albeit limited) of the critical account when he says, " . . . one can indeed point to nothing which the voice 'positively' recommends and imposes." However, to rest content with this "nothing" is not sufficient. For it is understood in relation to "assured possibilities of 'taking action' that are available and calculable." It *may* disclose nothing of *that* sort, but that is because it takes it out of "things" and confronts us with our existence. This latter is *not* just a "negative" affair.

Surely, however, this trivializes the critical interpretation of conscience. Even the "everyday" experience of conscience is more than a mere *not* this and *not* that. When it leads one to ask, "What should I do?" it is already more than *negative* criticism. But does this not just expand the sense of critique to the point at which it becomes identical with Heidegger's account

of the disclosive power of conscience? If Heidegger insists on the unambi-
guity, the direct, unmediated nature of the call of conscience, then this
cannot be what is going on. But suppose he were to say that he has never
denied the confusion that may *result* from the call, only that its *direction* is clear
(the direction/content distinction again)? If it is permissible, as Heidegger
allows, to include within a phenomenon those features that appear when it is
"fully disclosed," then our response would be that the question "What
should *I* do?" is not a question asked by and addressed to an authentic self,
but, precisely, leaves unresolved the question of whether there *is* such a self,
whether/how it can be distinguished from the everyday self, and so on.
Clearly the idea that I may have authentic possibilities, possibilities that are
most my own, is one that, for all its radicality to a world-absorbed "they-
self," is also *reassuring*, for it precisely allows the possibility of a *direction*
distinct from a content.

Further evidence of the divergence between Heidegger's position and
that being elaborated here, though not entirely unproblematic evidence, is to
be found in his account of Dasein's authentic existence as *resoluteness* (*Entschlossenheit*), which while embracing "anxiety" nonetheless seems to involve
an overcoming of the ambiguity, the doubt, and so forth that we have earlier
pointed to. Resoluteness is characterized in many ways. It is the name for
one's authentic liberation from the "they." And yet it is essentially in-the-
world, not some "free-floating 'I.'" It involves a kind of decisiveness about
one's mode of Being. But it is essential that one understands this aright.
Resoluteness not only makes choices, but resolves, in the sense of focusing,
the potentialities for Dasein's Being. Heidegger captures it as "the disclosive
projection of a determination of what is factically possible at the time." Hegel
has written that we truly exist only if we are prepared to be finite and not
while away our time contemplating possibilities. And resoluteness surely
captures this thought. Resoluteness is also meant to carry with it that sense of
finitude found in the idiom of "guilt" and "conscience" that has already
emerged.

Clearly this is an important beginning to an answer to the question
about the temporality of conscience. It will be taken up again very shortly in
discussing authenticity proper. Before that, just one small point, which will
again probe the relation of ambiguity, indecision, and so forth to authenticity. Heidegger writes rather interestingly:

> To resoluteness, the *indefiniteness* characteristic of every potentiality for
> Being into which Dasein has been factically thrown, is something that
> necessarily *belongs*. Only in a resolution is resoluteness sure of itself. The
> existentiell indefiniteness of resolution never makes itself definite except

in a resolution; yet it has, all the same, its existential definiteness. (H298)

Heidegger is saying that it is precisely our existentiell (preanalytic) *indefiniteness* that resoluteness *resolves*. Does this not, again, suggest that resoluteness is brought in to satisfy a desire? Indefiniteness is *unsatisfying*. Does it contribute to a temporal understanding of conscience? Surely not. The meaning of resoluteness is simply that what conscience disclosed must be made good, ultimately in relation to the future. But all the problems about the idealization of what "conscience" discloses arise again for "resoluteness." What can it possibly mean to have one's ownmost possibility of Being (completely?) disclosed? Such doubts, Heidegger would say, are just the symptoms of the "they's" failure to understand (see, for example, H296). But that is mere assertion.

ANTICIPATORY RESOLUTENESS AND TEMPORALITY

It has been suggested, without offering much proof, that the concept of resoluteness reinscribes, without itself *resolving*, the problems discerned in Heidegger's earlier description of what it is that conscience discloses. That points in the direction of a vital task—the temporalization of our understanding of conscience. My reasons for insisting on this temporalization are not, however, the same as Heidegger's, as will now become clear. For Heidegger, conscience points in an unambiguous direction—toward my authentic potentiality for Being. This formula can be seen as the expression of a *desire* that such a possibility be available, a *desire* that can be explained without any consideration of its possible satisfaction. And it is characteristic of such desires that their character (of being a desire) is revealed in the course of time, when the exhilaration of the moment fades and the need to give body to such a vision reasserts itself. Resoluteness would bring to conscience a certain constancy and determination that would allow such a grasp of one's authenticity for Being to be retained. But it would be a mistake to think of this constancy as occurring *within time*. Rather, it effects a certain transformed understanding of time itself.

Heidegger has already taken the first step toward such a transformed understanding in his discussion of Being-toward-death, where death was seen as something to be *anticipated* as the possibility of one's nonexistence, rather than, as expected (*Erwarten*), as an event sometime in the future. Clearly *anticipation* (*Vorlaufen*) is a mode of temporalization that *builds in* an understanding of Dasein's *finitude*—that one of its possibilities is not-to-be at all. So Heidegger suggests that *resoluteness* is understood more fully when understood as involving "anticipation" in this sense, and that it is as "anticipatory resoluteness" that "temporality gets experienced in a phenomenally primordial way in Dasein's authentic Being-a-whole." And here is announced the pivot of his whole book, and of my interest in it:

> The primordial phenomenon of temporality will be held secure by demonstrating that if we have regard for the possible totality, unity, and development of these fundamental structures of Dasein which we have hitherto exhibited, these structures are all to be conceived as at bottom "temporal" and as modes of the temporalization of temporality. (H304)

This is the pivot of the book not only because it explains the need for a temporal rewriting of the earlier, more structural description, a rewriting on which the book turns, but also because it articulates the degree to which time as temporality does not merely penetrate Dasein's existence but is also fatefully bound up with it. The question that will continue to guide the reading of Heidegger here is whether there is not room for disagreement with Heidegger about the interpretation of time and temporality even while accepting, indeed gratefully accepting, his demonstration of the need for illuminating it via an existential analytic.

CONSTANCY

The existential account is all very well, but Heidegger insists, to ensure that anticipatory resoluteness not be construed as a mere abstract conjunction of ideas, that it be borne out existentielly. Heidegger rightly stresses the methodological importance of this move. It has already been suggested that he may not always be quite as successful in articulating the two levels as one might like.

This question will be pursued at the point where Heidegger himself endeavors to explain its importance (section 62), focusing on the issue of *constancy*. Heidegger has already helpfully explained how the concept of care transforms our understanding of the self. No longer can we think of it as a kind of self-subsisting substance that endures.[16] It will be just this undermining of the idea of the self as a fixed point, and time as a secondary modification, that will be reflected in such formulations as "temporality temporalizes"—for the subject "is" itself temporal. This will shortly be pursued in more detail.

What the self-as-substance did however achieve was a sense of permanence and constancy, not surprisingly, because time was "external" to it. With the concept of resoluteness, Heidegger attempts to reestablish the value at least of constancy. He must do so, it would seem, for the effect of "the call of conscience" to be more than a "pin-prick."

This desire to *maintain constantly* a strong sense of that potentiality for Being disclosed by conscience is, as has been said, a philosopher's desire. (It has also been suggested that Heidegger needs, but fails, to retain a clear grasp of the distinction between the existentiell and the existential levels of analysis of conscience.) Here is proof that such retentive constancy concerns him *as a philosopher*:

> The Interpretation of the ontological meaning of care must be performed on the basis of envisaging phenomenologically in a full and constant manner Dasein's existential constitution as we have exhibited it up till now. (H303)

(And compare " . . . the unwavering discipline of the existential way of putting the question.") (H323) It is perhaps worth commenting briefly on how he does this.

It can be said without criticism that Heidegger's style at times exhibits a very considerable density. This density is a rough-and-ready name for the way he tries to maintain a constant vision of "Dasein's existential constitution." There are of course many other aspects of his style, as will be shown later. Some of those that contribute to this density are: condensation, compounding, and repetition and reworking. They are the consequence of adopting a method that circles ever deeper rather than proceeding through a series of easily stateable logical steps using a taken-for-granted language.

Condensation: Without the handle of "care" by which to refer to "existentiality," "facticity," and "falling," the text would be largely unreadable.

Compounding: Consider the work of preserving and displaying theoretical insight accomplished by all the articulated expressions (such as In-Der-Welt-sein/Being-in-the-world) and by such expressions as "anticipatory resoluteness." These are vital ways of expanding the units of discourse, and so increasing what can be thought in one go.

Repetition and reworking: Being and Time is itself built on a reworking/rewriting principle in the movement from Division 1 to Division 2. Heidegger's methodological scrupulousness means a great deal of repetition of material as similar points are made in different terms or from slightly different perspectives. One of the key patterns of continuity, which is also a reworking, and which *maintains* the past, is the very common move of offering one account as a mere filling out, clarification of, deepening of, a previous one. Everything gets reworked, but more primordially. All this is mandated by and also required for his hermeneutic method of investigation.

How does Heidegger elucidate the existentiell dimension of anticipatory resoluteness, which he rightly says is essential methodologically, to avoid the charge that he is engaging in "mere" theorizing? The existentiell dimension is necessary for the existential analytic, and it concerns all those modes of existence in which, while there is understanding, (self-)disclosure, and so forth, the philosophical implications of these are not, as such, being drawn out. But if the existential requires the existentiell, the existentiell, *to appear as such*, has to be displayed, and section 61 involves a series of ways of displaying at an existentiell level the various intimate connections between resoluteness and anticipation, with death as the ultimate horizon.

The complexity of the relationship between existential/existentiell is enormous. Heidegger's method is one that gives the existential level a certain autonomy at the level of thought, but insists on the need to ground any results of that thought on the existentiell level. Thus resoluteness is said to have to "take over" Dasein's "guilt," and this is only achieved when Being-guilty is seen not as some act-related state but as a *constant* condition. This *constancy* itself has to be understood in relation to Dasein's potentiality for Being "right-to-its-end." This is as far as the existentiell account goes. But existentielly, of course, this "right-to-its-end" has to be understood as a *toward* relation, and ultimately as "anticipation of death." This result of existential analysis must then be validated *existentielly*, and that is what section 61 is all about. And it is here that the question about *constancy* is raised. Heidegger wants to *affirm* the idea of constancy, but translate it from its connection with things present-at-hand into a sense appropriate to Dasein.

And that sense is surely this: Dasein, understood in terms of its Being, *is* only in terms of its potentiality-for-Being, that is, its Being is to be able to be in various ways. And one of the vital dimensions on which its Being can be assessed is that of the degree to which Dasein has taken on board, appropriated, these truths. The constancy that attaches to one's "Being-guilty" is thus not the constancy of a state of an object but of an existential condition of a Being whose Being is essentially bound up with its potentiality for Being.

Is it then still possible to ask the question (or has it already been answered) of the precise temporality of this Being-as-potentiality-for-Being?[17]

It is sometimes easy to feel that Heidegger's basic move is a reversal. Kant's reversal was to say that if we are in time, time was first in us. Heidegger, on the other hand, thinks through and deepens the metaphor of containment, and any reversal would have to be reformulated. Death, for example, is "for-us," before it is anything "in-itself." But this "for-us" is to be understood not as a reference to "subjectivity" but to Dasein's Being. And Heidegger's discussion of resoluteness turns out to be a discussion of the modalities of our appropriation of our Being-toward-death. What is difficult to accept—and it is equally difficult to decide the status of that difficulty—is that questions about the continuity, constancy, and perpetuation of a certain potentiality-for-Being can be answered without resort to ordinary temporal categories. Resoluteness as "anticipatory" emphasizes the sense of its Being-toward-death, toward the ultimate possibility. But ordinary questions about how to understand the temporal status of resoluteness (for example, as an act of resolution, as a *momentary* insight, as, once achieved, a permanent level of Being) surely remain. Heidegger's strategy seems to be (a) to defer these kinds of questions to a later discussion of the temporality of everydayness and (b) to suggest a priority of potentiality-(for-Being) over actuality that would undermine any attempt to give weight to these "ordinary questions."

Consider in this light certain of Heidegger's formulations in this section. First, it is clear that "resoluteness" cannot be thought of within the temporality of an act. He writes, "By 'resoluteness' we mean 'letting oneself be called forth to one's ownmost Being-guilty.'" This "letting-be" signals the refusal of "act-psychology." What is this Being-guilty? A "potentiality-for-Being"? And how can it be constantly guilty?

It is

> . . . not just an abiding property of something constantly present-at-hand but the *existentiell possibility of being* authentically or inauthentically guilty. In every case, the "guilty" is only in the current factical potentiality-for-Being. (H306)

Insofar as Dasein is understood as a potentiality-for-Being, it has no "is" independently of this potentiality for Being. Authentic and inauthentic existence are modalities of the "is" dependent on whether the full understanding of this potentiality-for-Being has been appropriated. (Anticipatory) resoluteness seems to be the name for that appropriation. But again what of its own temporality? Heidegger writes:

> The explicit appropriation of what has been disclosed or discovered is Being-certain. (H305)

> The primordial truth of existence demands an equi-primordial Being-certain, in which one *maintains oneself free* for the possibility of *taking it back*. (ibid.)

Heidegger is trying to describe a certain resoluteness with regard to a situation that *maintains the possibility* of determining the situation differently, without lapsing into an indeterminate or irresolute indecisiveness. His solution is a kind of redoubled resoluteness.

> ... this holding-for-true, as a resolute holding-oneself-free for taking back, is *authentic resoluteness which resolves to keep repeating itself.* (H308)

It is this resolution to repeat that seems to bear the weight of the notion of *constancy.* But can it possibly succeed? There would seem to be two immediate difficulties. The first is that any problems attached to resolving would apply equally to resolving to resolve. The second is that unless the idea of resolving builds in its own success, it is surely open to all the vagaries of time—bad memory, the dimming of vision, falling back into inauthenticity, and so forth. And the only way it could guarantee its own success would be by referring to an act of disclosure, which would be judged, for example, in terms of its luminosity rather than its being sustained through time. And surely no reference to potentiality-for-Being is going to improve matters. A concept such as constancy involves an achievement in time and cannot be reformulated in terms of anticipations, possibility, or potentiality for Being without losing that sense. The *desire for* constancy can of course be so reformulated, but that is a different story.

These doubts are, perhaps, ill-conceived. Do they not represent a falling back to an understanding of, for example, anticipation that treats them as psychological acts, albeit of a privileged sort? Surely anticipatory resoluteness is neither an act, nor a momentary state, but a certain mode of Being *continually revealed* in one's thoughts, behavior, and so forth or one revealed whenever it would be relevantly brought into play. That formulation, however, merely throws back the question of the *basis* of this continuity. It cannot be habit, for that would reduce Dasein to being the subject of properties like some sort of present-at-hand thing. It must surely be something like a power of ontological disclosure that can both open up the widest horizon (and deepest dimension) for the appreciation of and determination of the situations one finds oneself in, and can also see that this is directly applicable to all situations whether or not one does so apply it. This second clause is fundamentally only a weaker version of the redoubled resoluteness looked at earlier. It too leaves unanswered the question of how one moves—existentially or existentielly—from revealing experiences (beginning with and developing from the call of conscience) to a life-permeating transformation. It is claimed here that Heidegger can only multiply the levels and complexity of anticipation and desire. In the case of death, anticipation is not (from this point of view) problematic. For death as the *limit*, as a *positive* negation, is a *single thing*. But the anticipatory transformation of the rest of my life,[18] the projection of the decisiveness of anticipatory resoluteness on to the series of events that will trace my future, cannot be based on the supposition that the appropriation of death be distributed in dilute form, as it were, over every future event.

By making this contrast, the suggestion that the anticipation of death is both possible and legitimate and that it is a proper modalization of resoluteness has been left intact. It is clearly vital to discuss this briefly, for failure to understand these matters can lead to what Heidegger calls "the grossest perversions." Chief among these is the suggestion that anticipatory resoluteness is an attempt to escape, to "overcome" death.

THE APPROPRIATION OF DEATH?

I take it the remark of Heidegger most vulnerable to such an interpretation of anticipatory resoluteness is one quoted earlier:

> When in anticipation, resoluteness has *caught up* (eingeholt) the possibility of death into its potentiality-for-Being, Dasein's authentic existence can no longer be outstripped (Überholt) by anything. (H307)

The importance of the misreading that Heidegger sees as the grossest perversion is that it raises again the question of whether Heidegger is not (and if so, whether he is right to) taking for granted the ideas of unity, identity, and wholeness of a life, and (merely) giving an existential underpinning (perhaps the only possible one) to that wholeness. The importance of this for Heidegger has already been pointed out. Without Dasein's Being-a-whole, he cannot make the move from an existential analysis of Dasein's Being to more general illumination of the question of Being. But a requirement of method cannot dictate what must be the case in any absolute sense. This is not to begin to reject Heidegger's understanding of death, simply to say it is *possible* to deny it with no other immediate consequence than that Heidegger's own wider project would, in his own view, be broken-backed. Heidegger's view seems to be based on the idea that a *complete* understanding of Dasein's Being requires an understanding of Dasein's *Being-a-whole*, attested in an existentiell manner. With such attestation there is no possibility that the "wholeness" one obtained would be a mere theoretical picture. In this light, is Heidegger right to deny the charge of trying to "overcome" death?

As has been suggested already, Heidegger seems to perform a certain inversion on death. When he first introduces the question, he insists that death is "not to be outstripped." Now he says that with the anticipation of death, it is Dasein's authentic existence that "can no longer be outstripped by anything." It might seem as though authentic existence has displaced death in this way. But the sentence clearly suggests that it is by virtue of having incorporated death (as possibility) that authentic existence has this status, not by contrast with it. Is there not, nonetheless, a sense in which by anticipation the fear, even terror, of death has been muted? And is not that an "over-coming"? The text must be attended to more carefully:

1. Anticipatory resoluteness is not a way of escape, fabricated for the "overcoming" ("überwinden") of death . . . (H310)

2. . . . it is rather that understanding which follows the call of conscience and which frees for death the possibility of acquiring *power* over Dasein's *existence*, and of basically dispersing all fugitive self-concealments. (ibid.)

First, Heidegger somewhat loads the question. The words "fabricated for" suggest he might think his intellectual honor was at stake. It *could* still be that some sort of escape was the functional consequence of the concept of anticipatory resoluteness even if there was no fugitive intention. It is quite correct to say that at the level of everydayness, anticipatory resoluteness offers no escape from death, in the way that, say, the elixir of youth might, if regularly imbibed. But in that quite *literal sense*, nor does the promise of life after death where that life is spiritual rather than material. However, Heidegger's point would be that this religious solution *does* deny that the event of physical death is a true end, and this finality Heidegger does accept.

The crucial issue in assessing Heidegger's position is whether there is not a sense in which grasping the possibility of *anticipating* death, rather than merely expecting it as an event, is not to transform one's understanding, and perhaps dispel one's *fear* of death. For on Heidegger's reading the fact that death, from one point of view, may come at an inconvenient time—too soon, too late—peacefully or painfully, is unimportant when compared to the question of whether one's existence when alive is illuminated by the possibility of one's death. Clearly if by "overcoming" (*überwinden*) were meant either the *denial* of death or the lifting of its burden, then anticipatory resoluteness does not "overcome" or escape death. But it is nonetheless the case that anticipatory resoluteness is an *appropriation* of death, such that its primary significance is no longer that of something *absurd*, something radically *external*, something that radically subverts the human endeavor—which is surely Sartre's position. Sartre takes Heidegger to task for just this.[19] The question, however, is how one understands the entity to which death has been appropriated—whether the appropriation of death completes a process of ontological closure, or whether it demonstrates, as powerfully as anything could, the impossibility of such a closure.

The second half of the sentence just quoted makes clear how it is meant to achieve the latter. For it portrays death not as something safely internalized within some enlarged concept of existence, but as itself holding sway over Dasein's existence (as the possibility of the "impossibility" of that existence). If "overcoming" death means a kind of cosy tranquilization, then Heidegger surely does not do this. And when one reflects on the matter, the fact that Heidegger certainly does direct our attention from the mere event of our biological demise is both thoroughly justified and in no way a lessening of our concern with death. Heidegger transforms our understanding of death, but the new shape it takes is no less powerful.

A discussion was promised of both the possibility and legitimacy of Heidegger's account of the "anticipation" of death. Heidegger's own defense against the charge that this is a form of escape has been extended in a way that makes some contribution to the question of its legitimacy. (Though even so, it only denies one basis for its legitimacy.) But of course the question of its *possibility* is not thereby decided. Something could be legitimate were it possible, but not in fact be possible. The mode of achievability of anticipatory resoluteness has already been questioned, and it would perhaps be worth extending this.

The comments that follow carry a risk of misunderstanding Heidegger to the extent of arguing what is his own position against him.[20] The reader must judge.

Anticipatory resoluteness is a condensation of an account of how the recognition of one's finitude can effect an intensification of existence. That account is presented through a range of concepts: conscience, guilt, the call, anxiety, the "they," for which Heidegger gives existentiell foundations and to which he applies existential clarification. Insofar as anticipatory resoluteness is a condensation of such an elaborate account it will possess all the strengths and weaknesses of the terms involved, and their articulation. The problem is surely that while one can give existentiell illustrations of these concepts, it is quite another matter to demonstrate their necessity or completeness. That is simply a formal point. More concretely, it might be suggested that (a) the English (or modern) reader's difficulties with the idea of Dasein's fundamental Being-*guilty* are not merely difficulties stemming from an inauthentic *refusal* of the thought, but from a deep suspicion of its (theological) roots in a particular tradition; and (b) even if in every case Heidegger will perform a shift of sense on the terms he employs (so that "conscience" cannot be identified with the result of wrongdoing, for example, the "the call" is not audible, nor in words, nor *from* anyone else), it is surely *still* the case that what he discusses is a deeper sense of *these particular everyday words in the German language*. What else could he do? it might be retorted. The point however is not to suggest an alternative, but to feel one's way to discovering the limits of his thought. What if anticipatory resoluteness was *unthinkable* in some other natural language? (The philosophical privilege of the German language is hardly an acceptable position.)

But quite apart from these gestures to radical cultural diversity,[21] there are surely other difficulties with anticipatory resoluteness that center on whether the disclosive orientation it captures is in any way *prescribed* by our finitude, even when understood as our ultimate potentiality-for-Being. Heidegger, I would argue, follows Hegel in giving determinateness a privileged value for finite beings. In Entschlossenheit one focuses, one determines, one resolves. For Hegel, the alternative was to waste away in dreams and fantasies. For Heidegger, the alternative is being lost in "the they." It would of course be a separate project to compare the two thinkers on this matter, but they surely share a basic assumption, which is that the proper response to finitude is commitment—albeit retractable—rather than a mere wavering in indecisiveness. But finitude could equally teach a *positive* indecisiveness in which the failure to come to a decision would be no mere wavering, but, perhaps, a deep suspicion of the frame of reference in which the demand for a decision is posed. It might be said that the very concept of Entschlossenheit rests on just this rejection of an *everyday* either/or. And yet it still echoes the ordinary sense of decisiveness, resoluteness, at the same time as it insists on the radical distinctness of the authentic and the everyday levels. In such a distinctness, ambiguity is necessarily, methodologically, denied. (Indeed, such ambiguity was diagnosed in section 37 as itself a form of everydayness.)

Consider what Heidegger says about wanting-to-have-a-conscience: " . . . it brings one without illusions into the resoluteness of 'taking action.' " (H310) Now Heidegger is at this point denying that anticipatory resoluteness is an unworldly attitude, so the reference to "taking action" should be set into this context; action is not being stressed, it is simply the focus of his remark.

But what sort of actions are being recommended? Surely taking a stand, telling the truth, and honoring one's commitments are the models here. And yet the suggestion that once one has cleared away the "incidentals" of one's everyday existence, freed oneself from illusions, these authentic possibilities of honest straightforwardness will remain is surely itself illusory. Built into the whole idea of anticipatory resoluteness is a denial of ineliminable ambiguity, faith in a space of purity—Heidegger's position rests on a radical incompatibility between resoluteness and illusion. But what if illusion was in principle not separable from truth? Indeed, why does not Heidegger stress, as he does elsewhere with other forms of disclosure, that it is accompanied by a necessary concealment?

The notion of anticipatory resoluteness may not have a *moral* connotation to it, but it is hard to deny a certain seriousness of tone. I will endeavor to bring this out by some comments on Heidegger's dismissive remarks about skiers.[22]

HEIDEGGER ON SKIING

Heidegger describes most favorably the peasant who silently puffs on his pipe and is in tune with his world. He contrasts this humble, honest belonging to the shallow pleasure-seeking of the skiers who drive up from the city. It will be left to Sartre, who almost certainly did not read this piece of Heidegger's, to give an existential interpretation of skiing; Heidegger's thought seems wholly negative. What is characteristic of the peasant's world? The regular beat of the changing seasons, the repetition of simple pleasures, a life in which little changes. . . . And the skier's? One response to Heidegger would be this: first he knows what the peasant may not—that the peasant's world is not the only one. The visitor to the provinces may, of course, learn nothing about himself, but the peasant cannot even begin the journey. And what of the skiing itself? It is an elaborate act of pleasure, one that shows no return. For the peasant this would be wasteful. What use has the peasant for the brightly colored ski fashions, the waxed fiberglass skis? And why do the skiers just go up and down? The peasant can see no point in this Sisyphean activity. And think of the skiers' conception of time. For the peasant every day, in its season, is much like the rest, a sequence of reassurances. He is a creature of tradition, repeating what he has seen done by his elders. The skier is different. His element is *speed*,[23] technique, control, risk, a play on the edge between gripping the snow and falling, perhaps suffering injury or even death. For Heidegger the skier makes the world of the peasant into a spectacle. But this *reduction* of the skier and his world is itself the consequence of reducing the skier to a spectacle.

How do these doubts relate to anticipatory resoluteness? If it is allowed that skiing for pleasure has bound up with it risk, excitement, and intensity in playing at the very limit of control, might it not be that this is an alternative way of appreciating finitude in the project of the intensification of the present? It is one in which the future is anticipated, but always in the form of danger, unlike that of the peasant for whom the main danger is to become lost in the illusions of the city. For the skier, the question of his Being-a-whole is not posed. What matters is whether he can cross a certain slope, execute a certain turn at a certain speed, gracefully, skillfully, in control. In an important sense, the whole problematic of "ownmostness"[24] and authenticity dissolves, and not in passive pleasure but in active boundary play.

Heidegger's dismissal of the skiers has been taken seriously because it can be seen as a symptomatic refusal to contemplate a radically different mode of Being-toward-death, a refusal that undercuts the *necessity* Heidegger gives to his analysis. However, the *idea* of anticipatory resoluteness remains, and even if it represents a questionable ideal of authenticity, it does offer something enormously important, perhaps the answer to the unresolved question of its temporality. For when Heidegger discusses its relation to care and the temporal foundations of care, he provides us with something of a model for understanding, at least one important possibility of existential temporality—a complex articulation underlying the structure of care ("ahead-of-itself-Being-already-in [a world] as Being-alongside entities encountered within-the-world")—and its role as the basis of "projection." And it is his development of the idea of "temporality" from the structure of care that must now be pursued.

CARE AND TEMPORALITY

One can see at a glance that "care" has a tripartite structure to which past, present, and future can easily be made to correspond. It is hard to imagine that this consequence did not influence Heidegger's original characterization of care! But the relationship among the three parts has to be shown to be other than a simple juxtaposition. And to this end Heidegger links authentic care with anticipatory resoluteness and elaborates the multiple projective overlapping of modes of temporality that are involved in anticipatory resoluteness. In this multiple projectivity, temporality itself is disclosed.

As is true of the whole of chapter 3, the section 65 in which this is carried out bears repeated reading, and it is hard to do it brief justice. Simply illustrating and commenting on certain of Heidegger's formulations will show how anticipatory resoluteness is deemed capable of effecting this temporal condensation and intensification, and hence how temporality itself appears.

The scene can be set by noting the primacy of the future for Heidegger. This devolves from the primacy of Being toward death, and from the kinds of existential move required to give sense to Dasein having a relationship to its own wholeness. The future is conceived of not as a reservoir of new "nows" but rather as what makes possible "the coming in which Dasein, in its ownmost potentiality for Being, comes towards itself." But how, then, do the other temporal ecstases come in? Heidegger argues (H325–7) first that Dasein's futurity and pastness are intimately connected, and that ultimately the future has priority.[25] He does so beginning from his understanding of Dasein as Being-guilty, as "being the thrown basis of nullity." This sets a condition on any authentic anticipation (future-oriented) that Dasein is first recognized as an "I-as-having-been." And at the same time, one's rootedness in the (absolute?) past of one's own nullity can be existentially manifested only in our relation to the future. ("Only so far as it is futural can Dasein *be* authentically as having-been.") Having a past is one thing, but for one's past to be part of one's being, one must first exist, projectively—that is,

toward the future. And yet anticipatory resoluteness situates itself, in situations, by "making-present." This "present" he treats as what is released by
the future as it becomes a future that "has been." And he uses these
interlacings to define temporality.

> This phenomenon has the unity of a future which makes present in the
> process of having been; we designate it as *"temporality"* (Zeitlichkeit).
> (H326)

Heidegger will then proceed to spell out the temporal map underlying the
structure of care in a way that is fairly obvious.

Let us now comment on *temporality* as Heidegger has just introduced it,
for we are approaching crucial questions about the connection between time
and Being. Heidegger rightly and at all costs wants to avoid any understanding of the articulated structure of time that would give to its discrete
components some sort of independent status. But there are easier ways than
the one Heidegger pursues to achieve this. One could, as Kant and Husserl
did, simply talk about time as subjective. However, Heidegger's critique of
the "subject" and "subjectivity" precludes that option. For him, temporality
emerges as an articulation of modes of Dasein's Being, itself thought in terms
of the structure of care, and of the "authentic" care he calls anticipatory
resoluteness. So all the ontological delicacy that attaches to his discussion of
Dasein's Being is equally applicable to his account of temporality.

When temporal predicates are applied to things, the question traditionally posed is whether to count the resulting "objects" as real. Do future or
past things possess Being? When they are applied to Dasein, they modify—in
necessary and fundamental ways—its Being. They answer the question *how*.
And the sign of this in Heidegger's formulations is the expressive "as."
Dasein is an "I-am-as-having-been," "the future *as* coming towards," "as-
it-already-was," "*as* authentically futural," and so forth.

Heidegger has already discussed the "as" structure explicitly;[26] it
appears as the structure of interpretation. In the as-structure interpretation
lays out the meaning of its subject matter: "the 'upon which' of a projection
in terms of which something becomes intelligible as something." (H151)
This account suggests only one of the aspects of the "as" that are important
for understanding how Heidegger uses it in introducing temporality. Here
the "as" represents a dimension of explicit disclosure. But it does not merely
function in interpretation. The as-structure is, as it were, part of the structure
of Dasein's Being. Equally, it locks temporality into the question of Being.
Heidegger has attempted to displace the traditional problematic of the self as
substance with the explication of the structure of care, with temporality as its
fundamental structure. And problems that have always accompanied Heidegger's talking about Being and Nothing arise again for temporality. And Heideg-

ger attempts the same formulation. "Temporality" is so much not the name of an entity that it seems to Heidegger misleading even to allow it to function as the subject of a predicative "is," let alone the "is" of existence. He is happy only with the explicative verb form "temporality temporalizes (itself)" (*Zeitlichkeit zeitigt*) (H238)—to which "das Nicht nichtet im selbst"[27] should be seen to correspond.

Heidegger's formulations here have important implications for any ultimate assessment of his success in having found in temporality an adequate basis for his account of Dasein's Being and, equally, for an assessment of his account of temporality itself.

"Temporality," he writes, "temporalizes possible ways of itself. These make possible the multiplicity of Dasein's modes of Being and especially the basic possibility of authentic or inauthentic existence." (ibid.) It is when he proceeds to a discussion of temporality as the "ekstaticon" that the radicality of this really emerges.

TEMPORALITY AS THE "EKSTATICON"

In an important sense the whole of Heidegger's discussion so far has been concerned to elaborate a nonmetaphysical conception of temporality. But it is only too easy to *say* that by going existential one somehow transcends metaphysics. In the last few sections the precise way in which the investigation of Dasein will yield the required result has become clearer. His persistence in posing questions in terms of *Being* has more and more convincingly opened up the possibility of a thinking that does not center itself on things, on substances, on the present-at-hand. With Heidegger's discussion of the *ekstases* of temporality, he accomplishes at one and the same time both a grounding of Dasein's Being and a transformation of our understanding of time.

On the brief account just given of the "as" of temporality, it would seem both to fill out, and yet in the end to constitute, Dasein's Being. The relation of temporality to Being might be said to have something of the structure of supplementarity as Derrida describes it.[28] Here, however, with Heidegger's discussion of the "ekstases," the Derridean parallel is surely "*différance.*" And Heidegger himself focuses on more of the little words: "to," "towards," "alongside," "being," and so forth with which the various modes of temporality are articulated. They each mark a primitive break in self-identity, for which Heidegger uses the Greek-derived word "ekstasis." In clearly Hegelian language, he describes temporality as "the primordial 'outside-of-itself' and for itself." And he continues, "Temporality is not, prior to this, an entity which first emerges from itself; its essence is a process of temporalizing in the unity of the ekstases." (H329)

In other words, primordial temporality is difference rather than identity. It is not something "from which . . ." but the "from which" itself. Heidegger calls it a process, and yet it is not one process among many. It is prior to all entities, and all entity-dependent forms—such as "events," and "processes" ordinarily conceived. These "ekstases" are fundamental modes of "activity," and the need for such formulations as "temporality temporalizes" becomes clearer. Temporality *is* only in its temporalizing. That is not just one of the things it does.

Heidegger's discussion of temporality, and its awkward and tautological syntax, forms one of the strongest arguments for a continuity of project between *Being and Time* and his later writings, including "Time and Being," where all these same difficulties reemerge. (*Ereignis* is an "event" but an extraordinary one; one cannot say simply that it "is.")

One of the key consequences of this exposition of primordial time is that it means that we can treat our ordinary understanding of time (". . . a pure sequence of 'nows,' without beginning and without end") as a "levelling-off" of the ekstatical character of primordial temporality. And in particular it obscures (a) the primacy of the future "ekstasy," and (b) the fact that primordial time is *finite*. These claims are interesting and important for different reasons—the first because it raises a rather interesting methodological problem. Heidegger later came to reject the primacy of the future.[29] How is this possible?

When referring to his claim that it is only as futural that Dasein can "be authentically as having been," it was not in fact possible to find any very convincing arguments for the dependence between the two being unidirectional. Dasein's finitude has a clear future flavor to it, but when thought of as the "possibility of my impossibility," that flavor is somewhat dissipated. I am finite backward too, after all. Dasein's projectivity, again, even if one thought to apply it primarily to the future, might equally well be said to depend crucially on my thrownness, and on what in each case I *bring* to bear on the situation, out of the past.

But, more pointedly than these three ways of bringing a certain reasoned doubt into the "primacy of the future" thesis, there is the methodological question—How could Heidegger have got it wrong? What scope is there for *error* in existential analysis? And if acknowledged here, where else might it reside unacknowledged? Paradoxical questions arise at this point: for Heidegger, interpretation rests on a forehaving, foreconception, and so forth. Even if their disclosive adequacy is questioned *in this case* it still seems to confirm the primacy of the future! Disclosure-in-advance seems to be the condition on which the status of the future would be judged. If this circularity is not to be a defect, it must either be that this hermeneutic priority of the future is quite legitimately self-justifying, or that the priority of the future disclosed in advance is itself of a different order. And to explain how the priority of the future could be dropped, the same alternatives arise. It may be that the thought responsible for the fading of the future is that disclosure cannot be adequately thought on strictly temporal lines, and with the advent of time-space comes a falling away of the future. The horizonal features of spatiality displace the primacy of the future. If such speculation has any basis, it will run parallel to doubts about the possibility of a distinctly primordial time, and consideration of this topic must await later discussion of *Time and Being*.

The second claim—that primordial time is finite—is interesting for a different reason. On the one hand, it undermines our confidence in the ordinary concept of time as infinite in both directions. (Heidegger argues that this *infinite* time is the product of a leveling off of finite time.) But on the other hand, there is an important sense in which the two concepts of time are compossible and have their own spheres of valid application. For Heidegger agrees that time goes on after my death. Infinite time and finite temporality are in a sense compossible because incomparable:

> The question is not about everything that still can happen "in a time that goes on," or about what kind of letting-come-towards-oneself we can encounter "out of this time," but about how "coming-towards-one-self" is, as such, to be primordially defined. Its finitude does not amount primarily to a stopping, but is characteristic of temporalization itself. The primordial and authentic future is the "towards-oneself" (to oneself) . . . (H330)

In fact, there is a tension between the claim that time and temporality are derived one from the other and the claim that the two have their own legitimate domains. Heidegger is arguing uphill for the legitimacy of conceiving primordial time as finite, and in doing this he both wants to acknowledge that time *appears* endless while insisting on a place for a temporality that is not. And a legitimate place for that temporality is ensured by arguing that infinite time requires it. His argument is basically that the ordinary understanding has not *understood* infinite time. And for this understanding we need to go to primordial temporality, for which the distinction between finite and infinite is significant. But, as the penultimate sentence just quoted makes clear, finite and infinite are not really in opposition, because finite does *not* mean "stopping."

This makes Heidegger's position both easier to accept and more difficult to understand. Consider the last sentence in which Heidegger is perhaps jokingly alluding to Husserl's "to the things themselves." What this brings home to the reader is that the idea that for Heidegger existential time is concerned only with the humanly meaningful organization of time is quite mistaken. This may be our loss. What he *is* committed to, it seems, is a sense of time or temporality in which the guiding thread is the transformation into a temporal framework of the problem of selfhood—of self-gathering, returning to self, self-fulfillment, and so forth.

The question this raises is whether there is room for, and whether Heidegger can himself provide room for, an understanding of human temporality that is harnessed *neither* to the question of selfhood nor to the simplicities of "the ordinary understanding of time." If not, one may come to suspect that his guiding question, the question of Being, has functioned not only to illuminate but also to limit. These questions must be pursued in relation to his discussion of Dasein's everydayness, historicality, and the phenomenon of within-timeness. When Heidegger introduces these themes he sketches out the kinds of relationships he will establish among them. Both everydayness and the within-timeness of things will be seen to be dependent on primordial time, and historicality will modify primordial temporality itself. Equally, in section 83, he points forward to yet another level of analysis "in which the concept of Being is discussed in principle" that is not in fact provided but only, again, suggested at the very end.

4 *Time and Temporality*

In a later chapter (see Part Four, chapter 5) a variety of temporal structures are discussed, those that borrow their organization from the texture of language, some that mingle representation with experience, and others that may seem free of all representation. It will be argued that the understanding of time as structured in various explicit ways does not in itself reek of authenticity or inauthenticity. Heidegger does give everydayness both a derivative and yet a positive status. But will it supply the horizon of intelligibility for the outstanding issues alluded to? Or will it insist on reducing their pertinence to the problematic of everydayness? The next task will be to assess the status Heidegger accords "everydayness." What is at issue is the precise way in which the sphere of everydayness is demarcated, and the insight it provides, by contrast, into the values constitutive of authenticity.

We all have what might be called an everyday understanding of "everydayness." And to a large extent Heidegger relies on this, both in the sense that he himself does not make clear its precise meaning, and in the sense that it names a prephilosophical or reflectively inexplicit mode of existence, but one that for all that can serve as a starting point.[1] His "preparatory analysis of Dasein" relied on it, and so does much of what he wants to say about temporality. The chapter he devotes to discussing its links with temporality (Part 2, chapter 4) gives more *content* to it, but opens as many questions as it resolves about *what it is*. A number of claims can be made in a preliminary way. Everydayness is not the same thing as inauthenticity. The latter is something like the "self-satisfaction" of the former, the belief in or assertion of its adequacy. Everydayness is a *way* of existing characterized by evasiveness, a reduction of existence to *familiarity*, the domination of a kind of shrunken horizon of complacency.

The very word "everydayness" (*Alltäglichkeit*) suggests, correctly, a link with time (a link that would be lost if, for example, one thought to equate everydayness with common sense.) At the end of this chapter, and indeed the previous one, Heidegger anticipates his further discussion of this connection by pointing to the need to discuss Dasein's historicality, and his within-time-ness. But here the inner connection is best amplified by his account of daily life, section 71, as the breeding ground of existential complacency. Everydayness both has as its domain of application one's common, daily, repeated, unsurprising mode of existence, and also derives from this repetition a certain limitation of vision. Dailiness constitutes a frame of reference by which possibilities are determined in advance, and in which the boredom and dullness this generates seek constant diversion. The concept of everydayness therefore already embodies an interpretation of time. In consequence, the relation between everydayness and temporality with which Heidegger here deals is more intimate than one might imagine.[2]

Let us first chart the course of Heidegger's discussion of everydayness. Everydayness is a mode of Dasein's Being-in-the-world, and as such exhibits a disclosedness distributed across the structural limbs of care. The temporality of everydayness can most economically be described by following the sequence these limbs offer: understanding, state of mind, falling, and discourse. With the exception of discourse, each will be claimed to exhibit the dominance of one particular ekstasy in such a way that the tri-ekstatic structure of care and of temporality is confirmed.

Heidegger has also discussed our general Being-in-the-world in terms of our distinctive relation to things ready-to-hand and present-at-hand (*Zuhanden, Vorhanden*) and has argued for the derivativeness of the latter from the former. Insofar as these can be seen to exhibit a temporal dimension, he will try to show that any objectified or objectifying (for example, scientific, theoretical) understanding of things merely "present" will be derivative from the time of our practical involvement in the world, that of "circumspective concern." Here it becomes clear that everydayness, for all its deficiencies in relation to the possibility of authentic temporalizing that can develop out of it, is nonetheless the true, positive ground of our temporal being, and the presupposed basis of any representations of time. Heidegger's discussion of the temporality of the scientific neutral projection of being as "present" leads him to question yet further how that thematizing or objectifying is possible. His answer is that it rests on Dasein's transcendence. And this *too* has to be understood temporally. (The same claim [later withdrawn] will be made for Dasein's "spatiality" as well.) Dasein's transcendence, not distinct from that of the world, is accomplished by temporality in relation to the things we encounter. Temporality provides the ekstatic horizonality of that encounter, articulating a dimension in much the same way as the *syntactic* articulation of language allows the transcendence of mere things ("present-at-hand") in the world. (Indeed, in tense and mood the temporal and linguistic modes of articulation coincide.)

Heidegger's discussion of the temporality of everydayness clearly demonstrates three things: (1) the immense resources of temporal significance buried in the structure of our daily existence, (2) the possibility of understanding more theoretical and representational ways of thinking of time as founded on or derived from these resources, and (3) (a corollary to [1]) the *power* of a temporal hermeneutic, by which we can come to understand the deeply *layered* significance of existence in temporal terms. This chapter (Division 2, chapter 4) alone is a strong argument for the claim that it is indeed structures of time that can give "content" to the "is" of human existence that would otherwise have to be defined negatively (man is *not* a thing and so on). To say that man is a *temporal* Being does not immediately resolve the question of how to talk about his Being in a nonobjectified way, but it does hold out the prospect of such an understanding. The problem of talking about Being and of talking about Time are the *same* problem, and the analysis of the temporal structures of Dasein's Being shows, in an exemplary and privileged way, how this can be so.

Appreciation of the general aims and accomplishments of chapter 4 does not, however, absolve one of critical responsibility. And any departure from its precise conclusions will of course require that appreciation be tempered by a critical appraisal. I begin by looking at his discussion of the temporality of disclosedness.

First, a brief word about disclosedness (*Erschlossenheit*) and its fundamental importance for Heidegger. It unites two vital features of Dasein's Being, as Heidegger presents it. On the one hand, it captures the way Dasein is not a closed-off substance but is rather articulated in various dimensions of relatedness. This disclosedness could be called an *essential articulation*. At the same time, these dimensions of disclosure constitute the *truth* of Dasein, the modes of its openness, and are presupposed by any other sense of truth as propositional, representational, and so forth. If disclosedness is fundamentally temporal, then this essential outside-of-itself of Dasein will also have a fundamentally temporal sense, and Heidegger will have accomplished, or taken vital steps toward accomplishing, the task of substituting for any substantial atemporal sense of self, a radically temporalized interpretation. The systematic working out of the temporality of disclosedness follows the structure of care, which will cover understanding, state of mind, falling, and discourse. Rather than providing a survey, the attempt will be made to (a) draw out the analytical distinctions Heidegger makes; (b) comment on the interpretive achievement at a general level; and (c) pose specific difficulties with particular analyses. These specific difficulties will add up to (and indeed are motivated by) doubts about the drive to formal and systematic completeness exemplified here, which is itself predicated on the persistence in this work of foundationalist thought, undertaken in all seriousness.[3]

Consider, first, that mode of disclosedness, that existentiale, to which he

gives the name "understanding." This is *not* the name of a cognitive achievement, as one might think, but a primitive condition for a whole range of
capacities. Abstractly and primitively, Heidegger defines it as "to be projecting towards a potentiality-for-Being for the sake of which any Dasein
exists." He goes on: "In understanding, one's own potentiality-for-Being is
disclosed in such a way that one's Dasein always knows understandingly
what it is capable of." (H336) Those two remarks point, respectively, to what
might be called the temporal/structural question and the ontological question. The former will be discussed in a way that may be seen as exemplary for
the other dimensions of care.

THE PRIVILEGE OF THE FUTURE

There are three main analytical claims: that this projective understanding unifies, in its temporalizing, all three temporal ecstases; that it nonetheless does this in such a way as to give the *future* a dominant role; and that
each of these can occur in an authentic or inauthentic manner.

Suppose one uses the expression "ahead of itself" as a neutral term for
Dasein's futurity. This can take the authentic form of *anticipation*, in which
Dasein "comes towards its ownmost potentiality for Being," or inauthentically, in which Dasein "awaits this (potentiality for Being) concernfully in
terms of that which yields or denies the object of its concern." (H337)
Expectation of future events occurs within the horizon of awaiting. Again the
crucial distinction between the authentic and inauthentic forms is that the
inauthentic understands Dasein's potentiality for Being in terms of *beings* and
loses the distinctive sense of Dasein's Being.

The primacy of the future to understanding can be portrayed by
reference to his discussion of the present. It too has an authentic and an
inauthentic form. The inauthentic Present is called "making present," a
"being-*alongside* the things with which one concerns oneself." An example,
one presumes, would be being engaged in an activity to which one was
paying attention (listening to birdsong, writing carefully, neatly arranging
books, and so forth). The authentic present, on the other hand, he describes
as the "movement of vision," and here the primacy of the future stands out
more clearly. This "moment of vision" actually occurs in the present (and
indeed what it discloses may not survive the passage of time), but more
important it brings to light the *possibilities* latent in a situation, in a "rapture
which is *held* (ideally) in resoluteness." The moment of vision occurs in the
present, but *points* to the future in its projection of possibilities.

Heidegger also distinguishes between forgetting and repetition of the
past, but these will be passed over for the moment.

The significance of the primacy of the future and this account of the authentic present should not be underestimated. For if Heidegger is right in supposing that the history of philosophy has been characterized by a metaphysical privileging of the present, and of presence, then there are two different directions from which this privilege could be unseated. The primacy of the future suggests that no attempt to find some foundation, or ground, or privileged point in present experience, intuition, and so forth will succeed because the meaning and value of what that "present" reveals will always be subject to a reference to the future. But second, Heidegger's distinction between an inauthentic and an authentic present is such as to correct any remaining interpretation of the present as a "now." For the now-present belongs to the specific mode of temporality known as within-time-ness. And the moment of vision, both because it is an ecstatic rapture and because it is subject to the future, ". . . in principle can *not* be clarified in terms of the 'now.'" (H338) The privilege of the present can never be the same again.

AMBIGUITIES OF AUTHENTICITY

In mentioning the authentic/inauthentic distinction, what was called the ontological question has already been broached. It can now be spelled out. Heidegger distinguishes knowledge of facts from understanding one's potentiality-for-Being. He writes:

> In understanding, one's own potentiality-for-Being is disclosed in such a way that one's Dasein always knows understandingly what it is capable of. It "knows" this, however, not by having discovered some fact, but by maintaining itself in an existentiell possibility. (H336)

But this formulation is surely problematic. It has already been suggested that Heidegger's account of the structure of existence is an attempt to give voice to, to make sense of its distinctive "is" (and thence to that of Being in general). There are two ways, however, in which this could be achieved. The distinctiveness of the existential "is" would be seen in *each* of the elements of the temporal structure so distinguished. Or, it would appear in the unity of that structure. Heidegger has made so much of the unified structure of care, and of the triadic unity of *every* mode of temporalizing, that it might be thought he had opted for the second alternative. But the problem of the "is" has been shown to be merely dispersed among the different existentials. The assertion that gives us trouble is this: "Dasein always knows understandingly what it is capable of." And our difficulties with this go deep into the heart of the idea of authenticity. Heidegger seems to us not just to be treading a very narrow line, but from time to time to be falling off it on to the wrong side. Authenticity is a *way* of being, and as such, to give it *content* is both a temptation and a danger. The temptation is one that language keeps offering us—to say something about it. The danger is that *what* one says will be radically misleading. Does not Heidegger run this risk with this remark? What *can* it be to always know what one is capable of? Heidegger's gloss is that Dasein "maintain[s] itself in an existentiell possibility." But surely this language, and indeed the whole problematic of authenticity, betrays its own phenomenological ground—which is that of the *problem* of (one's own) Being, of a sense of not being wholly identifiable with one's everydayness, of having to define one's Being without guidance, in the midst of anxiety, and the like. Authenticity, then, is surely closer to a *not knowing*, a self-questioning, a desire to know oneself for which there is no guarantee of satisfaction. One tendency of Heidegger's thought is surely such as to convert this self-questioning into a prescription of a positive existentiall possibility. Such a conversion would suggest that the problematic of authenticity was, after all, in the service of a reestablishing of the self at a higher level, rather than the continual reaffirmation of its epistemological fragility.

These doubts could be summarized as the suggestion that Heidegger may have transformed a negative phenomenon into a positive one. But the same point could be formulated somewhat differently. Heidegger, it might be said, has made a condition of possibility into a discrete phenomenon. Could this not be claimed for his general account of disclosedness, and perhaps even of his descriptions of the temporal ekstases? Take understanding, again, as an example. Excluding from understanding any sense of projection that would involve either factual knowledge or prediction or planning, surely there is nothing left that could ever be entirely separated from newly conceived modes of worldly involvement. These may indeed be so conceived as to embody a certain self-understanding or to exclude another inadequate one, but the idea of a knowledge of one's "potentiality-for-Being" that was not either a negative revulsion at a particularly anonymous or unchosen mode of involvement, or a new commitment with the new "content" this would involve, is surely a fiction.

Heidegger's discussion of the temporality of disclosedness is not, of course, confined to understanding, and a somewhat more succinct indication must be given of the kind of specific things he has to say about state of mind, falling, and discourse. The question here is what does this temporalizing rerun add to our understanding of these existentials?

ARE THERE DOMINANT EKSTASES?

In the case of the temporality of state of mind—mood—the most striking claim is not that temporal predicates are essentially (and illuminatingly) attached, but that there is again a dominant ekstasis, and that it is the past.

Heidegger takes as exemplary the cases of fear and anxiety. Contrary to the commonsense view that fear is primarily oriented to the future, he argues that it rests on a clinging to what one has (or is) *already*, in the face of some disturbing possibility. After all, he says, we could be faced by something threatening without feeling fear. What *one already is* must, as such, be threatened. Now there is real merit in this demonstration that the future is not alone in determining the temporality of fear, but surely his argument for the priority of the past is unsatisfactory. For it could, symmetrically, be said that one's sense of what one already is (such that it might *be* threatened) could actually emerge only at the point of being threatened. Might it not be fear that crystallizes the "already"? And certainly *without* the threatening future, one might just be complacently absorbed in the past and present. If the threatening future is so clearly what distinguishes fear from a sense of security, how could one fail to credit the future, yet again, with the privilege? A similar point could be made for anxiety. And Heidegger's interpretive efforts on other moods seem to us to fare no better.

Two comments are called for on this whole question of a privileged ekstasis.

First, even when less than totally plausible, the interpretive effort involved is not wholly unfruitful. It provokes a searching scrutiny of the less obvious temporal dimensions involved, for example. And it enables the claimed unity of temporalizing in each case to take an ordered form. For instance, he claims that "... anxiety ... must ... come back *as* something future *which* comes towards (zukunftiges)" (H343) The "as" and the "which" rest upon the *determination* of the "come back" as the focus for modification. Might the need for diagnosing a dominant ekstasis in each case be a grammatical one?

Second, the particular case of "mood" requires a comment similar in spirit to our doubts about the *positive* presentation of Dasein's "potentiality for Being." And again, it is a doubt that would accentuate a particular side of this double reading of Heidegger. Moods are understood from the outset as "bringing us back to something," (H340) and in explaining how hope, which *seems* primarily futural, really is not, he talks of it as "hoping for something for oneself" (—an "already" self!). (H345) In each case, the philosophical value of moods is that, far from being mere emotional clouds on the clear sky of the intellect, they reveal us to ourselves in our thrownness. Now this *thrownness*, literally understood, clearly involves a reference to the past, but more abstractly, if what it means is the "that it is" of Dasein, then it is not specifically the past that is privileged. Moreover, it can seem gratuitous to bring in one's *self*, and one's "having been" self in particular, (a) as if there were not moods in which the self is not entirely left behind, or marginalized, and (b) as if we could not, for example, hope for *others'* happiness (which would mean that that one does *not* always "hope for something for oneself."

The claim being made here, broadly speaking, is that Heidegger's discussion of the temporality of mood, while correctly eschewing any reduction of mood to twinges and tweaks, or to mere emotional disturbances, seems nonetheless to understand them as disclosive of a possibility of authentic selfhood. This seems to us to rest on a dubious privileging of the past in this case, and a challengeable optimism about giving a positive characterization of what anxiety discloses. Even more succinctly, Heidegger is *using* temporality in the restoration of a sense of human self-identity, one better left an open question.

It might be said in reply that it is anxiety above all else that puts in question all our worldly attachments, all our commitments, and any worldly sense of selfhood, and anxiety is just the opposite of complacency. This reading would make Heidegger into the hero of nonidentity, of fragmentation. A double reading is required, however, because of the ease with which what anxiety discloses is named and pressed into service.

Heidegger's discussion of disclosedness makes a further contribution to a reevaluation of the status of the present in his discussion both of the temporality of falling and of discourse. The temporal significance of falling (which can have only an inauthentic form) lies in the way it represents an entanglement, for example, in mere curiosity, in the affairs of the present. Discourse, however, is somewhat more interesting.

It has always been clear that "discourse" was different in type from the trio just dealt with (understanding, state of mind, and falling). They by themselves disclose the "there" of Dasein, and it is this that discourse then articulates. Linguistic articulation is only a possibility for discourse, and as such it does not have a primary ekstasis of temporalization. However, in its familiar form of spoken and written language, he claims, discourse is typically dominated by "making-present," addressing itself to the significant world about us. Heidegger makes this claim very briefly largely because much has already been said about discourse in Division 1 (section 34). But in all its innocence, it is actually a very important claim. For it takes a very clear position in debates about the respective privileges of reference, meaning, and play in language, one that would seem to give circumstantial application (here and now reference) the privilege. And yet it is not difficult to argue that reference is possible only if language is constituted by a structure (or play) of differences, or differential functions that do not themselves carry any primitive referential power.[4] It is only on such a position, it has been suggested, that it is possible to take account of that literature and those literary effects that break with referentiality. This is perhaps too large an issue to be explored here, but it is not without relevance to his next move.

THE INTRINSIC TEMPORALITY OF LANGUAGE

Heidegger offers a further very condensed discussion of the *intrinsic* temporality of language, a theme developed further on pp. 335–360. He distinguishes (a) the fact that discourse can discuss time, temporal processes, and so forth, (b) that it takes place in time, and (c) that it is temporal "in itself." And it is of course this last that is the most revealing, and is itself revealed in the fact of "tense" and "aspect." Tense and aspect in discourse themselves require a grounding "in the ekstatical unity of temporality." The time of language, in other words, is derivative from the time of existence.

Now this is a crucial claim. It means that the reason for the inadequacy of ordinary temporal concepts for understanding the time of language is much clearer. It further means that existential problems that take an explicitly linguistic form (for example, how to rescue the "is" from the status of a mere copula) can be more appropriately dealt with. It is also crucial in that it might conceal a more complex relation between the temporality of disclosure and existence than it suggests. (This will be discussed in greater detail on pp. 335–60.)

This is an opportune moment simply to lay down some questions. Would the dependence of the temporality of discourse on that of existence survive a radical interrogation of the foundationalist model implicit here? If one came to doubt that "in any discourse one is talking about entities" (H349), would one not have released discourse at least from being informed (and constrained) by a particular rhetorical mode? If the link between discourse and existence is retained, what would be the effect on one's understanding of the temporality of existence if one's understanding of the temporality of discourse were, on independent grounds, to change? The possibility of some of these questions bearing fruit should be assessed in the light of the critical approach persistently taken here to a whole succession of Heidegger's formulations. The claim will be made that Derrida's writing supplies the theoretical justification for just such an attempt at a transvaluation of the relationship between discourse and existence. And most important, it does so without compromising Heidegger's claim that ordinary concepts of time are inadequate to comprehend either. As was indicated at the outset, Derrida will claim that the concept of time itself is circumscribed by metaphysical motifs, a claim that will be disputed.

A number of critical claims have been made in the course of an appreciative discussion of the temporality of disclosedness. Doubts have been voiced about the possibility of giving a positive significance to authenticity, and it has been suggested, in the specific case of understanding, that it might better be understood negatively, or as a condition of possibility. Reservations have been expressed about Heidegger's claim that each of these modes of disclosedness has a dominant ectasy. And it has been suggested that Heidegger's account of temporality for all its revolutionary impact is still subservient to the problematic of selfhood and the project of restoring a sense to the self. But a general comment on this section was promised.

Heidegger has been engaging in a hermeneutic of temporality. And notwithstanding the doubts here expressed, there is no doubting the success with which he has demonstrated that the most fundamental phenomena of human existence (or at least those he has chosen to focus on) are susceptible to, and are greatly illuminated by, being rethought in temporal terms. Of course the possibility of doing this rests heavily on the way he has transformed our ability to articulate the temporal dimensions of existence in the form of iterable *modifying phrases*, thus supplying what I have elsewhere called a "virtual temporality" for every apparently simple position, orientation, projection, and so forth. And without meaning to be patronizing it is Heidegger's example, quite as much as the particular results he arrives at, that is so enormously compelling. His account of fear, for example which as a state of mind is meant to be based on Dasein's having been (rather than, as is popularly held, on the future), was criticized as unconvincing. But whatever the truth of the matter, the temporal multidimensionality of fear is incontestable, and Heidegger has made the search for such dimensions into an intellectual habit (in the best sense).

THE TIME OF THE WORLD

It might however be argued that the temporality Heidegger has displayed is limited to Dasein in its self-understanding, or to a "subjective perspective," or at least that the existential dimension is being considered very much from the point of view of Dasein, and with little reference to the world. And interesting as this might be, the place of time cannot be so limited. Surely Heidegger needs to offer *some* sort of account of what is thought of as the time of the world. In Heidegger's view such an account is only *now* possible: the temporality of "concern" (our involvement with the ready-to-hand) can be understood only on the *basis* of (in the language of) the more general temporality of disclosedness already developed. And if his earlier analysis is correct, that the present-at-hand is derivative from the ready-to-hand, it is *only now* that the questions of the temporality of the present-at-hand, of the theoretical attitude, and of the transcendence of/by the world that makes it possible can be posed. In short, Heidegger believes he can follow through, in a logical sequence, the derivation structure of temporality—the theoretical being based on the practical, the practical on the disclosedness of the "there" in general, and the theoretical also presupposing the possibility of transcendence, which possibility itself is to be understood temporally. In following this through, the discussion of Heidegger's account of the temporality of everydayness will be completed.

Central to Heidegger's being able to open up this new dimension of temporal reflection is his original conception of Dasein as Being-in-the-world. And for Heidegger, it is the ready-to-hand that is primitive here. His *general* argument for its fundamental temporality goes like this: We relate to things with which we are concerned not singly, but within a framework, which he calls "the equipmental totality." And that relation is one in which we projectively disclose (understand) the involvement of one item of equipment (a typewriter) in an equipmental context.[5] Heidegger uses the expression "letting be (involved)," which will emerge later and more prominently in such essays as "Gelassenheit."[6] Even here there is as much importance in the word "be" as in the "letting." What is at issue is the constitution of the specifically ready-to-hand mode of *Being*.

Now in every case of understanding, it is clear from the previous analysis of the temporality of disclosedness that its structure is fundamentally temporal. His way of showing this is to insist that the relation of involvement (which he dubs a "towards-which") between an individual item and that context of use in which it is located is one which, as an example of projective (that is, prethematic) understanding,[7] must itself be understood not as an "analytic" relationship but as a dynamic, or better, temporal one. And the two-way relationship (between part and whole) is characterized by the terms awaiting and retaining.

> The *awaiting* of what it is involved in, and . . . the retaining of that which is thus involved, make possible in its ecstatical unity the specifically manipulative way in which equipment is made present. (H353)

And

> . . .[this] making-present . . . makes possible the characteristic absorption of concern in its equipmental world. (H354)

In almost Husserlian language, Heidegger insists on the debt of the present to past and future projectivity.

At another level, one that necessarily *compromises* the authenticity of such a mode of involvement, the possibility of the kind of absorption and loss of reflection that goes with it requires "a specific kind of *forgetting*." And the interest of such a comment lies in the fact that it shows how—given the Heideggerean problematic of authenticity—the complexity of temporality is not confined to the possibilities of overlapping or modifying ecstatic projections, but that any one such complex can itself be the result of or made possible by an essentially temporal move at another level—in this case, forgetting makes possible "the unity of a retention which awaits," that is, absorbed involvement.[8]

With this analysis of the temporality of involved concern, Heidegger can and does proceed to show how it allows him to make sense of those forms of "circumspective concern" with the ready-to-hand which arise when things break down, are damaged, go missing, surprise us, or in some other way "obtrude" from their inconspicuous context of use. Each of these succumbs to, is illuminated by, an analysis in terms of retaining/awaiting.[9] They constitute an important step closer to the emergence of the theoretical attitude. (section 69b) The move Heidegger makes here is to claim that the theoretical attitude is the result of "changeover" from that of practical manipulation. What is particularly interesting is his denial that this "change-over" is simply the result of the suspension of the practical attitude. When that happens, "our concern" then diverts itself specifically into a "just-looking-around."

There is surely little doubt that it is Husserl and his account of the suspension of the natural attitude that he has in mind here. Heidegger argues, in fact, that the active circumspection (for example, checking equipment) often born from our suspension of practical activity is, again, not theory.

However, it is his contention that the theoretical (scientific) vision is derivative from the circumspection that can arise out of practical involvement. Such circumspection, based on Dasein's basic care structure, takes as its object the world of equipment, and although it involves a break with unreflective involvement, it also "brings the ready-to-hand closer to Dasein" by interpreting it within if-then frameworks of utility.

There follows an account of what Heidegger calls "the genesis of theoretical behaviour," one supposed to demonstrate the temporality underlying this behavior, and, if the title of the section is to be believed, of the transition itself. Heidegger's line of thought *seems* to involve the filling in of various steps, and the implications of these steps, from practical involvement to theoretical detachment. Circumspection, for example, is a making-present that brings to light the if-then structure of Dasein's deliberation. But what is thereby linked together in if-then relations must *already* have been understood *as* this or that. And understanding something *as* something, in which the future and having-been are drawn into the present, provides circumspective understanding with a temporal foundation.

Now, although it is far from clear in his exposition, I take it that Heidegger is saying that "understanding-as" is the closest we get, on the side of the ready-to-hand, to a radically new mode of relating to the world as

present-at-hand, between which two there is a "new way of seeing," one that
marks the gap between the hammer being heavy and the hammer possessing
weight as one of its properties.

This particular move, even in the elaborate version Heidegger gives us,
is surely disappointing. The reader may in some sense have been prepared
for it, but it seems to be more descriptive than explanatory and, moreover, to
little involve anything significantly temporal. It is important then to realize,
as Heidegger insists, that even if it is possible to have the ready-to-hand as
such, the "object" of a science (for example, economics), the fully scientific
approach involves a completely new projection of the Being of the present-
at-hand. And it is here that the *temporal* foundations of the theoretical,
scientific attitude are truly found, for (taking *physics* as the paradigm):

> In this projection something constantly present-at-hand (matter) is
> uncovered beforehand, and the horizon is opened so that one may be
> guided by looking at those constitutive items in it which are quantita-
> tively determinable (motion, force, location and time). (H362)

Heidegger is saying that the physicalist model—in which space and time are
mere qualitatively determined dimensions—has an existential ground. That
this is fundamentally temporal might be thought clear from the reference to
this as a projection (which takes one back to understanding, the temporality
of which has already been shown) and to its being "a priori."[10] But for
Heidegger, this peculiarly *scientific* kind of making-present, a unique form of
disclosedness, grounded in Dasein's Being-in-the-world, has its temporal
foundations best illuminated by an account of the temporality of the *transcendence* that its thematizing presupposes.

A superficial comparison of *Being and Time* with *Time and Being* might try
to develop the idea that the latter abandons the possibility of a purely
temporal perspective in favor of time-space. Heidegger's attempt in section
70 to subordinate Dasein's spatiality to its temporality might be thought to
support this view. Both readings would be mistaken, and it is Heidegger's
account of the transcendence "of" the world that makes this clear. The key
word will be "horizon," recycled from Husserl.

A brief word: the "of" in "transcendence *of* the world" is a subjective
genitive. It refers to the "world's" transcendence. Equally, it refers to
Dasein's transcendence (qua Being-in-the-world). What is transcended?
Both Dasein and the world, or Dasein as being-in-the-world, are transcendent in relation to "things," conceived of as just there, present. But how?
Heidegger's answer, in a phrase, is Dasein's "ekstatic-horizonality," and the
claim is surely a brilliant one.

The world, understood as a unity of significance, involving such relationships as "in order to," "towards-which," and "for-the-sake of," is essential to Dasein's being and vice versa. And to each of the temporal ekstases already isolated corresponds a "horizonal schema":

> *The existential-temporal condition for the possibility of the world lies in the fact that temporality, as an ekstatical unity, has something like a horizon.* (H365)

And the distinct horizonal schema (future ["for-the-sake-of-which"]; having ["in-the-face-of-which"]; present ["in-order-to"]) constitute a unitary framework within which things in the world can be encountered.

Two comments on this. It can helpfully be contrasted with the Kantian treatment of time as the form of inner intuition. On such a view, time would be subjectively real, but in some sense an *imposition* on our experience. Heidegger, borrowing the word "schemata" from Kant's "Schematism of the Understanding," shows how it is possible (and sketches the possibility in principle) to see temporality not merely as a subjective condition but as constitutive of the world itself. This account, of course, both rests on and further consolidates the original insistence on understanding Dasein as primitively Being-in-the-world. Second, that these ekstases constitute dimensions of significance of the world is surely incontestable. But that they are the last word on what makes transcendence possible is less clear. The other obvious candidate is language, or discourse. At every level—the individual word, the assertion, the complex articulation of tenses—language exhibits a transcendence in relation to *what* it discloses. If it is thought of as constitutive (in whole or in part) of the world, then its candidacy is ensured. But the status of language, certainly in *Being and Time*, is limited. Propositional truth rests on a prepredicative disclosedness, discourse articulates pretheoretical understanding, tense reflects Dasein's primordially ekstatic conditions. Language *seems* derivative, and at best a "supplement." And yet (a) one cannot fail to notice that Heidegger later comes to credit language with a much more originary role; and (b) even in *Being and Time* the argument is borne on the shoulders of tiny but vital words ("in the face of," "in order to," "as," "is," and so forth [*vor, um-zu, als, ist*]). Minimally, one might suggest that his consideration of the temporality of transcendence at least raises the question of the role of discourse here, and of the inner relation between language and time.

CRITICAL REFLECTIONS

What is the status of Heidegger's discussion of the relation between everydayness and temporality, and what conclusions can be drawn from it? Above all else, one has to recognize the tension in the notion of everydayness.

For it names *both* the basic and normal condition of Dasein, one on which the possibility of there being a temporality at all depends ("temporality is essentially falling"). (H369) It is, in other words, an inestimable source of temporal significance. (Indeed, he writes, ". . . at bottom, we mean by the term 'everydayness' nothing else than temporality." [H371]) And it is also a way of being characterized by evasion, suffering, (self-)dispersion, and "enigma after enigma." If Heidegger *equates* everydayness and temporality, is it enough to say that authenticity is a *modification* of everydayness in order to make sense of the idea of "authentic temporality"? The notion seems to be all too easily squeezed out. Heidegger's own remarks suggest considerable general uneasiness. He doubts whether "the explication of temporality . . . so far . . . is sufficient to delimit the existential meaning of 'everydayness'" (H370)—and his final chapters on historicality and within-time-ness will attempt to make this good. But he repeats a sentiment that has become increasingly frequent in this chapter—that the clarification of the meaning of Being in general is the ultimate aim of the book—one that, as will be seen, remains unachieved at its conclusion.

My own response to this vital chapter must now be formulated. A number of critical themes have already been drawn together—dissatisfaction with the role played by authenticity in its *negative* description of the phenomena of everydayness (for example, as a *forgetting* of one's ownmost potentiality-for-Being), with the attempt to exclude representation from his description of primordial temporality, and with the attempt at assigning dominant ekstases to the different modes of disclosedness. And a persistent interest has been taken in the relegation of the structures and functions of language to a secondary status in relation to time. What if they could not be separated? (Heidegger's discussion of the "as" [H360] could serve as a point of departure.) But there are one or two particular criticisms so far unvoiced that it would now be appropriate to bring forward.

THE TEMPORALITY OF PLEASURE AND DESIRE

One could ask, first of all, whether Heidegger's whole existential frame of reference, for all its internal subtlety, does not reflect a rather traditional and unexamined view of the nature of everyday life—which is essentially that of purposeful activity, means/ends relationships, and so forth. As a hermeneutic of the temporality of concern, it is surely unrivaled, but the *place* of such concern is surely not established beyond question. There would seem to be two "existentials," perhaps of greater importance than he would allow, not covered by such an account, existentials that have their rightful place in "everydayness"—those of *pleasure* and of *desire*. And what is important about them (and in this they are doubtless not unique) is that they may be thought to suggest modes of temporality somewhat disruptive to Heidegger's overall aim.

To demonstrate this convincingly, even supposing that these terms (pleasure, desire) could be innocently deployed without a glance at their philosophical (and wider cultural) history, would be an enormous task. It might, however, still be of some value to stimulate a rereading of Heidegger's account by *suggesting* certain limitations.

Pleasures could perhaps be distinguished into those that are ontologically conservative (that reassure) and those that are ontologically subversive (by exhilaration, by rupture, by challenge, and so on).[11] Suppose one were to hold on to the second, and add to it a third pleasure, which one could begin by calling ontologically innocent—say, the taste of honey.[12]

One could further distinguish two types of desire—that desire which seeks to fill a lack, to remedy a deficiency, and that which proceeds from an original exuberance. Again, the latter will be held on to.

There is no doubting that in each case, the overlapping modifications of Heidegger's temporal language would be appropriate, often revealing. But in each case, Heidegger could surely treat their unrelatedness to a self only *negatively*. And yet their significance is in no way confined to that of a "forgetting" of my potentiality for Being. They *cannot* be regarded as deficient modes of my coming-toward-myself. The "taster of honey" may be tasting, enjoying, a *liberation* from the whole question of his real Being. And the man of jouissance or of exuberant desire has found a mode of action that is not a mode of self-enactment. The clue to the break with Heidegger's account that these each involve is the positive disappearance of a projected horizon of the future. The future is *allowed to be* absolutely other.

THE ORDINARY CONCEPTION OF TIME

To the extent that Heidegger's account of existential temporality is successful, it also poses a problem: How is one then to think about *time*? For there is a clear sense in which Heidegger, like Husserl, has bracketed out the question of time to deal with temporality more effectively. By "time" is meant something not immediately obviously just a mode of Dasein's Being but one that could capture the thought that events take place "in" time, that certain operations "take" time, that times can be shorter or longer than one another, and so on. In our ordinary way of thinking about time, even if not treated as a thing, it seems to have some independence from our individual Dasein, it seems to serve as a dimension in which things out there arise and decay. The book is entitled *Being and Time*, so what can now be said about it?

If Heidegger is right, it is only now that he can properly answer the question. For his answer will consist of a complex derivation of our ordinary understanding of time from those most primitive modes of our temporality. It is complex because, as will be seen, it involves a number of stages, and because it involves a certain reflexive redoubling of temporal structure. At a vital moment, the full existential horizon of temporality is *forgotten*. And forgetting is itself a mode of temporalization. The ordinary conception of time, as a series of "now" points, dominated by the leveled-off present, is an inauthentic temporalization of (primordial) temporality. The import of this analysis is twofold. On the one hand, it serves to explain the origin of the unreflective references to time that we make in countless everyday expressions. On the other hand, it also enables Heidegger to offer an explanation of the source of what in his view has become the standard philosophical treatment of time, from Aristotle onward. The opportunity will be taken later to discuss the way Heidegger amplifies his views of the traditional philosophical treatment, in his *Basic Problems of Phenomenology*, and how this affects Derrida's discussion of a seminal footnote on Aristotle in *Being and Time* in his essay "Ousia and Gramme." The question will be the one with which this treatise began: Whether and how far Heidegger still shows allegiance to the traditional metaphysical understanding of time. I shall first consider the derivation of within-time-ness and of the ordinary conception of time.

In his discussion of circumspective concern he had argued that it was "grounded in temporality . . . in the mode of a making-present which *retains* and *awaits*." His first move is to say that to each of these there corresponds what we could call a mode of temporal reference. These can be thought of as represented by the words "then," "once" (as in "once upon a time"), and "now" (*dann, zuvor, jetzt*). To each of these temporal references there corresponds a horizon, for which, again, we have simple words: "later on," "today," and "earlier." And implicit in our use of words like "once," "now," and "then" is a further elaboration, such as "once when," "then when," and so on. In other words, they play their part in a relational structure of datability. And in such a use the primacy of the "now" already becomes clear. For the references "once" and "then" refer to a past and future "now" respectively. One can see, then, in the primitive activity of temporal reference that he calls datability (no calendar dates need be involved) the emergence of the privilege of the now as "making-present."

This set of corresponding triads of words was obviously not intended by Heidegger as an explanation of the possibility of giving dates, because he immediately asks the question, implying that he has not yet answered it, of how this datability is grounded. The commonsense answer supposes that there *are* points of time, and that these temporal references simply indicate them. But if what is sought is an explanation of such expressions as "now that," "once when," and so on, this will not do, for such "points" would have to have a significance that they are, as such, incapable of bearing. Datability, then, has to be grounded in something by which significance can be bestowed. Heidegger's answer is that expressions like "once when," "then when," and "now that" reflect not just an interpretation of what I am concerned with, but also a self-interpretation. Dasein is expressing itself as "Being-alongside the ready-to-hand." Moreover, it does so in the form of a "making-present." This is quite obvious. We say, "now," "then," and the like within a certain interpretive context in which we are involved (for example, recounting one's life, reminding the other of a promise, remembering a scene, and so on), and one does this from the point of view of and with reference to "the present." It might be asked why, if we always mean "now *that*," "then *when*," and so forth, we do not always spell out the tacit elaboration of significance. Heidegger's answer employs the reflexivity noted earlier. We do not always spell it out because although the ekstatic character of temporality is built into each act of dating, it equally gets covered up in each "making-present." "Making-present" hides its own ekstatic character in the "now."

In addition to these cases of temporal reference, which go to make up datability, Heidegger shows how Dasein understands itself not only in relation to temporal points, but also periods. He explicates phenomenologically Dasein's spannedness, stretchedoutness, through such simple phrases as "during," "meanwhile," "until-then." And he could have talked of the continuous present tense. It is this spannedness that makes it possible for Dasein to "take" time "off," to "spend" time, to "have" time to spare, and so forth and for every temporal reference to have its own span (some "nows" are longer than others!).

But this spannedness, stretchedness, of Dasein cannot be understood in terms of a sequence of nows. It does not, for one thing, offer the guaranteed (if spurious) continuity of such a sequence. So what is its basis? Clearly a linguistic phenomenology is not enough. Heidegger's answer is that our ordinary stretching along, our ordinary ability to think of ourselves as living in spans of time, is an inauthentic falling away from that projected stretching ahead of oneself opened up in the "moment of vision":

> One's existence in the moment of vision temporalizes itself as something that has been stretched along in a way which is fatefully whole in the sense of the authentic historical *constancy* of the Self. (H410)

The person who says he has no time for this or that has (inauthentically) lost himself in objects of immediate concern, and is someone for whom the moment of vision is no longer operative. In other words, Heidegger interprets the possibility of *periods* of time in terms of *ekstatic* temporality, transformed and covered over. But is this not confined, as formulated here, to each individual, one at a time? Where do we get the idea that we share the same time? Heidegger's answer is that the significance of expressions like "now" and "then" is one that rests on our public Being-in-the-world. We do not always mean precisely the same thing by "now," even when we say it together, but the everyday disclosedness of Dasein is something all the more shared for being everyday. Indeed, it is one of the distinguishing features of this public time that we think of it just as something "there is," *in contrast to*, say, the authentic understanding of time in relation to one's ownmost-potentiality-for-Being.

But in a sense this only raises more problems. For one might well think that what had been offered was only an *explanation* of how "public time" arises rather than an account of what "it is" that has so arisen. And Heidegger insists on the need to return to such a question. One might, for example, still wonder whether this "public time" was something we imagine, project, impose on the real world, or whether it was really there. In his view, these sorts of metaphysical questions get confronted seriously only by detailed phenomenological work.

The publicness of time cannot be dependent on the business of applying numbers, the quantification of time, for there is a reckoning with time that is altogether more primitive, and that is quite sufficient to establish a public time. Moreover, it is one in which the question of whether time is objective or subjective seems out of place. The difference between day and night, and the height of the sun in the sky, supply us all in a publicly available way with a grid for the assigning of times, one that applies equally to the natural and the social worlds. This is the first clock, on which all subsequent clocks, or at least all subsequent use of clocks, ultimately depend. Heidegger insists, however, that "temporality is the reason for the clock."

His sometimes complex formulations have as their goal (a) to demonstrate the dependence of any sense of time as an autonomous thing, on its being "public," (b) to show that its being public is dependent on our utilization—in a shared world—of natural and manufactured clocks for measurement and dating, (c) that such utilization has to be understood in existential terms, and (d) that these terms are ultimately temporal in their significance.

What is the relation between this public time and the "world"? What can be made of the expression "world-time"? Heidegger presses forward his demonstration that our use of simple temporal words has a built-in worldly significance. Time references often have a built-in element of appropriateness or inappropriateness. We talk of "time for . . ." ("Time for tea"), and again, a reference to the present betrays the structure of awaiting-retaining that extends its ultimate reference beyond the present. But this significance structure that we are drawn into is nothing other than the world itself. This public world-time is not something in the world, but part of it, it belongs to it. Perhaps the relation could be summed up like this: concern structures the world and is rooted in temporality, and the temporal way it structures the world appears as time.

Heidegger, as mentioned earlier, has claimed the peasant's world as his inspiration, but he does not want to restrict his discussion of public time to the way we can make appointments by measuring the length of our shadows, however significant such primitive time reckoning might be. He is interested in the existential significance of the fact that we have moved on from there to the increasingly sophisticated use of devices to tell time and, as he might have added had he been writing a few decades later, to the regulation of clocks by the vibration of a cesium atom, even more reliable than the sun in its travels.

A PHENOMENOLOGY OF MEASUREMENT I

At this point, Heidegger attempts, in effect, a phenomenology of the measurement of time, a move that justifies the subtle way in which Derrida handles Heidegger's negotiation of a relationship with the traditional Aristotelian view that connects time with measurement.[13] The connection is undoubtedly there, and the question is whether phenomenology, and in particular this existential phenomenology, can transform our understanding of that connection in a significant way. This question is discussed explicitly later.

There is more to telling the time than looking at a clock. We can see the hands, their spatial positions and the divisions on the face of the clock, but we do not see what time it is unless we already bring an understanding of time to bear on our looking. For, "when we look at the clock and regulate ourselves *according to the time*, we are essentially *saying 'now,'*" (H416) and, as Heidegger says, this "now" has already been understood in terms of "datability, spannedness, publicness, and worldhood." Heidegger is surely right here in supposing that our use of clocks already presupposes temporality. Think of what it is like to wake up after a doze and wonder after looking at one's watch whether it is morning or evening. It would not make sense to say that one *knew what time it was* but did not know whether it was morning or evening. And yet one might know that it was seven-thirty.

VERSIONS OF "PRESENT"

What is the ontological significance of our ability to tell the time? Heidegger's answer to this begins with a paragraph that deserves a certain unpacking (H416–17). For those who rely at all on the English translation of *Being and Time*, it becomes essential at this critical point for our understanding of the significance of a phenomenological reading to return to the German original. A number of translation choices that are justified independently come together to wreak confusion. Etymological connections are lost when they are important, and parallels appear which were never intended. In the sentence that ends ". . . an entity which is present-at-hand (vorhandenen) for everyone in every 'Now' (Jetzt), is made present (Gegenwärtigen) in its own presence (Anwesenheit)," (H418) the word "present(-ce)" occurs three times, in various forms, in English, where three quite different words are employed in German. And the sentence beginning the paragraph makes it far less mysterious how Heidegger connects "making-present" with "retentive-awaiting":

> Saying "now," however, is the discursive Articulation of a *making present* (Gegenwärtigens) which temporalizes itself in a unity with a retentive awaiting (behaltenden Gewärtigen). (H416)

To these illustrations one should also add the words "the Present" (*die Gegenwart*) (literally, the "waiting-toward") and "in the past" (*gegenwärtig*). The complexity of the interconnections among these words (to which in all forms the translators allude in their footnotes to H25) raises all manner of questions it is hard to know how to begin to answer. Is Heidegger's whole problematic, or parts of it (for example, the etymological suggestion of *futurity* in the literal sense of *Gegenwart*), dependent on the peculiarities of the German language? (If we tried to do the same thing in the same way in English, might we not conclude that the present was a pre-esse, an "is before," with roots in the past?) Still, one might at least confirm the importance of the connections between time and Being by such etymologizing. And Heidegger might well argue that the ontologico-etymological parallels among par-ousia, an-wesen, and pre-sent (L. *praesens, praesentum,* prop. pres. ppl. of *praesse*, to be before, to be at hand [*SOED*, p. 1573]) were obvious enough. There is a remaining difficulty. If for Heidegger the historical "translation" of Greek temporal terms into the Latin language represented a covering over of the temporality of time,[14] then the attempt to recapture Heidegger's claims in a term (present) with Latinate roots may be a further historical reversal. The same would, of course, be true of French. While one might occasionally find if not solace, at least an explanation for the difficulty of understanding Heidegger in such a consideration, one's hope must be that even if such a problem did exist, it could be overcome by (a) the awareness of it, (b) having a multilingual basis of comprehension, and (c) the use of subtle syntactic forms to compensate for the unwanted implications of the terms we must use (for example, the verbalizing tautology "das Nicht nichtet," "the present presences"). Clearly Heidegger thinks one even needs them in German.[15]

It will be appreciated that a *presentation* of Heidegger's phenomenology of measurement—which, as has been suggested, comes to a head in H416–17f.—is far from easy. Although *this* is written in English, there is a sense in which it is an Anglo-German hybrid that is getting written *in* English. It might therefore be helpful to be reminded of one or two of Heidegger's fundamental claims about what history has done to our understanding of time, claims first made in his famous section 6, "The Task of Destroying the History of Ontology."

Heidegger believes that it is only in ontology that the fundamental questions can be posed, that the possibilities of ontological thought, once open to ancient thought, were closed off by their particular formulations and need to be reopened. The key to this de-struction is that it was always in relation to Time that Being was interpreted. For them,

> ... the meaning of Being [was] ... παρουσία or οὐσία which signifies in ontologico-Temporal terms, "presence" ["Anwesenheit"]. Entities are grasped in their Being as "presence": this means they are understood with regard to definite mode of time—the "Present." (H25)

There are two points here: first, "presence" is an ontological term; "present" is a temporal term. The claim is that the temporal has determined the ontological. Second, a definite "mode of time" has been so privileged, that is, the present, rather than the past or the future. Heidegger's aim is to reevaluate both of these connections. For the reader the fundamental interpretative question must be: Does Heidegger accept the determination of Being by time as "presence" (and simply want to give a new sense of "presence"), or does he want to challenge the relationship more fundamentally? Derrida, will claim the former,[16] but it might be more accurate to say that with Heidegger the question becomes undecidable. The best discussion of this question can be found in David Krell's *Intimations of Mortality*. He shows that even while Heidegger is elaborating the ekstatic analysis of existential temporality, he is committed to the environmental Praesenz of the world. The question remains as to whether the ekstatic analysis merely articulates this Praesenz, or whether Praesenz is something deeper, more fundamental. (See the discussion here on pp. 28–9) The undecidability touched on here is the source of the double strategy of deconstruction by which this book is organized.

Discussing the waning of the importance of "dialectic" after Plato, Heidegger continues:

> . . . Aristotle "no longer has any understanding" of it, for he has put it on a more radical footing and raised it to a new level (aufhob). Λέγειν itself—or rather νοεῖν—that simple awareness of something present-at-hand in its sheer presence-at-hand, which Parmenides had already taken to guide him in his own interpretation of Being—has the temporal structure of a pure "making-present" of something. Those entities which show themselves in this and for it, and which are understood as entities in the most authentic sense, thus get interpreted with regard to the Present; that is, they are conceived as presence (οὐσία). (H25–26)

Both our fundamental mode of awareness and the things of which we are aware, both poles of intentionality, to use Husserlian language, are indebted to an emphasis on a specific temporal mode—making-present (*Gegenwärtigens*). The possibility of a phenomenological investigation seems here to reside in a certain receptiveness to etymology.

A PHENOMENOLOGY OF MEASUREMENT II

The next task is to elucidate Heidegger's phenomenology of measurement. A brief explanation will first be given of what Heidegger is doing, followed by more detailed comments.

What *is* Heidegger doing? He is giving a temporal interpretation to the apparent immediacy with which we "tell the time" and understand and measure intervals in time. The sentence in which he begins to explain this has already been cited. When we say "now," we are giving voice to a "making-present" (*Gegenwärtigens*) that temporalizes itself in a unity with a "retentive awaiting" (*behaltenden Gewärtigen*). (H416) In other words, the immediacy of the "now" dissolves in an activity in which a bringing about of the present, making-present, presencing, presentifying . . . grasping/ affirming something as present is itself "performed" only in conjunction with an orientation that points both forward and backward. What we relate to in this measuring activity is something Vorhanden, something just "there" ("present-at-hand") as it seems. But our relating to it involves more than just passive receptivity—a "making-present." There is a sense in which the measuring of an interval involves a double relatedness to the "standard"—by which he presumably means the "unit of measurement"—say, the way the clock is *divided* into minutes, hours, and so on, and to the actual *extent* of the interval. Each of these is confronted as something present-at-hand, and measuring consists of a "making-present" of what is thus Vorhanden. We may not think of ourselves as relating to Time as such when we measure it, but the fact that measuring thinks it is dealing in a public way with a present-at-hand multiplicity of "nows" provides a basis for its subsequent theoretical interpretation.

It might be said (and many, including Bergson, *have* said it) that the measurement of time involves a (misleading) spatialization, but Heidegger's response is that the "space" supposed to be responsible for this is itself dependent on temporality, and so we are not dealing with a misplaced transformation of modes. Measuring is not a reductive spatialization but a specific making-present.

What does all this have to say about time's ontological status? Clearly "time" is made public by the institution of measurement, and things can have "times" attached to them as soon as they are seen to be "in" this public time. However, this time in which it is possible for things to be "in time" is neither objective nor subjective (that is, in Heidegger's language it is not Vorhanden [present-at-hand] in any way). It *could* be called "objective" (though it is no "object") insofar as it is the "condition for the possibility of entities within-the world." It is certainly not just (with Kant) the form of inner intuition; it is encountered in the world just as immediately. Again it could be called subjective, but only as a condition of "the Being of the factically existing Self"—not as some psychical faculty or component. What then is the status of this world-time, this time-within-which? It cannot now be thought of, if we have followed Heidegger's argument, as a Dasein-independent framework. Rather it is a product of "the temporalizing of temporality." Keeping this in mind allows one to see through these false ways of construing time as either "objective" or "subjective." Within-time-ness is both a legitimate and important phenomenon, but it also hides its own temporality in its immediacy. Through it, the everyday concept of time becomes intelligible.

Heidegger's moves here are surely commendable. He is giving a plausible account of how "time" can be thought of as fundamental without adopting the usual course of calling it either "objective" or "subjective," while yet managing to keep in the air at the same time the fact that its status is distinct from and makes possible particular things ("in time"), its publicness, its role as condition of factical selfhood, and its relation to the (significant) world.

ARISTOTLE AND THE ORDINARY CONCEPT OF TIME

With this account of what it is to be "in" time, of the time "in" which things are to be found, he is now in a position to offer an explanation of how our "ordinary conception of time" (as a series of "now-points) has arisen. Heidegger explicitly and for the first time acknowledges that what he has been offering is ". . . nothing else than an existential-ontological interpretation of Aristotle's definition of 'time'" (H21) (which was ". . . that which is counted in the movement which we encounter within the horizon of the earlier and later"). It is not clear from this context whether he means that it just so happens to have this connection with Aristotle or whether it has been his underlying aim all along to provide such a phenomenological deepening. But it is fairly clear from the early remarks recently cited (H25) that he has seen a close connection between the ordinary conception of time and Aristotle's view. But the claim is not that the ordinary concept of time is somehow derivative from Aristotle's. Any appearance to this effect is explained by the fact that Aristotle and philosophers after him have interpreted time on the same basis as is shared by the ordinary (prephilosophical) concept of time, namely, our commonsense horizon (or lack of it).

There is clearly a considerable gap between what Heidegger has called "world-time," which possesses "significance" (for example, time we were finished, time to get up), and "datability." His explanation for this is in terms of "covering over" and "leveling-off." In fact, these terms seem to walk the line between description and explanation somewhat uneasily. For at one level they are purely *descriptive* of the difference between the "now" to which significance relations are essentially related and the cleaned-up and isolated "now" that stands in a series, without suggesting how the change has been effected. Heidegger's answer to this question is essentially that everyday concernful living in the world gets lost among the objects of its concern and fails to see the horizonal structure that makes this "content" possible. So we treat each "now" as something "present-at-hand." The ordinary conception of time as *infinite* is then only symptomatic to such a view. For once all relations between "nows" have been reduced to those of *sequentiality* there is nothing to put an "end" to the series. Apart from the leveling brought about simply by the myopia of everyday concern, the ordinary conception of time as infinite also serves to underpin our flight from our own mortality,[17] this leveled-off, infinite time belongs to everyone and no one.

Heidegger's symptomatic reading is elaborated by two further claims: (1) that the expression "time passes away" should be seen as a clue to our recognition that our own passing away has not been entirely forgotten, and (2) that the irreversibility we attach to this ordinary concept of time should be seen as the legacy of *primordial* time poking its way through.

Does this expose an error? Heidegger claims that the ordinary concept of time has its place, its natural justification, as belonging to "Dasein's average kind of Being." Error arises, one might say, only when it "errs"—when it strays beyond its horizons of legitimate application. Heidegger is offering diagnosis, interpretation, not critique. It is the relationships of derivation, dependence, grounding, between temporality, world-time, and the ordinary concept of time that are quite crucial.

> From temporality the full structure of world-time has been drawn; and only the interpretation of this structure gives us the clue for "seeing" at all that in the ordinary concept of time something has been covered-up, and for estimating how much the ekstatico-horizonal constitution of temporality has been levelled-off. (H426)

Starting from the ordinary concept of time (as of course Heidegger has *not*) so much is simply invisible and unthinkable—not only "temporality" but the primacy of the future for "ekstatico-horizonal temporality." There is a radical difference between the temporal elements of ekstatic temporality and those of the ordinary concept of time, even (and precisely) where they seem to map on to one another. What has "the moment of vision" to do with a "future-now"?

It is Heidegger's contention, of course, that it is the neglect of this "ekstatic-horizonal temporality" that characterizes not just the ordinary concept of time at some preliminary commonsense level of reflection, but also the key philosophical accounts of time offered after Aristotle, for example, by Hegel and Kant (and, he would add, Nietzsche). Certain remarks to which attention has been drawn in passing suggest that Husserl too would be included in this list. And near the end of *Being and Time* (H433f.), Heidegger explicitly discusses the difference between his own views and those of Hegel, and in a long footnote (H432 n.xxx) discusses Hegel's indebtedness to Aristotle. This important question, raised again by Derrida in his essay "Ousia and Gramme"[18] and by Heidegger's *Fundamental Problems of Phenomenology*,[19] is discussed separately on pp. 251–263. It supplements in a more historical way the discussion here of the possibility of a nonmetaphysical concept of time.

FROM TEMPORALITY TO TIME: SOME CRITICAL THOUGHTS

Heidegger's discussion of the genesis of the ordinary conception of time has been presented largely without criticism, so as not to detract from the course of the argument. But passive acquiescence is neither possible nor desirable.

One can accept, broadly speaking, his account of the dependence of theoretical concepts of time on one that is more tacit or primitive. Heidegger's main insight here is the double sense of everydayness—both as the *source* of significance and of its concealment. There is, however, surely room for doubt about his explanation of the ordinary concept of time.

He makes no reference to the possibility that this abstract succession of "nows" might have its basis elsewhere—for example, in the development of the number system, which exhibits the same unending seriality (in both directions, if we allow negative numbers). This is of course debatable.[20] But the following two arguments might be adduced in its favor.

First, the idea of an infinite series of dimensionless, meaning-less "now" points is one it would be rash to attribute a priori to primitive people (people lacking, inter alia, a number system of the sort to which we are accustomed). But the scope of *Dasein* is not restricted to modern Western man. Indeed, one might have thought that, as Durkheim said of primitive people, they might display the essence (Being) of men all the more clearly.

Second, the move from the practice of measurement to the conception of time as an infinite series of nows is not that convincing, taken as an argument. Quite why the "now" should come to be seen as present-at-hand, shorn of significance, is left unclear. This is not to deny the value of a phenomenology of measurement, but it might suggest an acknowledged role for "structures" and "representations." Surely something ought to be made of the fact that when we measure temporal intervals, we mark the *limits* of these intervals, and *in themselves* these point instants *have no meaning*. If we think of time as something *in which*, then the dimensionless time of the "limit" can have no meaning of the sort that requires the scope for an event to occur. However, if *inner* meaning has been eroded ("shorn"—Heidegger's word is *beschnitten*), we can *still* have outer relations—of serial order—and we have in the number series a model of just how that is possible.

This last remark, however, marks the point of difficulty without resolving it. Brouwer, for instance, might argue that any idea of succession we found in the number series would have to be derived from our experience of real succession (that is, that of time). But if Heidegger is right, our fundamental intuitions are *not* of empty intrinsically meaningless points succeeding one another, even without bringing in the polyecstatic nature of temporality. They rather consist of meaningful periods, in which the "units" of significance, if one can so call them, would be existential. The time it takes to tie a shoelace would be more primitive than "the instant." This primacy is not chronological, nor logical, but hermeneutic—that is, making reference to "that in terms of which something can alone be fully understood."

But if it is central to the ordinary conception of time that its elements be meaning-neutral, and their succession be endless, then what is in question is something from which *representation* cannot be excluded. A structure of *external relations* is being *brought to* our awareness of temporal succession, and surely cannot just be derived from it. The argument would be this: *meaning-neutrality* and *infiniteness* (just to begin the list of formal attributes of such a series) could not be thought to be *discovered* true of the series and its elements nor could they plausibly be thought to be derived from each individual "now." But whatever the source of this ordinary concept of time, its status is an a priori one in relation to any part of the series we may be confronted with. And in each case, grasping its infinity is cashed out in a confidence that one will always be able to continue counting, measuring, and projecting time in any direction. Again, one might say what is at issue is a *concept*, and as such something with a representational content. Heidegger *presents* the ordinary concept of time as a product of a certain covering over, leveling out. But surely it has a positive content—a content manifest in the clarity of its structure, one that compensates to some extent for its apparently "negative" origins. If one were to take a piece of paper and cut out of it a perfect circle, it *could* be said that one had just *cut off* certain bits, but the circular line one followed or the template one cut around gives positive guidance to the scissors. By analogy, one might suggest that the existence of the number system might supply a model from which could be drawn the ordinary concept of time. This is not a proof, rather a possible solution to a deficiency.

There is, however, a danger of offering a facile objection to a more interesting error, and one should at least spell out the kind of response open to Heidegger. Suppose he were faced with the suggestion that the idea of infinity be understood as a rule—for example, to any series of "nows" another "now" can be added at either end. This, in other words, is an a priori concept guided by an interest in formal properties, not derived by a negative process from a certain mode of existence. Heidegger might reply, this recursive rule is fundamentally a *temporal* projection, one that applies equally to past and future. So even if it is admitted to be an a priori structure, only one further step is needed to *confirm* its fundamentally *temporal* status. (Even the words "a priori" give that away).[21] That must be true. But it still seems that what is confirmed is an intimate fusion of the formal and the temporal, structure and time. One cannot specify the rule that gives infinity to the series without involving both senses.

If this is right, and the ordinary concept of time were to be inhabited or informed by a structure imported from outside, then the possibility of a very different explanation of its origins could be opened up. One might wonder whether the philosophical theorizing about time had not so much taken for granted an historically invariant "ordinary" concept of time as it has privileged that mode of representation—the number system—associated with the birth of systematic measurement. Heidegger's account (particularly in 11.4 and 11.6) should then be seen as an attempt to incorporate one language of time into another. The other is that found in *ordinary language*, in tense, in certain indexical expressions, in numerous *little* words (and indeed in suffixes and prefixes).[22] If the argument has a linguistic level, this however is itself treated as underpinned by the existential. Discourse merely *discloses* existential articulation. Were this view to be questioned, however, were one to come to believe that no concept of time could be kept free from the determination of this or that language, model, schema, or mode of conceptualization, then a third term—"language"—would have to be added to *Being and Time*. And in effect, this is what Derrida, building on Heidegger's own later work, has done.

5 From the Earlier to the Later Heidegger (and Derrida)

The discussion in the last three chapters could be likened to a critical trawling along a path laid down in advance by the internal development of *Being and Time*. The key question determining the shape of the "net" had to do with the way Heidegger's ultimate concern with the question of Being opened up the possibility of a radically new existential sense of temporality, and yet at the same time remained subordinated to values of wholeness, unity, and closure. Heidegger's requirement that Dasein exhibit itself as a "whole" stems not from the "facts" of human existence, but from his wider project of making the existential analytic open the way to an illumination of the question of Being. To be specific, I take issue with his attempts to exclude "representation" and "otherness" in general from an account of the existential significance of death, I deny that Being-toward-death can serve as the foundation for Dasein's wholeness, I question Heidegger's claim that what conscience discloses it does so unambiguously and directly, and argue that it is its negative, critical role that is important. What is traced in detail is a *double movement of opening up and closing off*, one indeed that Heidegger's very account of disclosedness would have anticipated. However, just as something of the Hegelian dialectic can survive loss of confidence in the Absolute as telos, perhaps much of the Heideggerean account of existential temporality still retains its plausibility, and in two different ways. First, it is possible repeatedly to discount the positive content of his moves toward authenticity, wholeness, and resoluteness, and to take them in their negative, critical aspect. And second, one can read *Being and Time* as a witness to a "metaphysical" desire, the elimination of which—either for Dasein, or for the reflective philosopher—still remains problematic.

In making these points, it is not implied that Heidegger just stood still after *Being and Time*. He himself clearly found the transition from his existential analytic to an account of Being as such to present more than a temporary difficulty. In his writing after *Being and Time*, he moves in a direction that gives to *this* book its double trajectory. My interest in developing an account of temporal structures, informed and guided by those of "textuality," could be said to be grafted on to a version of existential analytic purged of the problematic of Being. And yet in the account of time as an opening on to radical otherness, drawing on Nietzsche, there is some convergence with the path Heidegger takes from *Being and Time* (1927) to *Time and Being* (1962). I shall now try to sketch this move.

In *Being and Time*, while the orientation of the book as a whole is ontological, directed toward the renewal of the question of Being, the actual content, as has been seen at length, deals more specifically with the existential structures of human being, of Dasein. In the second half of the book an adequate account of these structures is shown to be possible only by their interpretation within the horizon of temporality. Temporality is understood as a quite general existential condition, illuminating Care, Dasein's basic existential structure. It is through Care, for example, that the different temporal ekstases—being-as-having-been, being-as-making-present, and being-ahead-of-oneself—are united. The time we associate with the clock, time as measured, is not basic, but dependent on human temporalizing. Time is not only existentially primordial, but also primordially existential. Time has long ceased to be a sequence of nows. It has become a basic existential condition.

As has always been demonstrated, there seem to be all sorts of virtues in this position, not least of which is that Heidegger has given us a phenomenological account of temporality that is not merely that of an internal time-consciousness. While Heidegger was undoubtedly indebted to Husserl, no adequate account of human temporality is possible unless it takes on board the fact of our existence as self-questioning, moodful, embodied, concerned, and historical beings. And in these respects, Heidegger marks a certain advance on Husserl. But in Heidegger's view it would be a great mistake to think of this existential analysis as an end in itself. The existential analysis of Dasein, Being-there, was intended as a gateway to Being itself. Without this wider gloss, philosophy would risk being confused with anthropology. Husserl, who had previously leveled this charge against Heidegger, had himself fought long against such dangers. But whatever doubts Heidegger was later to have about whether it was philosophy he was doing, he was certainly not offering just a general account of man, a new "humanism."

While an existential account of temporality seems extremely fruitful, Heidegger only scratched the surface of such an account in *Being and Time*. And yet in his later work he moves not only away from a phenomenology of Dasein, he moves away from an account of Time as related to Dasein at all.

For all the brilliance of *Kant and the Problem of Metaphysics*,[1] in which he draws out the latent thematics of finitude in Kant's first *Critique*, there is very little mention of man as a being-in-the-world. If Kant's central aim, as he claims, was not epistemology (any more than Heidegger's own in *Being and Time* was anthropology), it is still true that Kant poses the problem of metaphysics within the framework of a theory of knowledge. And central though time is to this account, its role is cognitive, both in relation to the transcendental imagination and to the three modes of synthesis. Heidegger's account of temporality does not substantially change here. Rather, the framework of *Being and Time* offers a way of reading the Kant book, and as a result of that reading *Being and Time* can be seen as an advance on Kant's first *Critique*.

Suggestions of a break or reversal in Heidegger have been exaggerated. But there is an implicit devaluation of *Being and Time* in the direction taken by his later work, one that affects his account of temporality. The account of time in *Being and Time*, is not peripheral to his development.[2] Its "ekstatic temporality," when detached from any particular existential content, can be seen to serve as a kind of optics or framework within which Being can be determined in various ways—as "presence," for example. For time has ceased to be an objective succession in terms of which independently identifiable beings could be ordered. It has become an essential feature of the "constitution" of such beings, to use a phenomenological expression.

Heidegger has always avoided the hint of subjective activity that the term "*constitution*" might be thought to imply. The concept of the *subject* as active origin becomes increasingly important for him to avoid. So time cannot be something we add to raw data to create temporally complex objects. Rather, through the opening, the time-space, as he comes to call it, Being gives itself, or *it gives* (*es gibt*), or there is an *event of appropriation* (*ereignis*).

The central point of all his meditation on the *es gibt*, and on *ereignis*, is the repudiation of the categories of activity and passivity, subject and object, in understanding Being, for the good reason that these categories always presuppose Being in some way or other. But no one, surely, comes away from *Time and Being* with any real sense of having grasped at last what Time is all about. Time is here not only assimilated into time-space, but even that general dimensionality has lost most of its content.

This seems to be inevitable, given Heidegger's project. But it either casts doubt on the value of that project, or on the concern to elucidate the nature of time. Might it not be that "time" is inextricably bound up with metaphysics, as Derrida has suggested? So that one ought to expect that increasing clarity about time would see it dissolve away as some sort of metaphysical illusion.

The relationship of historical determination between Time and Being goes in each direction. In the first direction, a modality of Being determines our conception of time. This can be found in summary form at the end of Heidegger's *An Introduction to Metaphysics*:

in the beginning of Western philosophy the *perspective* governing the disclosure of Being was time, though this perspective *as such* remained hidden, and inevitably so. When ultimately *ousia*, meaning permanent presence, became the basic concept of Being, what was the unconcealed foundation of permanence and presence if not time? But *this* "time" remained essentially undeveloped and (on the basis and in the perspective of "physics") could not be developed. For as soon as reflection on the essence of time began, at the end of Greek philosophy with Aristotle, time itself had to be taken as something somehow present, *ousia tis*. Consequently time was considered from the standpoint of the "now," the actual moment. The past is the "no-longer-now," the future is the "not-yet-now." Being in the sense of already-thereness (presence) became the perspective for the determination of time.[3]

The essential argument here is that while Being had since the pre-Socratics been understood through time, this temporal condition was not reflected on until "time" had been taken over by physics. And this way of conceiving of time made its earlier role in interpreting Being impossible to formulate.

This fascinating claim opens up the possibility of rethinking the original phenomenon of time while at the same time making it clear how difficult that would be. It is the project that Heidegger takes as his own.

In the reverse direction, Heidegger offers us an analysis of four ways in which Being has been *determined*, in particular oppositions, by time. It is worth, again, briefly quoting *Introduction to Metaphysics*, at a point at which a long and detailed discussion is being summarized:

over against *becoming* being is permanence
over against *appearance* being is the enduring prototype, the always identical
over against *thought* it is the underlying, the already-there
over against *the ought* it is the datum, the ought that is not yet realised or already realised
Permanent, always identical, already there, given—all mean fundamentally the same: enduring presence, *on* as *ousia*.[4]

So clearly, in Heidegger's view, it is a temporal modality— "presence"—that each of these determinations of Being shares. And together these constitute much of the fabric of metaphysics. For while the fact that being can be *determined* in these ways presupposes a distinction between Being and beings, this distinction, which is absolutely necessary for Heidegger, is itself covered over and hidden. To raise again the *question* of Being is to render problematic the very meaning of Being and the possibility of such determinations. And this means to ask about "presence," to ask about primordial temporality.

But will it further the understanding of Time to follow Heidegger down this path? Am I not being *lured* by a phrase into a discussion with a quite different focus? Heidegger's concerns are with Being, with an original disclosure of Being, such as can give man a new historical orientation . . .[5] Surely anything he might say about time would be by the way.

And yet Heidegger *is* making a move that seems essential for any account of time that does not merely reflect our ordinary prejudices. He is trying to reawaken an experience not only of Being but also of Time, an experience both nonstandard and fundamental. The odds are against such an enterprise:

> an age which regards as real only what goes fast and can be clutched with both hands looks on questioning as "remote from reality" and as something that does not pay, whose benefits cannot be numbered.[6]

Nevertheless, the prospect is held out of an understanding of time more basic than that of a succession of "nows," and that can be developed in relative independence of the existential analytic of Dasein in *Being and Time*. And for this it is necessary to return to "presence" again (see pp. 241–3.)

This is a particularly difficult and yet important question. That it is difficult will soon become apparent. It is particularly important because of the weight attached to the concept of "presence" by Derrida, and because of the ambiguous distance from Heidegger this involves him in. Derrida regards an appeal to (a) presence of one sort or another as the hallmark of the logocentric or metaphysical tradition, and he is not sure that Heidegger, on whom he draws for this analysis, actually escapes from this tradition:

> to the extent . . . logocentrism is not totally absent from Heidegger's thought, perhaps it still holds that thought in the epoch of onto-theology, within the philosophy of presence, that is to say, within philosophy itself.[7]

This will be returned to later. Meanwhile, it provides a way of approaching Heidegger's account of "presence" (*Anwesenheit*). For it is true, Heidegger does not seem to claim that the interpretation of Being as presence is *wrong*, but rather that we need to rethink the meaning of this "presence." We need to reexperience it.

An excellent account of what the author calls "the presencing process" is offered by Werner Marx in his book *Heidegger and the Tradition*,[8] and I shall draw substantially on this in my remarks that follow. Marx tries to capture

Heidegger's account of the "presencing process," or what could be called, "presencing," by focusing on three of the key Greek words that Heidegger rethinks or reworks: *physis, aletheia,* and *logos.*

As has been seen before, Heidegger in *Being and Time* had already talked about presence in the form of the Vorhanden, Being-present-at-hand, a type of Being contrasted with the Ready-to-hand (*Zuhanden*)—also translated as "tools" or "equipment"—by its inert, static presence.[9] The most obvious example of this is Descartes's conception of a *res extensa,* a merely extended thing. This reading of presence would weight it on the side of the object. But in his later work, Heidegger came to understand presence as characterizing at the most general level the relationship of man to the world, and to himself, or what Werner Marx calls the "subject-object" relationship. Heidegger calls presence "the constant abiding that approaches man, reaches him, is extended to him."[10] It has taken a number of different historical forms:

> . . . presencing shows itself as the *hen*, the unifying unique One, as the *logos*, the gathering that preserves the All, as *idea, ousia, energeia, substansia, actualitas, perceptio, monad,* as objectivity, as the being posited of self-positing in the sense of the will of reason, of love, of the spirit, of power, as the will to will in the eternal recurrence of the same.[11]

But to see what presence really involves it cannot be identified with any of these historical embodiments. Heidegger's way of probing the meaning of presence is by way of what could be called a speculative etymological phenomenology. For Heidegger, certain words preserve fundamental experiences, and these experiences can be wrested from them sometimes only by violence. He often translates *physis*, when used by the pre-Socratics, as *Anwesen*, an active presencing, or making present. To *physis* as *Anwesen* he attaches a number of distinct phenomenal characteristics: (1) *Creative occurrence.* The emergence of something out of itself, illustrated in the overwhelming powers of nature or the fruits of the earth. (2) (subtly distinct) *Originating.* Bringing out, exposing, producing. (3) *Appearing.* Showing itself, coming to light (even while pointing back to the darkness).

All these together, or each of them at different times, go some way toward capturing the early Greek sense of *physis*, a "concept" that so interpreted is not really a concept at all but rather a pointer to a productive flux, governed by no restitutive necessities or teleological principles. This is the first way in which presence can be unpacked.

Next there is *aletheia*. Marx relates it to *physis* in the following way:

> the basic traits of the presenting process thought as *physis* were manifested in its "self-emerging-prevalence" which as a coming to be, an appearing, a coming of the fore, is a stepping forth into the *light*.[12]

What is being offered, in other words, is something like a primal scene as the underlying link between *physis* and *aletheia*. In the "Essence of Truth"[13] it is understood as openness, in "Letter on Humanism"[14] a clearing. So *aletheia*, usually translated simply as *truth*, is for Heidegger the "openness characteristic of the occurrence of presence experienced by the early Greeks." But Heidegger is not suggesting that *aletheia* offers some absolute transparency, some perfect illumination, or unconcealment. That would point toward the idea of an unmediated presence, in which truth, for example, would just be a matter of contact or correspondence. For Heidegger, *aletheia* is light always struggling with the powers of darkness, lodged in a field of strife. Openness is always threatened and defined against what threatens it. Plato had already forgotten this.

In all too summary fashion, one must also mention *logos*. Through his interpretation of Heraclitus, Heidegger has developed a reading of *logos* that rounds out our understanding of "presence."[15] Presence as *logos* is expressed in a number of ways: (1) a laying down, a laying out, a laying forth; (2) a depositing in unconcealment; and (3) a gathering of all that strives apart, out of its dispersion, a collecting together.

But *logos* does not mean order, for as with *aletheia*, and indeed *physis*, there is no finality about this "gathering," no necessity. There is no relief in this *logos* from struggle. "On the contrary, presence must be kept in contention as a 'conflict' a 'primeval strife,' if it is not to lose its creative power, if it is to remain an occurrence in the original sense."[16] *Aletheia* is no final illumination. And *logos* is no privileged static order. The huge importance of this latter claim is that *logos* itself has no ground but is a "groundless play" indifferent to our search for principles and foundations. So the *logos* for Heidegger can offer no absolute grounding, and this in his eyes is enough to dismantle the pretensions of ontology and theology.

Clearly there is a great deal more to be said about the *logos*, especially about its intertwinings with language. But already it is clear how Heidegger can be said to have returned from "presence" understood on the basis of some preconstituted concept of time to a more original process of presencing. I shall shortly make explicit some of the elements of this analysis, elements that Derrida picks up in his continuation of the general thesis about the history of metaphysics as the history of Being as "presence."

But first at least an attempt must be made at an assessment of this account of presence as an account of, or as suggesting an account of, a primordial temporality. I would suggest two meanings of "primordial temporality": (1) a phenomenon that, although not itself temporal, is nonetheless specifically related to the temporal and makes it possible: and (2) a basic form of temporality on which other more developed and familiar forms depend.

The issue in either case would then be whether or not the phenomenon in question was itself *temporal*. But this cannot be decided without some standard of what counts as the "temporal." And the trouble is that the assured constitution of the object of inquiry cannot be presumed in this case. The question of what "temporality" means is not independent of the question of whether presencing (*Anwesen*) can be treated as primordial temporality.

Within the Heideggerean framework one would not expect all that is true of Time, as commonly understood, to be true of primordial temporality. One would not for example expect primordial temporality to involve a sequence of nows. But are there not some characteristics it must have?

Primordial temporality—understood through presencing—has, I would suggest, the following important features:

1. It is neither the temporality of a subject nor of a preconstituted object (this theme reappears in the language of the *es gibt* and *ereignis* in *Time and Being*).
2. In a corollary way, its description has to constantly *resist nominal* forms. Heidegger is far happier with verbal forms. His use of language here is often itself a kind of presencing. Presencing itself appears, emerges, is shown, drawn together, through verbal forms.
3. This temporality is weighted on the side of the creative, the emergent, while not ignoring its relation to what lies hidden.
4. Temporality is understood in oppositional terms. Its essential moment is that of struggle, conflict.
5. This temporality is void of any supervening principles of order. It has no purpose, its form is not even linear, it has no dialectical powers of resolution.
6. It is not a neutral base, but always has a "color." It is always *something* that emerges, or is given, or is appropriated, or it always occurs in some way or other.
7. The purity of temporality itself seems to recede in this account. *Aletheia*, for example, is an opening, a clearing—more of a spatial concept. One might be forgiven for thinking of presence as a primitive spatiotemporality. And indeed in *Time and Being* Heidegger talks of a primitive time-space, as if time by itself were a later abstraction.

If this characterization is at all fair, it begins to seem as though there is no clear answer to whether *presencing* so described is itself a form of temporality, or rather an "event" or "process" underlying temporality. On the one hand, it would seem it could be equally construed as in part a primitive spatiality, and on the other hand, it clearly does describe an event, presencing, that does not cease to be temporal. But it does not seem to be all there is to time. Might it not better be thought of as a theory of irruptions in time, not of time "itself?"

So it might be argued that Heidegger's account has been overestimated. Surely it is not an account of primordial temporality but rather of a phenomenon—"presence"—that is both more than a temporal one (involving something like a general man/world relationship), and also only handling directly one aspect of the temporal. (What about the past and the future?) This question asks more generally, whether it is at all proper to try to understand Time on the basis of an analysis of a particularly important role that interpretations of time have played in the history of philosophy.

Insofar as these interpretations seem to hold enormous sway over the ways we think of time, one might well conclude that these interpretations of time could at least be profitably drawn up into a more general theory of time. Surely time has all sorts of descriptive features independent of any value that has been accredited to different temporal modalities.

Two disconcerting replies can be made here. First, it might be said that if there are these descriptive properties, they would hardly be the concern of a *philosophy* of time, philosophy being concerned with theories, rules, principles, but not mere description. The answer to this of course is that philosophy can indeed be descriptive and that phenomenology is a prime example of this possibility. The second reply is in many ways more worrying. Derrida has suggested that the concept of Time might be essentially linked to the metaphysical tradition, so that to try to think of time *outside* that tradition would be a vain undertaking. This position of course undermines the project of a *philosophy* of time from the outset, insofar as it undermines philosophy.

As a way into Derrida, this account of Heidegger's discussion of temporality and presence is not complete. Mention ought properly to be made of Heidegger's emphasis on the ontological difference between Being and beings, his essay on the "Onto-Theological Constitution of Metaphysics,"[17] of the *Question of Being*,[18] of his *Nietzsche* books,[19] of his essay on Anaximander,[20] and of his *Basic Problems of Phenomenology*.[21] But what I would like to do now is to evaluate, in the light of Heidegger's account of presencing, the account Derrida gives of the role of time both in metaphysics and in particular in the theory of the *sign*. After this I will explain what I take to be the *limitations* of Derrida's approach to an understanding of Time.

Derrida shares with Heidegger the general analysis of the history of philosophy as the history of presence. In addition to Heidegger's own arguments, Derrida makes use in particular of Husserl's account of time-consciousness,[22] and of Saussure's account of language as a system of differences,[23] as providing ammunition for the deconstruction of the value of *presence*. The use of Husserl is ironic in that it appears in the course of an internal critique of Husserl's theory of ideality, of intuition, and of the sign. Husserl is committed to an account of ideality that refers to a particular type of consciousness, one that has an essence directly presented to it. Derrida, as has been seen, uses against Husserl his own theory of time-consciousness, in which the present is never simply distinguishable from the past and future but overlaid with protentions and retentions. As such this present is never pure, says Derrida. And yet such purity is required for there to be a pure intuition. Husserl, he suggests, for all the superiority of phenomenology over previous philosophy, is still a metaphysician,[24] in that he is committed to crediting presence with a value and status that it does not have and cannot sustain. Again, he argues against Husserl that *there are no purely expressive signs*, no signs that would or could simply convey a meaning expressed through them. All signs function as part of a signifying system in which it is the play of differences between signs that gives each of them meaning. Derrida is committed to the view that all meaning, all signification functions like this, so the idea of coming face to face with an essence, or the meaning of a sign—a true presence—is an illusion, or better, a fantasy. Or rather, the possibility of an experience one might describe in such a way is dependent on relations outside that experience that destroy its sense of magical purity.

The belief in a fullness of expression, by the voice, of a meaning, which has given human speech a privileged value in the history of philosophy, is based on the idea that through the voice the other is somehow directly given, or revealed. But if speech is equally a system of signs, then it too must be governed by the Saussurean principle of difference. And this time he operates an internal critique against Saussure.

For Saussure, inconsistently as Derrida claims, reviles *writing* as a mere external copy of speech, crediting speech with a privileged primacy, a thesis Derrida dubs phonocentrism. If one accepts that the properties of detachability and arbitrariness found openly in writing are in fact typical of any systematic use of signs, it is not too farfetched to generalize the concept of writing and to consider every use of signs a *writing*. This is like the move that generalizes the term "language" to the language of art, of the body, and so forth. Derrida instead, and with more persuasive intent, extends the term "writing," and even to include speech. Whatever the problems this brings, it has, for him, the advantage that he seems to have cleared a site in which one might be able to think in a way free from the metaphysics that seems to appear whenever one talks about the subject, consciousness, meaning, experience. . .

Derrida is taking up the task that Heidegger laid at our feet—that of thinking if not beyond metaphysics, at least along different lines. But there is surely another topic that could have been added to the list—time. According to Derrida, as we have often remarked, "the concept of Time belongs entirely to metaphysics and designates the domination of presence."[25] And yet Derrida makes considerable use of temporal considerations in setting the stage for what he calls his grammatology.[26] After all, it is the pervasiveness of the illusory temporality of presence that makes his project of deconstruction plausible.

As was mentioned earlier, one of the fullest accounts he gives of what we will provisionally call the *primitive* role of the temporal is to be found in his essay "*Différance*."[27] It is worth exploring the role *the temporal* plays in developing this vital "term" because it is via *différance* that Derrida's consistent distance from philosophy can be measured.

After a few precious pages in which he makes it clear that *différance* is really a thoroughly undecidable term—neither a word nor a concept—he nonetheless tries to explain its manifold "meaning" by a little semantic elaboration. The French verb *différer* apparently does service for both the English verbs *to defer* and *to differ*. Both of these senses are embraced at once. Signs are quite traditionally thought to defer their meaning, they stand in for something not immediately available. Derrida calls this *temporizing*.[28] And the sense of *différer* that means to differ neatly expresses the Saussurean sense of the differences, dissimilarities, between signs that allow them to mean anything. This could be called a *spacing*.

So far so good. But care is needed here. First, the traditional understanding of the sign as deferring its meaning treats the sign as secondary and provisional in relation to some primary meaning. And this is a thoroughly retrograde view. Rather, the deferment, the temporizing aspect of *différance* involves an other that may never have been present, and may never become present. In other words, what is deferred is not to be thought of as a present just temporarily absent, but as outside the whole system of presence and absence.

The spatial aspect of difference—differing—also requires more thought. Briefly it must be remembered that the principle of difference applies not just to the signifier but also to *what is signified*, for example, meanings.

So what about *différance*? Can it be said to produce differences? Only if one is aware of the dangers of hypostatizing *différance* as some sort of real source or origin. If such a *productive play* can be grasped without attributing to *différance* the status of an originating presence, fine. If not, some other way of putting it is required.

Derrida tries to handle this problem by the term "trace," which, were it possible, would be an effect without cause. The trace, if anything is, is an element in the signifying process. Each trace relates to a past element, and to a future element; it is separated from other traces by an interval, and in such a fundamental way that this *spacing* can be said to be constitutive of its present. In summary, to use his own words:

> It is this constitution of the present as a "primordial" and irreducibly
> non-simple and therefore in the strict sense non primordial synthesis of
> traces, retentions and protentions . . . that I propose to call protowrit-
> ing, prototrace or *differance*. The latter (is) (both) spacing (and)
> temporizing.[29]

The possibility approaches of comparing this account with Heidegger's
account of presence. But it is worth searching just a little further for an even
more appropriate place to turn off the track.

> Presence . . . is no longer . . . the absolutely matrical form of being
> but . . . a "determination" and an "effect." Presence is a determination
> and effect within a system which is no longer that of presence but that of
> differance; it no more allows the opposition between activity and pas-
> sivity than between cause and effect . . .[30]

What Derrida is saying here is this: the *phenomenon of presence* can be
reconstructed within a framework that makes no use of its basic terms and
assumptions. If it did not risk a lapse back into metaphysics, *presence*, on this
reading, could be said to be merely a *system effect*, and not the ground, source,
origin, foundation . . . of meaning or anything else.

Surely this is something of a writing experiment (along the lines of a
thought experiment). The term *différance* is, as he puts it, an *assemblage* of the
basic themes of a number of thinkers from Hegel to Bataille, and it seems to
be possible to put it to work to construct an account of how presence could be
produced nonmetaphysically.

How, then, does this square with the Heideggerean account of presence?
I will briefly compare these two accounts with respect to two questions:
Where does each of them leave time? and Where does each of them stand
with regard to metaphysics?

The second of these questions will be taken first. Derrida is not at all
sure where he stands in relation to Heidegger's account of the ontico-
ontological difference, the difference between Being and beings, one obvious
point of correlation with Derrida's *différance*. Certainly, he says, *différance* is in
part "the historical deployment of Being."[31] But equally, he would be
happier if *différance* were thought of outside of all considerations of Being
and/or its relation to beings. Certainly Heidegger himself came to be less
interested in the ontological difference and more in Being as such. But
Derrida seems to want to say that, even then, metaphysics has not been
thrown off. It is not enough, it seems, to awaken the question of Being, as
Heidegger had done. It must in sone way be actively forgotten.[32] Derrida
suggests that the ontico-ontological difference might be a particular determination of his more original *différance*.

To return now to the first question, there are some fascinating parallels between Derrida and Heidegger as far as time is concerned. In particular:

1. In both cases, presence is underpinned by something not itself a presence. Whether or not it is christened primordial temporality, it seems to have some quasi-temporal properties, and it certainly undermines the privilege of what it underpins, the ordinary metaphysical concept of time as a series of nows.[33]

2. Both Derrida and Heidegger talk about this process as one of *play*, a movement in which principles of order, finality, origin, have fallen away.

3. Finally, both suggest in a way quite distinct from relativity physics that fundamentally temporality and spatiality are not distinct. Derrida finds these linked in *différance*, and Heidegger in his time-space.

When all is said and done, how convincing is this general thesis about the interplay among the history of philosophy as metaphysics, the interpretation of Being, and the value of presence?

As far as its consequences for philosophy are concerned, it can be immensely fruitful. One is thrust forcibly against a whole series of questions that lurk on the boundary between philosophy and non-philosophy. One is driven to asking questions of a basically Nietzschean sort about the theoretical investments involved in the tools of philosophical thinking—in particular those systematic and pervasive investments that seem to be self-concealing. And there is no doubt that a value, in part temporal—which has been called "presence"—can be seen to unify, if loosely, a number of key philosophical concepts and values.

What of the suggestion that a new way of thinking, or a new writing is required? Has not a radical break in the history of philosophy often been advertised in these terms, to the point that it is almost a genre?

Much of the problem lies in the temptation—to which I perhaps succumbed in discussing Nietzsche—to offer a description of a primal scene, either of presencing or *différance*, in which certain ways of thinking about space and time are legitimized. For even as description of this scene recognizes, say, the enormous wealth of language or the variety of ways of Being, it cannot help unifying this variety, comprehending it, and anticipating its diversity. And the consequence is a novel sort of reductionism. Moreover, the unifying drive that each of these positions reflects makes one wonder whether the understanding of *time* has not been sacrificed in an ostensibly greater cause.

The conclusion to be drawn from Derrida is that in the only place at which the question of time can properly be posed—within the discussion of *différance*—it is welded to a spatiality and has no genuinely independent status.

Both Derrida and Heidegger emphasize and draw attention to the fact that a philosophical study of time is systematically beset with the problem of taking into account the role time has already played in constituting the terms and standards within which the discussion is to take place. I take it this is what Derrida meant when he said that "In a certain way it is always too late to ask the question of time. The latter has already appeared."[34] And it is not merely a question of *discounting* the work already done by time in these terms and standards. If Heidegger and Derrida are right there would be *no philosophy left* if the interpretation of time as presence were put aside, for that has determined the interpretation of Being since the pre-Socratics, and *that* has shaped all or almost all of philosophy.

Surely one general thesis to be drawn from their work is that whatever the consequences for philosophy in general, an investigation of the role of Time in philosophy spells the end of the philosophy of Time.

In words like *différance*, Praesenz, and Rapture (and, indeed, Husserl's *Fluß*), temporality seems to disappear at the very point of its fundamental illumination. This is where my first strategy leads to. If, that is, our critical resources are focused on the fundamental unit of time, traditionally understood as the now-point, the moment, it reveals itself to be about as "simple" as a bud about to burst.

But there are surely limits to what can be gained by this route. Focusing on the moment, or on the phenomenon of presence, or on the primitive phenomenon of temporal transition does indeed disturb sedimented complacencies of unity, simplicity, and identity, but equally it can be a distraction from the task of charting the complexities of articulated temporal structure, and this second strategy plays just as vital a part in the general project of deconstructing our ordinary concept of time.

Part 4
Time Beyond Deconstruction

1   Derrida's Deconstruction of Time and Its Limitations*

In this chapter the discussion centers on the possibility of an adequate account of the temporality of language. I claim that Derrida's contribution here is essentially preparatory and that he does not (and perhaps cannot) himself offer such an account. The force of this demonstration for the argument as a whole rests of course on the generalizability of these claims about the temporality of language to "time" itself. My general line here (developed more explicitly in the next chapter) is that this expansion of scope is achieved when language (understood in its widest sense as significant articulation) is seen as the very source of that wealth of temporal structures whose description is only begun in this book. I argue that Derrida's specific sterility here is a consequence of his one-sided development of a quite traditional philosophical opposition between identity and difference, and particularly of his inability to shake free of what is for him a necessary strategic detour through transcendental argument.

The fact that these remarks will be fairly straightforwardly philosophical does however present in itself a kind of theoretical problem. For Derrida does not pretend to, or pretends not to, seek the approval of philosophy. His aims lie elsewhere. He writes in *Of Grammatology*, "To make enigmatic what one thinks one understands by the words 'proximity', 'immediacy', 'presence' (the proximate, the own, the pre- of presence), is my final intention in this book."[1] There may be a certain amount of irony in these words, but they should not be dismissed. Nor should their implications be ignored. Derrida is trying to produce an effect. His writing is governed by a strategy, not by, say,

* This chapter is a very slightly revised version of an article, "Time and the Sign," published in the *Journal of the British Society for Phenomenology* 13, no. 2 (May 1982). It was presented in an earlier form to the International Association for Philosophy and Literature at the University of Maine at Orono in May 1980. I am indebted to Andrew Benjamin for his helpful criticisms of that version.

the ideal of truth. Indeed one reply to my claim that his position is one-sided and undialectical is that he is not actually taking up a position at all. A position implies a location within a space, and Derrida is a frontiersman. To accuse him of theoretical overkill is to forget this strategic status of his writing.

But this strategy involves and invokes numerous philosophical arguments and numerous stands taken on other philosophical positions as well as on philosophy itself. Derrida's writing is armed to resist traditional philosophical reappropriation for as long as possible, and in trying to open him up, I have all the hesitation he has about trying to bite into Hegel. One can come to a sticky end. Without wanting to dismantle these defenses in detail, I will suggest just a couple of ways of getting a foot in the door, ways of dealing with his most general defenses against philosophy, so that a philosophical approach to Derrida will not seem futile from the outset.

The structure of this defense mechanism is classic. It could be called rebuttal by preemptive engulfment. Derrida knows what the essence of philosophy is, and he has taken its measure. Philosophy is the systematic deployment of concepts invested with a privilege that (after Heidegger) he calls presence, such as meaning, truth, experience, consciousness, the subject . . . the list is very long. In each case the same value is demanded, promised, preserved . . . of a ground, a foundation, a beginning, a point of privileged encounter.

Is there not at least a danger—to put it no more strongly—that the whole vast bulk of writing called philosophy, and much besides, might be thought to have been reduced to this one character? Are not all those enormous differences being treated as evidence of the Same? Surely this *closure* of philosophy, as Derrida calls it, is a gesture born and bred in the stables of philosophy itself? And yet on Derrida's own premises, premises independent of the thesis that philosophy is the development of presence, no writing, philosophical or otherwise, can have its *meaning* so summarily represented or extracted. Why does his circumscription of philosophy not commit him to an essentialism untenable on his own terms?

But preemptive engulfment is never a conclusive ploy, especially against philosophy. The second way I would demonstrate this, after which Derrida's writing should be sufficiently softened up to withstand a little philosophical probing, is by offering a rather non-Derridean characterization of his treatment of philosophy. In writing about the privilege of presence in philosophy, he uses terms like "security," "reduction of anxiety," and so forth. The value of presence is the value of a desire for such security, expressed in a variety of ways. Derrida's totalizing treatment of philosophy can be understood as a hermeneutic of Desire. More often even than philosophy invokes the value of a presence, Derrida interprets philosophy as the desire for such a first point or *arché*. This interpretation, albeit as a partial reading, is generally accurate, but surely not legitimate for Derrida. As I read Derrida, the real meaning of philosophy is this Desire. But there are no *real meanings* for Derrida, and if there were, to privilege such Desire would be an interference with play.[2]

With these two attempts to jam his basic defense against a philosophical intrusion, I have simply tried to create a little space in which to think critically and yet philosophically about Derrida. These lines of argument may be thought both unnecessary and unjustified. Unnecessary because Derrida is the first to admit there is no real escape from philosophy. And unjustified because of the precautions with which he surrounds his writing. The logic of precautionary gestures will not be spelled out here, the conclusion I would come to can be: Derrida privileges his precautionary gestures— graphic and propositional—in a way he cannot justify, for there can be no such privileged writing.[3]

I began with a claim about the absence of an adequate account of the temporality of language in Derrida, and how this defect is rooted in an undialectical promotion of difference at the expense of identity. I will first try to show how the specific temporality of language gets both partially uncovered and then covered over again. Fortunately the ground is still fresh.

The last section of "Linguistics and Grammatology" (referred to hereafter as LG), entitled "The Hinge" ("La Brisure"), provides most of the material, but the problem can be most sharply fronted by three brief quotations from the end of Derrida's long and difficult "Ousia and Gramme."[4]

> The concept of time, in all its aspects, belongs to metaphysics, and it names the domination of presence. (p. 63)

> . . . if something which bears a relation to time, but is not time—is to be thought beyond the determination of Being as presence, it cannot . . . still . . . be called time. (p. 60)

> . . . an *other* concept of time cannot be opposed to it [the whole historical system of metaphysical concepts] since time in general belongs to metaphysical conceptuality. (p. 63)

The territory here is Heideggerean. Heidegger's general thesis was that the history of philosophy could be understood only through the way it had interpreted Being.[5] And for Heidegger, since the pre-Socratics, Being had always been determined, in some form or other, *as presence*. But by the time this temporal determination of Being came to be reflected upon—by Plato and Aristotle—the understanding of Time had been taken over by physics. Time's ontological status receded from view.

It should be clear from this account that Heidegger leaves room for *reappropriation* of a more original time or temporality, a recovery of what was lost. And indeed he writes of *authentic* or *primordial* time in *Being and Time* (1927), and in *Time and Being* (1962) he refers to *true time*, to designate admittedly different readings of this original time.

For all his debts to Heidegger, Derrida not only thinks of the question of Being as a metaphysical residue, but explicitly questions this hankering after a lost primordial time. And here the quotations from Derrida come to life. There is no *alternative* concept of time to the metaphysical one. To think outside metaphysics, or to try to, the concept of time must be abandoned. The traditional way of understanding the concept of the sign, according to Derrida, is in part at least as a secondary, instrumental, and disposable representation of a meaning or referent that is not itself present, whose presence is absent and deferred. This concept of the sign invokes a value of presence that only the traditional conception of time can sustain. On Derrida's view, if that concept of time is rejected, there is none other available to take its place. It is that or nothing. One may not be struck for words, but they will not announce an alternative temporality.

It might be thought that some sort of compromise could be reached here. Could one not agree on a "new" sense of time and temporality? A sense perhaps fundamentally different from the old one? The reason for his intransigence here is revealing. Derrida has a thesis (that slips in and out of erasure) which provides a transcendental ground for time (and indeed space). Naturally this transcendental thesis is hedged by qualifications, for there could be nothing more metaphysical than such a claim. But in the first sentence of "The Hinge" it appears in all its birthday innocence:

> Origin of the experience of space and time, this writing of difference, this fabric of the trace, permits the difference between space and time to be articulated, to appear as such, in the unity of an experience. (p. 65–66)

There are three problems with this. Is it a transcendental claim? If not what sort of claim is it? Is there not a primordial temporality in *différance* or the *trace*?

These questions are only sharpened when one notices the extraordinary structural parallels between this description of the role of *différance* (let us say), the role Kant attributes (on Heidegger's reading) to the transcendental imagination in the first *Critique*, and the role Heidegger gives to what he calls the *es gibt* in *Time and Being*.[6] But whereas for Kant transcendental imagination *is* (sometimes "is rooted in") temporality of a primordial sort, and for

Heidegger what is given is true time, Derrida has no such *alternative temporality*. At least he says no such is possible. This is not to say of course that what seem like temporal characteristics are not central to his characterization of the *trace* and *différance*. The stumbling block is surely that Derrida takes seriously the traditional link between time, the logic of identity, and the law of noncontradiction. And it is not possible to characterize the trace or *différance* without violating these logical principles.[7]

I will now try to show the vital role of time in the development of the terms *trace* and *différance*, and how they in turn then absorb that very same temporality.

The term *trace* can be understood as a transformation of the concept of sign, a transformation in which the horizon of presence that governs the classic concept of the sign gives way to the "horizon" of *différance*. While a sign is classically understood as standing in for a meaning or referent that while not actually present, is potentially so, the *trace* on the other hand refers to a past that cannot fully be reactivated, even potentially, a "past" that can no longer be thought of as a past present, now dormant. The term *trace* is the result of depriving the concept of sign of its signifier/signified structure. This deprivation rests on the claim that the idea of the *signified* here is a metaphysical legacy.

With this term *trace* arrives a problematic temporality, which may no longer be a temporality at all. It might be thought, for example, that what was needed was some sort of phenomenological enrichment of the traditional concept of time as a series of nows. Would it not help in understanding the peculiar mixture of dependence and independence of what is *other than it* to consider the phenomenological account of temporality elaborated by Husserl? Here the present is from the very beginning permeated with past and future, as retention and protention. But the past and future are here each grasped as presents, albeit potential presents. That is, they are experiences that we once were having, or may soon be having. In this expanded living present, the value of experiential evidence, of "presence," remains and is indeed shored up. And it is this very value that the trace denies. There can be no phenomenology of the trace.

Now it is easy to see that an account of the temporality of language could not just make use of clock time, physical time. The words I am now writing or speaking, even understood from the point of view of simple succession, have no single order to them. They can be ordered as individual word units, or as parts of larger units of articulation such as sentences. I have elsewhere called this "nested articulation." And succession is by no means the key to the temporality of language. One might have thought that a theory of temporality that began by bracketing out objective, world time, as phenomenology does, would be exactly what was required. The force of

Derrida's criticism here (LG, p. 67) is to claim that it cannot account even for the most basic temporality of the trace, ultimately because its structure is anterior to even phenomenological temporality. Phenomenology was after all concerned with time-*consciousness* and Derrida's "trace" is an attempt to handle something like the "unconscious" of language.

The trace is said to have a structure. Loosely speaking it could be said to involve a certain sort of relating. Derrida was prepared to use the word "referring." If it breaks with the idea of time, it is, in part, because that relation neither requires nor establishes a dimensionality, any form of neutral extensiveness such as a concept of time always seems to require. To pursue this question, I will now turn to the related term *différance*. I draw on the versions found in LG and on the essay "Différance."

In this essay, Derrida puts the term through its hoops so that without defining it he can nonetheless show what use can be made of it and explain the historical context of its appearance as what he calls an "assemblage" (not a word or a concept). In LG he uses it to do the work for which a transcendental critique is traditionally reserved—to unsettle a naive objectivism—in this case that of Hjelmslev's linguisticism.

The term *différance* allows us another access route to Derrida's theory of writing, this time through Saussure. Derrida's relationship to Saussure has three key aspects. Derrida endorses in broad terms the priority of language over the language user. To make sense one has to make use of a preexisting language. Second, he shows that Saussure's account of the sign is at odds with his privileging speech over writing, and Derrida himself argues that the rupture with presence, with the transcendental subject, found openly in ordinary writing is characteristic of all signification, including speech. Even speech, one could say, is a kind of writing. Third, he appropriates Saussure's diacritical theory of the sign, the idea that language is just a system of differences. *Différance*, indebted to Hegel, Nietzsche, Freud, and others, is most directly drawn from Saussure. The sign has no meaning in itself, no semantic content. If it is thought of as having a meaning, then it is not a "presence," not a unity graspable in itself. For Saussure, both the signifier and signified have their identity only through the ways they differ from other signifiers or signifieds in the system. But how is this relationship of difference to be thought of? Derrida's development of the term *"différance"* brings out what looks at first glance like an original spatiotemporality of *différance*, and does so using the semantic duality of the French verb *différer*. It stands for both the English verbs "to differ" and "to defer." In differing there is a kind of "spacing" and in deferring a temporizing, that is, a delaying. *Différance* is meant to capture both. This "spacing" and this "temporizing" are each offered as the fundamental conditions of any signification at all. They are two ways of unpacking the idea that a sign is constituted not by a re-presentative relation to another presence, but by a deferment to and a spacing from what is absolutely other. Identity is the product of difference; presence is derivative from a fundamental absence.

So while Derrida makes important use of a temporal notion like *delay* or temporizing in his account of *différance* (a good example is Freud's delayed effect), this is not to be thought of as an operation that takes place *in* time, nor as constitutive of another sort of time. Somehow, *différance* is a "movement" that makes time possible. But it is not itself temporal. Does it then offer us an account of the temporality of language? If it succeeds at all, it does so by offering an account of the temporality of the sign, of its "presence." Derrida is, as it were, saving the appearances. The spacing and temporizing of *différance* makes possible the underwriting of the apparent simplicity of presence as a manufactured complexity.

> It is this constitution of the present, as an "originary" and irreducibly non-simple (and therefore *stricto sensu* nonoriginary) synthesis of marks, or traces of retentions, and protentions [terms used only provisionally] ... that I propose to call ... *différance* ... which (is) (simultaneously) spacing (and) temporization.[8]

So the argument is that *différance* precedes time understood through presence—for example, as made up of a series of now-points, or as the living present—in that it is only on the basis of *différance* that presence is possible.

Derrida was quoted earlier as saying that no alternative temporality was available outside time as presence. One of the important consequences of this is that he will not allow a kind of two-tier temporality, in which, say, an unconscious temporality, one governed *by différance*, for example, was woven on to an underlying phenomenological time. That would leave phenomenological time *in itself* untouched by *différance*. And of course it precisely is affected if its presence is a mere "effect" of *différance*. *Différance* cannot be used as a corrective; it transforms what it touches. And the implication certainly is that the very idea of an unconscious temporality is based on using *différance* in a merely corrective way, so it must be dropped. I think this is a mistake. An account of what we could call the unconscious temporality of language can be provided and can even begin to be constructed out of some of the material that Derrida himself has gathered.

To license my efforts in assembling a few hints toward a genuine temporality of language, I must first justify my misapprehensions about *différance*, albeit too briefly to be absolutely satisfying. I begin with a general reminder. Derrida's writing is governed by *strategy* and by the historical situation in which he finds himself. Preceded by and surrounded by a vast sea of philosophical writing unaware of its naive and uncritical commitment to presence, Derrida responded by writing a shadow text. By a shadow text I mean one that precisely in its intransigent repudiation of presence as a founding value makes a difference that a more conciliatory text would not have made. But this is just assertion, not an argument.

There is, after all, an important objection to be made to giving a transcendental status to the relationship between *différance* and presence, a relationship he characterizes by all the forbidden words—priority, production, origin . . . in and out of formal dress. He argues, correctly, that one has to go *through* transcendental types of argument rather than just bypassing them. But he retains the value of privilege and priority that only such arguments bestow. Remember his words: *différance* " . . . is a *constitution* of the present . . . as a synthesis . . . of traces." Or: "This writing of différance [is the] . . . *origin* of the experience of space and time." (emphasis added twice) In neither of these cases, and they are not unusual, are precautions offered. It will be said that he knew what he was doing . . . But is there agreement on what that was? In making "enigmatic" the idea of presence, Derrida cannot be indifferent as to the legitimacy of the means he employs. He is, I believe, engaged in high-grade mimicry of transcendental arguments. But he does not and cannot show that *différance* can be the origin of anything. Let us suppose that there are always scare-quotes or erasures *implied* in such terms.[9] But then surely the consequence is that they do not have the required force. They cannot establish the priority they claim.

In my view, undeconstructed presence, and various of the concepts associated with it, are ineliminable in principle. Derrida himself says something like this—that there can be no question of abandoning the concepts of metaphysics when deconstructing it. I mean it in a rather different way. At a somewhat rarefied level of argument, I have just concluded that Derrida could not claim a privileged position or priority for *différance* over presence without self-stultification. Is there room for the term *différance* without this privilege attached to it? I think there is, and it is at precisely the same level as there is room for "presences," "unities," and "identities" functioning without the metaphysical security that philosophy is apt to bestow upon them.[10]

There is in Derrida's work a self-confessed danger of inducing sterility. The language of *différance* not only legitimates a school of criticism, but it has terror potential in its exclusion of a whole range of important analytical concepts. More important for my purposes, it discourages interest in the temporality of language.

The level just referred to, at which presence and *différance* meet without privilege, I would call the level of human finitude. We *are* satisfied with partial answers to questions, incompletely fulfilled meanings, good enough cases of immediacy. We do have a tacit knowledge of contexts, we construe situations, we know what it is to see a piece of paper, to go for a walk, and so forth. Derrida would probably not dispute this. What I think he forgets is that it is *from these everyday securities*, and our tacit grasp, however vague and average, that we necessarily start when considering, say, the contribution of *différance* to their constitution. The always-already present is the condition for *différance* being thinkable. Derrida's way of handling this problem involves the use of the concept of *economy*. The play of *différance* is not infinite. Stabilities are the product of an economy of forces in a field. And that is importantly quite as true of the "terms" difference and *différance*. Difference and identity, or *différance* and presence, are each linked pairs, and it might be salutary to think of metaphysics as the result of a hypertrophic privileging of one of the terms in each pair.

The temporality of language can now be considered more freely. I have said that, in spite of his rejection (in LG) of an alternative temporality, he nonetheless supplies some material for this project and points the way. The best sources for this are to be found in his discussions of the linearity of the "vulgar conception of time"[11] and a similar account in "Grammatology as a Positive Science" (GPS).[12] And there is a surprise in store. For he refers in passing, at the end of the second source, to increasing "access to pluri-dimensionality" and to a "delinearized temporality."

It will help to be guided by a question: What is wrong with the picture of the temporality of speech or writing that represents it simply as a linear succession of elements? Saussure seemed to claim just this, at least about speech: "Auditory signifiers have at their command only the dimension of time. Their elements are presented in succession. They form a chain."[13] Now there are two fundamental things wrong with it. First, it is rarely if ever true, even superficially, of writing. So it is not generalizable to all signification. And second, it is only very superficially true of speech.[14] For Derrida it reflects the historical domination of the linearist conception of time, the "vulgar concept of time," as Heidegger calls it, of time as a series of now points. Its hopelessness as a model of linguistic temporality should be obvious.

Derrida describes this linear model of time in a number of places. Its various attributes include consecutivity, irreversibility, unidimensionality, homogeneity, and being "dominated by the form of the now and the ideal of continuous movement, straight or circular."

But if such a model of time has to be abandoned to account for, among other things, the temporality of language, Derrida's account of the "trace" and of "*différance*" does not offer us an adequate conceptualization of the alternative—and not just because they are not properly speaking "concepts," but because their power to articulate the complex temporal structures of speech and writing is restricted in scope. Even if *différance* were basic to all the other temporal structures of language, it need not hold the key to their complexity and variety. There is no reason to suppose that an adequate account of the temporality of language can be generated from an account of the temporality of the sign (or the "trace"). An adequate account, I suggest, would have to take note of four different levels at which language production is temporally involved: (1) ludic, (2) intentional, (3) structural, and (4) communicative.

The *ludic* level is that of the play of sounds and shapes that allows patterns of echo and repetition to be set up in discourse. The *intentional* level concerns itself with the way the construction and comprehension of discourse involves both protention and retention and complex layerings of each. The *structural* level deals with the multiplicities of orders of unit (and thus levels of successivity) in a text, the different series to be found in the same text (constituting character, theme, plot, and so on), the many continuities and discontinuities it enclosed, and so forth.[15] Finally, seen as *communication*, speech or text production enters into the whole field of interaction, interpretation, understanding, interruption, response, and correction.

This gesture toward the different levels at which the temporality of language can be elucidated is designed only to suggest one thing: that a pluridimensional and delinearized temporality—one that escapes the most obvious forms of the metaphysical determination of time—can be developed by a careful elucidation of the complex weave of temporal structure found in discourse. A deconstructive practice that obscured these different levels would perform a great disservice. And what is being suggested here is, if not *sensu stricto* a new *concept* of time, at least a new model of temporal structuration.

The obvious response to this from one who insists that the notion of time itself is inescapably metaphysical—a thesis being attributed to Derrida—is that each of these different levels is identified under classically "metaphysical" headings—how else can references to intention or to communication be understood? Do they not involve an appeal to *presence* in some form? And if so, does not the contribution each makes to what we have called the weave of textual temporality just reestablish the old concept of time on new ground?

I think one has to take seriously Derrida's own claim that concepts are only metaphysical when inscribed in a particular way in a text.[16] And I take it that the kind of textual inscription that counts here is one that gives a foundational value—the value of presence—to such concepts. But this is just what referring to a weave does not do. Consider "intention" first. To understand the temporal structure of "telling a joke," one needs to refer to the setting up of expectations, and the retroactive cancellation of these expectations in the punch line. This is not the whole story about the time of the joke, but it is an essential part of it, and it involves a reference to the intentional. Or again, consider "communication." One *could* think of communication as the establishing of some ideal co-presence in which minds, if only briefly, are unified. That would indeed be a metaphysical notion, as would the idea that it involves the transmission of some "meaning" from one person to another. But one does not have to abandon the whole notion of communication, and the effort to understand its more subtle temporal structuring, just because it is open to a metaphysical appropriation.

To summarize: Derrida's deployment of *"différance"* and "trace" so as to strategically displace "identity" and "presence" in a deconstructive ma-noeuvre would become an act of destruction if it were allowed to obscure the place at which "identities" and "differences" meet without privilege, with-out, that is, the relation between identity and difference being reduced to a foundational relation in either direction.

Derrida's claim that there can be no nonmetaphysical concept of time is founded, it is argued, on an illegitimate conversion of the status of the term *"différance"* into one with (quasi-)transcendental significance. Only then can he argue that as *différance* is "prior" to time (and its distinction from space), "time" is only a fundamental concept within a framework that denies *différance*, that is, within metaphysics. But if we refuse to allow *différance* to play this (quasi-)transcendental role, then the possibility of an alternative nonmetaphysical temporality is opened up. To this end Derrida's work supplies much of the motivation and some of the tools, but finally he obstructs the path.

2   *Derrida and the Paradoxes of Reflection*

Reading Derrida's work as philosophy, perhaps a foolhardy enterprise, one soon realizes that something like a paradox courses through it. Derrida is proposing an account of writing that refuses even philosophy (or particularly philosophy) the status of "purveyor of truth" in any linguistically naive way. For Derrida, linguistic naivety does not consist in forgetting, as Whorf might have said, that we are using a *specific* natural language when we make our universal pronouncements, and that these might not even be sayable in a language with different ontological commitments. It consists in supposing that philosophy could ever escape the condition of writing, which he thinks of as semantically ungrounded, as "originating" with what he calls the "trace" (which is not a trace of anything) rather than a meaning that can be grasped in itself. Writing is constituted by a play of differences, a constant deferring of the point at which its meaning could be cashed out.[1] For those who would pin down a text to what it really means, Derrida's account of the determination of meaning is like that of catching an infinitely slippery eel.

The paradox lies in the status of what he writes, and the fact that he too is *writing*. If what Derrida writes is true, it would follow that he and other philosophers ought to be read in a new way. But if what he says is true in the ordinary philosophical sense of truth (which he describes as metaphysical), then in fact it *cannot* be true, for there would then be at least one species of writing—namely, Derrida's type of metaphilosophy—that has escaped the universal condition of writing of never just being able to deliver the truth for consumption. But if the claim to truth is dropped, then how and why should Derrida's claims about language as writing be believed? Derrida has the problem of saying what he means without meaning what he says.

If Derrida were not aware of the form of this problem, my task would be lighter and shorter, but he is very much aware of this paradox and his highly reflective and self-conscious texts can in part be seen as responses to it. He talks of his writing as strategic (though without an end according to which the strategy could be more or less successful), he insistently admits that metaphysics in some form or other is inescapable, but that does not make all texts equally metaphysical, and he varies his style. The difference between *Speech and Phenomena* and *Glas*, for example, makes it clear that he is almost as concerned about his relation to his texts as was that master of authorial disguises Kierkegaard. It is finally worth pointing out, as further evidence of Derrida's concern with reflexivity, that when he deals with Hegel, one of the places he makes for is Hegel's preface to the *Phenomenology of Spirit*, in which Hegel expresses his distaste for prefaces.[2] It is after all inappropriate to offer the reader reflections on a work he has not yet read, and, more important, it is impossible to summarize a work that is inseparable from its actual detailed textual development. A philosophical preface is, as such, about the worst introduction to philosophical writing. Despite Hegel's attachment to the logocentric ideal of presence, in the shape of Absolute Knowledge, albeit with a long-delayed delivery date, Hegel's concern that the reader realize that philosophical work cannot be represented by summary forms, that one piece of writing cannot stand for another piece of writing, gives him a place in the fictional history of nonlogocentric philosophy, philosophy not detachable from its written form, even if Hegel still hung on to a telos.[3] If I am right, the way Derrida latches on to Hegel's most important textual reflections is symptomatic of the importance of reflexivity for understanding Derrida's own work. I hope that pursuing this theme will help us to answer the question of the status of his writing.

By the question of its status what is in play is its relationship to philosophy and possible future philosophical practice. I have already explained how this problem first arises. If we take his claims in a straightforward philosophical way, they can be reflexively applied to themselves to generate a puzzle about their own status. One way out of this problem is to abandon this naive reading. Derrida's own agreement with my suggestion about the inadequacy of a naive level of reading him is reflected in the apparatus of internal warnings, instructions, security systems, recommendations, and the general level of guardedness with which he writes. Whereas the initial first-level paradox depended on making his conclusions reflect on themselves, I will now try to show that a second level of paradox can be discerned in the internal reflections within the text itself. I will argue that the use of these strategies of textual reflexivity that can be seen as a solution to the first order paradox seems at least to realign Derrida firmly within the logocentric tradition he is criticizing, and moreover that it does this in ways that he did not anticipate and cannot find acceptable.

But the problem about the *status of his writing* can be put another way. Following Heidegger, he has a large-scale view about the pervasiveness in the history of philosophy of a single underlying theme. For Heidegger this was the forgetfulness of Being—or forgetting the question of Being. It took the form of substituting some specific *determination of Being* (such as Plato's

Ideas, or any other specific form of ontological commitment) for Being itself. Instead of a question we have a series of answers unaware that they are answers because the questions have been forgotten and are now hidden. Derrida agrees with this analysis of the history of philosophy as a series of "determinations of Being as presence" but does away with the reference to Being, adding that even Heidegger's "Being" is just another product of the philosopher's desire for "presence." The concept of presence is Derrida's term for the principle that unifies all epistemological and ontological touchstones—intuition, self-consciousness, direct awareness of the other, the revelation of meaning, and so forth.[4] Now if we deem metaphysical those philosophical texts that are organized by some appeal to the privileged value of presence, it raises the question as to whether there could be any nonmetaphysical philosophy—a question Derrida has himself also raised.

What Derrida opposes to the ideal structuring of a philosophical text by the appeal to the value of presence is the status of all texts (as well as speech, and even consciousness) as *writing*.[5] And he uses the term "writing" to capture a view of language as a dispersion of signs with no center, no point privileged by presence. The structure of presence could be regarded as the false consciousness of the metaphysical text, and it is reflected in the refusal of such texts to recognize their own status as writing (by the kinds of timeless ideals they endorse, for example). If we suppose that such a blindness was a constitutive feature of all philosophy and yet that it is inconsistent with the ideals of clarity, autonomy, and self-justifiability that philosophy holds, then Derrida's account of writing puts in question the very possibility of philosophy not determined by metaphysical and thus ultimately contradictory values.[6]

There are three more general external considerations relevant to assessing Derrida which I shall simply list.

First, it seems to be a lesson drawn from his analysis of the work of Husserl and Heidegger that the principle that effort is rewarded does not apply to attempts to escape metaphysical determinations. I have already mentioned Heidegger's case. Derrida thinks of Husserl too as someone who despite the best intentions built into phenomenology the most basic metaphysical motifs—intuition as a guaranteed basis for knowledge. And Husserl had made the avoidance of metaphysics a thematic aim of phenomenology! It follows that Derrida's own efforts to (more modestly) limit the extent of his involvement with metaphysical aims *need* bear no relation to his success. History teaches us that the road back to metaphysics is paved with the very best intentions.

Second, metaphysics, as Derrida understands it, is not simply to be found in the use of certain types of arguments, or the holding of certain sorts of propositions (those that cannot be verified, say). One is implicated in metaphysics as one might be implicated in an ontology,[7] even by the language one uses to escape it. The way Nietzsche describes *truth* applies in large part to the terminological armory of metaphysics, and hence its insidiousness:

> What therefore is truth? A mobile army of metaphors, metonymies, anthropomorphisms . . . illusions of which one has forgotten that they are illusions; worn out metaphors which have become powerless to affect the senses. . .[8]

Our language contains all the conceptual oppositions within which the history of philosophy unconsciously plays itself out. And that language is not isolable from the one used in what Derrida calls the deconstruction of metaphysics.[9]

Finally, perhaps a rather provincial problem in assessing Derrida. In Britain and in the United States, the thinkers who, in Derrida's view, have been most successful, and in the light of whose writing Derrida is best understood, are often as much the subject of philosophical suspicion as Derrida himself. And the success of Joyce, Mallarmé, or Artaud is no guide to judging Derrida, who never seeks a wholly literary evaluation.

I have called these considerations "external" because while they bear on Derrida's writing they can be grasped without special reference to his work and allow some advance scene setting. I have mentioned the names of a number of philosophers with whose work Derrida's is intimately linked. Further connections and comparisons will help in assessing Derrida's textual reflexivity in a historical context, but before such comparisons, I shall spell out what I mean by his strategies of textual reflexivity.

What is at issue here are the various ways in which Derrida explicitly focuses in his texts on the ways in which they break new ground, the remarks he makes about the complicity of any such critical writing with the metaphysical matrix that it is taking as its object, and his account of the strategic way in which, as a consequence, he is using terms he cannot avoid using. I shall argue that these strategies (including Derrida's claim that all writing in this area is a matter of strategy and risk) are not the only possible responses to the theoretical situation as he sees it, and that by the very reflective devices he uses he appeals to the most traditional metaphysical values, values he himself has analyzed. While one might be prepared to accept the metaphysical complicity of deconstructive theory on the strength of Derrida's explanations, if these explanations themselves are even clearer and more obvious examples of metaphysical thinking, there will surely be cause for a certain skepticism. I am saying that the general line of prudence with regard to metaphysics that Derrida takes is directly self-defeating.

The first level of reflexivity I would point to is that of Derrida's lexical innovations and appropriations. I am thinking of such examples as "trace," "*différance*," "brisure," "écriture," "presence," "grammatology," "supplement," and so forth. These terms have the function of occupying a space created by the first stage of a deconstruction, that of reversing the dominance relationship between two opposed concepts,[10] so that the value of presence is annulled. They are used as sticks can be used to hold open a crocodile's mouth. In each case Derrida not only offers examples of the use of these terms, but also covering remarks designed to prevent us from understanding them in traditional logocentric ways:

> "*différance*" is neither a word nor a concept
> a "trace" is not a trace of anything
> a "supplement" is neither an increase, nor an addition

These are cases of what he calls undecidable terms, which do not obey ordinary logic and grammar. And yet it seems as though they can be given work to do in a text, and they are repeated in like contexts. One way of ensuring that their special status is retained is to write the words "under erasure"—by first writing them and then crossing them out, and then leaving them in that state. Even when Derrida does not use this Heideggerean device, one is expected to remember the fact that these terms do not in any ordinary sense have a meaning. One might compare the way in which mathematical fictions such as the square root of minus one can function in equations without themselves being rational numbers. The question then becomes: Just how are these terms used? If one were just to read them in a new attitude, this appearance of something like a phenomenological shift might be thought amusing. But one is also supposed to accept accounts of the functions of such terms that sound suspiciously as though they are fitting into the very same patterns as the terms they have displaced. It is of interest that Derrida vigorously denies this, but such denials constitute a different level of textual reflection, to which I shall return.

Let me quote some remarks he makes about "*différance*" in his essay of that "name":

> What is written as *différance*, then, will be the playing movement that "produces"—by means of something that is not simply an activity—these differences, these effects of difference.

> *Différance* is the non-full, non-simple, structured and differentiating origin of differences.

It is because of *différance* that the movement of signification is possible . . .

. . . We shall understand by the term *différance* the movement by which language, or any code, any system of reference in general becomes "historically" constituted as a fabric of differences.[11]

There is no doubt that he can seem to be offering us a new transcendental argument, with *différance* as the ultimate ground.[12] But that would be a disaster if it were true because the term "*différance*" is set up in direct opposition to just such an appeal to a first principle. Derrida not only employs an army of scare quotes to prevent us understanding his elucidatory terms in normal senses, he proceeds to point out that to take these accounts with their normal metaphysical weight would be quite mistaken. He insists that he is using these quoted terms only strategically. But surely the scare-quoted terms do have to be understood in order to be able to grasp just how they situate a new sense for *différance*. And yet one is not allowed to do this. The tide however is pulling strongly, and lingering over these problems is not encouraged. Derrida already knows where he is going. He will refuse the identification of *différance* with Heidegger's Being, he will deny that there is any sense in which we can point to the essence of *différance*, and he will finally insist that *différance* is not a name at all. The reason for this is that what it would name, if it were a name, is "the play which makes possible nominal effects, the relatively unitary and atomic structures that are called names, the chains of substitutions . . ."[13] The reason it cannot name this is presumably that the condition for the possibility of names cannot itself be named.

I have a number of worries about this theological argument. First, I cannot see why, if there was such a condition, it could not for all that be named. After all it seems to be possible to describe it. My second worry is rather deeper. For it begins to seem as though Derrida's coyness about *différance* is based on his own mystical understanding of names. And I just do not see why one needs to go all the way through Heidegger to discover that a term can be functional without being a name, and that its function can consist in both the substitutions it allows and those it engages in, that is, its paradigmatic possibilities. Derrida is here following both Nietzsche and Heidegger in conceiving of language primarily at the lexical level, allowing no independent form to syntax. Are sentences just plays on words?[14] The whole problematic of language as representation seems to spring from such a limited conception. And I am not sure if one escapes it by meditations on the mystery of names, and the unnameable relation between names and things that Nietzsche called both metaphor and a lie. My concern over the term "*différance*" cannot be applied in quite the same way to Derrida's other terminological appropriations. Before making some general remarks about these, I would like to single out for comment one of his most powerful terms—"presence." When he explains the extension of this term, he does so in a way that reminds me again of the very sort of thinking that "the philosophy of presence" itself designates.[15] Derrida lists a whole set of themes which also appear as philosophical concerns and problems, which all exemplify the same privilege, albeit in different forms. Does not the deconstruction of the history of philosophy of presence already posit it as a *history*, as a series of expressions of the *same theme*? Does it not, in other words, at the very moment at which it discovers the pervasiveness of presence, thereby make "presence" present, display it as a unity of the series of its appearance? Derrida's list goes on:

> . . . presence of the thing to the sight as *eidos*, presence as substance/ essence/existence (*ousia*), temporal presence as point (*stigme*) of the now or of the moment (*nun*), the self-presence of the cogito, consciousness, subjectivity, the co-presence of the other and of the self, intersubjectivity as the intentional phenomenon of the ego, and so forth.

Derrida has had something like this point made to him before.[16] The condition of the unity of the list is surely the way in which each of its elements exemplifies the essence of presence. And this is one of a number of cases in which it is simply not open to Derrida to refer back to his claims that avoiding metaphysics is a difficult task, even an impossible one. Here, for instance, we are concerned with the very principle on which it is possible to identify what he has called metaphysics, that is, the condition on which even the sense of his doubts about total metaphysical hygiene rests. In other words, the very term "metaphysics," as Derrida uses it, is precariously balanced on the edge of essentialism. The consequence of these and other objections is to cast doubt on what Derrida thinks he is doing, not necessarily on whether it is worth doing.

More generally, I would point to the extraordinary care and detailed control that Derrida tries to exercise over all the terms he uses, the exceptional self-consciousness of his texts. Where there is a chance of misunderstanding, it is not uncommon to find at that very point, or a little later, a polite insistence that he be understood in the way he intends, that the logocentric meanings of words be bracketed out. I point to this because whatever he says about the merely strategic value of his writing, the function of control, of internal reflexive interpretations, "vigilance," ultimately responsibility, can be read as a transference to the field of writing of an ideal relationship between subject (writer) and object (writing) which is nothing short of metaphysical. It is as if having admitted the ineliminability of absence, alterity, otherness in writing, that all the old values of directness, contact, and—dare one say it—presence (the presence of the author to his reader) can be reinstated at another level. Is it a metaphor to call his writing constitutionally self-conscious? And whence does such self-consciousness draw its value?

The second level of textual reflexivity covers a number of different ways in which Derrida has spoken about the status of his own work, particularly in relation to philosophy, literature, and metaphysics. And it is worth stressing that any consideration of these reflections, to be accurate, must accommodate the fact of their own perpetual appearance within his writing, as well as from one text to another (this most notably in the published interviews *Positions* in which he talks about his texts).

It has been difficult to avoid trespassing on these themes already, but I shall try to make any repetitions clarificatory.

I have repeated the old charge that Derrida's writing is metaphysical, in his own sense of the term. I have also implied, and sometimes explicitly stated, that despite his own claims about the necessity of this state of affairs such assurances were not adequate to the cases I mentioned. But his remarks about the stickiness of metaphysics need fleshing out a little more.

In one of his papers on Lévi-Strauss, he summarizes this complicity strikingly:

> There is no sense in doing without the concepts of metaphysics in order to attack metaphysics. We have no language—no syntax and no lexicon—which is alien to this history; we cannot utter a single deconstructive proposition which has not already slipped into the form, the logic and the implicit postulations of precisely what it seeks to contest.[17]

And he cites the use of the concept of the sign to attack the metaphysics of presence (as in his critique of Husserl, for example), a concept that suffers from the same illness. Nietzsche, Freud, and Heidegger each in his own way tried to step outside the circle, and each was recaptured.

It is after this recognition of the trap set by the circle that the concept of risk appears. If hard work only digs one deeper in, perhaps what is needed is a different approach which may involve taking risks. And why not—it seems there cannot be anything to lose. After all, the risk is one that one could not help taking—the risk of having finally said nothing, or having repeated a movement that could have been foreseen. The method of deconstruction is considered to be the best way of actually going about the task with some limited chance of success.

What is this task? It is not simply the academic one of exposing for all to see the logocentric commitments of the great philosophers, but more important bringing about what Heidegger called a changed relationship to language. Indeed, deconstruction can accomplish both these ends. The change is not easy to spell out but one might call it the recognition of the materiality and vulnerability of writing, and thus the impossibility of those forms of discourse predicated on writing having some higher status, like the expression of truth.

What interests me at this point are the norms that guide the particular way he has chosen to bring about this changed attitude, as I suspect that they are such as to make it impossible to put in question the sort of writing in which Derrida himself is engaged. They take it for granted, even affirm its neutrality with respect to metaphysics. Derrida believes in the classical values of theory, rigor, system, precision, and control. He says that his texts "belong neither to the philosophical nor to the literary register," but they do belong to the register of theory and they do not question it fundamentally. And my remarks about lexical control and proprietorship apply even more importantly at this level. Derrida at one point in *Positions* says that it makes a difference how *rigorously* and systematically one takes up one's distance to metaphysics. But if this involves a distinction between good and bad metaphysics, it is a distinction that can be immediately deconstructed by refusing the value "good" (or "better") to that which is merely more rigorously controlled and self-conscious.

His writing is governed by rules, by norms, even, I would say, by a telos (such as: a changed relation to language). It is very unlike that Heraclitean play of differences to which he often refers. I will not repeat my question about the logocentric nature of these very principles of textual ordering. What does strike me is the poverty of Derrida's middle-range understanding of linguistic form. He is very interested in both the lexical and the systematic level but does not seem prepared to conceptualize the middle-range forms of syntactic construction, except in the case of his general account of the strategy of deconstruction.

One of the reasons for making these points about Derrida's residual choice of theory as a mode of writing is that it does seem to be a *choice*, it does not seem to be necessary; it is not so unavoidable for necessity to be an excuse. To show this, I will make remarks about Nietzsche by way of contrast.

One of Nietzsche's guiding threads was the finiteness of the circle of metaphysics, and the ease with which escapes can be just illusions. Heidegger takes over most of Nietzsche's history of metaphysics but finally locates even Nietzsche's *Wille zum Willen* within that very same tradition as— however concealed—a kind of "presence." And Nietzsche's account of language merely seems to be a radical sensationalist inversion of idealism. This is too fast a tour of a much more subtle diagnosis, but Heidegger is still too fast with Nietzsche. Nietzsche cannot be understood separately from his style. Many of Nietzsche's inversions, for example, can be read as a sustained rhetorical device. And then it would cease to be true that Nietzsche was locked into a simple opposition. One can only decide on the degree to which Nietzsche (or anyone for that matter) has evaded the framework they have tried to escape from, after one has decided that they are serious in what they say. And we know perfectly well that while Nietzsche was serious in what he sought to convey, he did not restrict himself to serious forms of expressions. Nietzsche is amusing, he exaggerates, he plays with his reader.

Nietzsche was, of course, well aware of what Derrida called the closure of metaphysics, its exhaustion, the fact that it had played itself out. Style was his way forward.[18] He chose a way of writing by which he could not be taken literally, referentially. If Nietzsche continued to contribute to philosophy, he did so by abandoning theory. However much Derrida may say he is not serious, many of the problems he sets himself arise from the ideal that may no longer be even intelligible in a strict sense, of being able to say what he means in a theoretical form.[19] Derrida's work is not marked by nostalgia, but a fixation on the importance of going through all sorts of traditional manoeuvres (which it would be wrong to think of simply as academic ones), many of which Nietzsche abandoned.

There may also be lessons to be learned from Kierkegaard's pseudonymous writings, which again use stylistic devices to effect a changed attitude to writing on the part of the reader.

At one point Derrida says that the passage beyond philosophy involves a new reading of philosophy, but unless that also includes an assessment of a philosopher's own understanding of the status of his writing, that new writing could be a form of blindness.[20] It is not possible for Derrida to dismiss such considerations as too much concerned with what the writer intended because he too is obsessed with having his intentions properly read.

Kierkegaard and Nietzsche developed real alternatives to scare-quotes. Quite how satisfactory they are as alternatives depends on how far one takes the activity of deconstruction to be a means to an end obtainable by other routes, and how far one takes it to be an end in itself. Derrida is not always wholly clear on this point.

There is another philosopher whose strategies one cannot ignore in trying to come to terms with Derrida. Surprisingly enough, I am referring to Husserl, the very object of Derrida's most sustained "philosophical" critique.[21] Yet there are some strange and unexpected parallels between Husserl and Derrida.

Both philosophers share the view that metaphysics can be escaped or its impact lessened only by the strictest control over one's writing. Husserl's phenomenological reduction was in part the internalizing of a rule that one should not read off ontological conclusions from one's experiences. It was a refusal to be committed to the ontological seriousness of ordinary everyday experience and language. Husserl's solution is to transform that experience so that it can support a new sort of seriousness, one which confines itself to essential structures rather than to empirical matters of fact. Husserl is constantly reminding his reader not to understand what he is writing in a psychologistic way, however tempting it seems, for that is quite contrary to his intentions/meaning. Many of Derrida's strategies seem very close to Husserl here. Like Husserl, Derrida knows he cannot avoid using ordinary language. But he can hope to prevent us taking it seriously. While Husserl tries to organize a new permanent way of avoiding taking it seriously (ontologically), Derrida, realizing that Husserl achieves this only by giving to language a new phenomenological seriousness, offers strategic uses of terms as an alternative short-life answer to the avoidance of seriousness.

But this appeal to strategy is only to a means. Finally Derrida wants to bring about a transformation of the available ways of reading and understanding language quite as global and total as Husserl, and he would like, ideally, to arrive at a point at which it would be no longer necessary to bracket out the logocentric meanings of words, to cross them out, to indicate that a certain expression is only a transitory evil.

Both Derrida and Husserl offer visions of "philosophical" work to be done, discover or anticipate; new sciences, logics, hold out promises to disciples. Husserl's dreams of the final scope of phenomenology are notorious. As far as judging the value of phenomenology was concerned, Husserl made constant references to the future work to be done. Doesn't Derrida repeat this? Grammatology, a logic of the supplement, even a parasitology are announced, but are they more than gestures in the direction of a new organization of a theoretical space?

It has been said of Derrida that his main strength lies in the fact that "he does not destroy anything or even refute—but simply offers a diagnosis."[22] There is a certain amount of truth in this. Derrida believes that the structures he discovers are actually there already in the texts, constituting them as the philosophical texts they are. Thus

> a patient reading of the *Investigations* would show *the germinal structure* of the whole of Husserl's thought. On each page the necessity—or the implicit practice—of eidetic and phenomenological reductions is *visible*.[23]

While *lisible* would be strictly more accurate a translation than *visible*, in the last line, the commitment to a realism with regard to the structures discerned is brought out well by the perceptual term. Derrida seems to be committed to a science of the structures of metaphysical writing in the same way in which Husserl was committed to phenomenology as a descriptive science of experience. The belief in the existence of structures of presence is a form of textual realism. On what other basis does he claim such a privilege for his readings? And yet such realism is utterly metaphysical!

Derrida cannot refute metaphysics for all the reasons that I (and he) have documented. He began at least by merely exposing it. And of course the parallel with Husserl's neutrality would begin to fade if the development of Derrida's thought were pursued. The use of "indecidables" is the kind of active intervention in a field which Husserl never contemplated. However, it will one day be seen that Derrida's debts to Husserl, through his transformed appropriation of some of the strategies of phenomenology, are quite crucial in understanding why Derrida is not more like Nietzsche or Heidegger than he is. We must not forget Husserl's passion for theory.

One wonders whether, quite apart from its philosophical perversity, the preparatory and internally directing care with which Derrida releases his texts into the scattering winds of history will ever have the consequences for which he must hope. It was Husserl, again, who insisted on the difference between his phenomenology and any ordinary philosophy, that it was not just a branch of philosophy, but the only serious substitute for it. But for all his training of disciples, they ended up betraying him, and his writings were absorbed into the public domain of philosophy.

Derrida talks of himself as having set up camp at a distance from philosophy from which he can still *communicate* with it. But if that communication is the condition of his writing on philosophy, it is also the source of the greatest risk—reabsorption, a condition we might describe as the mortality of the text, and a condition that Derrida's passion for acute textual control seems to be based on ignoring, or excluding.

Would he be entirely disappointed by such an absorption by the philosophy of the future? What would remain? When one considers, for example, the rhetorical exaggeration to be found in such phrases as "the originary violence of writing"[24] or "there never was any 'perception'"[25] one soon comes to realize that the exaggeration does have a function. A memorable phrase drags with it the special attitude needed for its comprehension. But the changes that one can envisage being brought about in philosophy cannot easily be determined in advance. In particular I can imagine two quite different "ideal" effects: (1) the disappearance of certain forms of writing and their "replacement" by others and (2) a change in attitude to the written word (such that writing was no longer understood as the articulation of meanings). I am not clear which Derrida would wish because the substitution of new forms of writing might well take place as the consequence of trying to maintain the seriousness of writing. Such changes would not be necessary, however, if one ceased to treat writing with that reverence.[26] The difference is between ceasing to use the word "God" because it is not a name, and continuing to use it without using it as a name, without that seriousness. If Derrida takes the former path, it seems to me that he follows Husserl in this respect. If he takes the second path, it is quite possible for logocentric texts, even authoritarian texts, to continue to be produced, with a kind of ironic subscript, as a new rhetorical form. Indeed such texts could even reappear, grounded by the self-conscious assertion of the ungroundability of their basic premises, as a new irrational voluntarism, asserted not justified, and logocentrically clean. Or, such texts could simply remain as rhetorical forms with the appeal to a ground appearing as an explicit textual structure, with a "centering"[27] as a formal device.

These various possibilities of learning from deconstruction without practicing it rest, fundamentally, on an exploitation of two factors: (1) Derrida's recognition that it is ultimately the strategies, articulations, and recourses of texts that are or are not metaphysical, not simply the concepts themselves. And that surely opens up an extraordinary range of possible textual ruses, from which it will become impossible to exclude questions of intention. (2) The sense in which the future of deconstruction depends on a steady supply of texts that at some not too deep level take themselves seriously, that continue to pursue a statable truth in a reasonably straightforward way. But (whether or not this is a result of deconstruction) there is no reason to think that philosophy will just stand still and offer itself for deconstruction. Not only could Derrida's defense of Nietzsche against Heidegger (using his "style")[28] be more generously extended to other philosophers who might similarly be thought to have sailed dangerously close to the edge of language and yet carried on writing,[29] but deconstruction could spawn a range of texts that knowingly anticipated the inevitable possibility of their own deconstruction. (Derrida's own texts, he agrees, have just such a status.) The difference such anticipation would make is not one for which there is yet an adequate name.

These deliberations on the power and limits of the strategies of writing are not just abstract conjectures; they license the postdeconstructive temporal musings being offered here. For it is possible at least to imagine a postmetaphysical pluralism about time(s) in which the demand for primitive "elements" would give way to proliferation without foundation. At least a short step is here being taken beyond mere imagination! On this view Derrida's "trace" and Nietzsche's "self-exceeding moment" are each cuckoo's eggs hatched in the nest of foundationalist thinking. There *is* no primitive unit, but if there were it would be like this . . .

3   *The Question of Strategy*

THE QUESTION OF STYLE

The distinction between the form and content of language, between the *how* and the *what*, is not only traditional but formative for philosophy. It is formative in that it implies their genuine separability and so authorizes focusing on one side, on the *what*, relegating the question of *how* to such "peripheral areas" as rhetoric, stylistics, and pragmatics.[1]

Despite the more recent interest of analytic philosophy (especially that of Austin and Searle) in speech acts, and of phenomenology in the noetic aspects of intentionality, this concern with the *what*, with the propositional content of language, has been the dominant tradition. The *how* of a philosophical text, which could be called its philosophical *style*, has too long been seen functionally as a means of conveying a content, or aesthetically as an end in itself. There have been exhortations to clarity on the one hand, and stylistics on the other. But might it not be that the "style" of philosophical writing is not treated with due philosophical seriousness when understood in either of these two ways, either as a means or as an end?

And in fact, whatever philosophy's dominant concern has been, its history does offer us examples of philosophers for whom style was more than a peripheral matter. Consider the examples of Plato, Hegel, Nietzsche, and Wittgenstein. Platonism is always represented as an otherworldly philosophy; knowing has in each case an ideal form as its proper object. And yet Plato so often uses not just the figure of Socrates but the form of his philosophizing, the dialogue. The Socratic maieutics is a dialectical art. This dialectic is a necessary preliminary to the earthly contemplation of the forms for those not trained in such vision from birth. On this view, Plato's use of the dialogue form is no mere means of exposition nor a gilding of the truth.

Hegel's use of the speculative proposition, as he called it, allowed what *he* called the dialectic to appear at the level of syntax, as well as almost every other level of his work, including that of the individual word. It is the German language (the only truly philosophical language, said Hegel) that we have to thank for allowing the form of his philosophy to so intimately reflect his method.

293

Of Nietzsche more will be said shortly, but two aspects of his "style" are worth mentioning here. First, he constantly changes his style in an attempt to avoid all appearance of building a system. System builders, Kierkegaard claimed, never occupy their castles but live in huts next door. As Heidegger put it, "What is important is learning to live in the speaking of language." Second, Nietzsche's *provocative* style makes it impossible to treat his work as a *picture* of the truth. In requiring a reaction from the reader to supplement it, it loses that "imaginary" status.

Finally, one ought to mention the style—questioning, fragmentary, "speculative"—of the Wittgenstein of the *Investigations*. This philosophical bricolage again demands a participation on the part of the reader that would have been redundant on the picture theory of language of the *Tractatus*.

Analogous remarks about the importance of style could be made about Hume, Spinoza, Whitehead, and others. For many philosophers, the problematic status of language is not confined to a localizable philosophical topic but invades the expressive medium of philosophy itself. He may merely believe in the importance of clarity, or he may try to make his style consistent with his philosophical views, or finally he may try to use the *way* he writes to convey something that cannot, or cannot as effectively, be said.

If this story is plausible, those who make a habit of impatience with Continental philosophy might begin to consider that very often a difficult style is not a gratuitous disfiguring of a more simply statable truth, nor a veil covering the shame of confusion, but a careful, serious philosophical choice. I would like to take the cases of Heidegger and Derrida as object lessons for this general thesis. I shall argue that the motivation for their somewhat different styles is an easily statable philosophical problem.

Consider the problem posed by a universal solvent. The problem is how to store a liquid that could dissolve any container one put it in. By analogy, a philosopher who wields a method highly critical of the history of philosophy has to be very careful of his own philosophical production lest it be reflexively destroyed by its own critical method. Both Heidegger and Derrida wield such methods, both are indebted to Nietzsche, and Derrida to Heidegger.

Some will feel that the interest in such large-scale enterprises should go no further. The original error can be seen in the use of such overgeneralizing critical methods, for they reduce the delights of philosophical variety to a gray nocturnal sameness. The short answer is that some games can be played for high stakes, and philosophy is one of them. Logical positivism and ordinary language philosophy were both based on the most general principles of method.

If one provisionally accepts that this problem of an undermining critical self-reference is a *real* one, the difficulties Heidegger and Derrida encounter are best illuminated by a brief reference to Hegel—another philosopher with an all-embracing critical method.

Hegel's dialectic, if one let it, would allow the limits of each previous philosophical position to be marked. If one thought of metaphysics as a kind of blindness to the finite and transitory nature of our conceptual schemes, the dialectic might seem to have a meta-metaphysical status. And yet closer scrutiny detects a teleological presupposition, which can itself be thematized and sublated.

In contrast, Heidegger and Derrida do not set out to rank the great philosophies of the past in some hierarchy of spiritual elevation—although they each have their heroes—rather they are concerned to reveal the metaphysical prestructuring of the texts they analyze and to avoid reproducing these moves in their own thought. Heidegger uses the terms "ontology" and "onto-theology," and Derrida the "logocentric tradition" as roughly synonymous with metaphysics. For Heidegger, metaphysics is characterized by a *forgetfulness of Being* (what it is to be) and for Derrida it appears as the *philosophy of presence*. These phrases will be expanded on later; for now it is important to realize that both of these characterizations of metaphysics appear as structural features of philosophical texts. Heidegger's reference to "forgetting" is not to some mental process, but rather to something like the conditions of possibility of a metaphysics that manifests itself in the structure of philosophical texts at a number of different levels.

To complete this preliminary account of the philosophical importance of their style, one further point must be made. Both Heidegger and Derrida associate the metaphysical history of philosophy with mistaken attitudes to language and misdirected attempts to use it, and each has something approximating a *theory* of language that renders suspect a number of traditional philosophical moves (such as definition). This theory of language is, as has been suggested, heavily indebted to Nietzsche, and a brief account of that relation would not be out of place.

Once it is accepted that there is a problem concerning the relation between metaphysics and language, the scope of that problem still has to be determined. This makes an enormous difference. If it is supposed that the problem is just a regional problem about the relationship between metaphysics and *its* language, and that its language could be isolated from the rest, one could at least in principle stand outside it, so to speak, and with a quite separate, clean, and hygienic language, handle the local involvement of metaphysics with its language from a distance. But if this was how one were to read, say, Heidegger's *Introduction to Metaphysics*, treating as specially philosophical such oppositions as Being and Becoming, Existence and Essence, Appearance and Reality, Fact and Value, and so on, it would soon

become clear that such concepts, let alone all their synonyms and deriva-
tives, cannot be thought of as specifically metaphysical, but are parts of
ordinary discourse. That would mean expanding the scope of our problem to
that of the ordinary discourse of the West and considering its relation to
metaphysics, both informing it and informed by it.

Nietzsche points to both these levels and to a third: that of language
itself. Language as such is, if not plain metaphysical, at least a dangerous
foundation on which to build anything, being a lie and a deception. For it not
only allows but requires that differences between particulars to which the
same term is applied are ignored. In summary, one can at this point
distinguish three levels of concern with the metaphysical adequacy of
language—that of the language of metaphysics, that of the permeation of
ordinary language with metaphysical concepts, and the problem of the
original lie of language as such.

If this last is considered more deeply, something apparently paradoxical
emerges. In "On Truth and Falsity in Their Ultra-Moral Sense"[2] Nietzsche
gives his answer to the perennial question as to how general terms can have
multiple reference. His answer is that they are lies or metaphors. But surely
to predicate "lie" or "metaphor" of a general term is to treat a word as a
proposition or a figure of speech, that is, as an entity of a higher order. And
even if that one is sidestepped, it is hard to see how such judgments could
ever get off the ground. The very word "lie" or "metaphor" would be a lie or
a metaphor, and so on. Clearly Nietzsche cannot here be taken *literally*. And
this is particularly ironic because what he is nonliterally trying to get across
is that the literal relation of *naming* should not be thought of as paradigmatic
for language. And in any case, should not be taken to be an external relation
between a word and something just taken off the shelf. Rather naming and
referring must be thought of as textual, or even intentional phenomena,[3] and
in any case not a direct word-thing relation.

On this analysis of Nietzsche's aim in using the terms "lie" and
"metaphor," it is clear that quite apart from the different levels at which
language and metaphysics are interwoven, Nietzsche's text offers us a quite
different direction to investigate—that of style.

Derrida is right in thinking that Heidegger was wrong in calling
Nietzsche the last metaphysician.[4] Through his style, or rather the variety of
his different styles, Nietzsche escapes this charge. All of Nietzsche's promo-
tions of laughing, dancing, leaping, and other ways of modulating a distance
from a surface are metaphors for the advocacy of style as a field for
philosophical innovation. Quite what it is capable of is one of the questions at
issue here. But when one grows weary of trying to eliminate metaphyscial
terms from philosophical and everyday discourse, might it not be a change of
style that is needed?

It is now possible to summarize the claim I hope to be able to make good: Heidegger and Derrida both employ highly critical philosophical methods, the generality of which threatens to encompass their own philosophical work. In both cases, their diagnosis of the sources of metaphysics is linked to views about language and what it can and cannot do. Their respective "styles" are deliberate and careful attempts to direct their own writing so as not to retread too many of the paths of metaphysics, and in this and other respects they are each following in the footsteps of Nietzsche. In support of this claim, I begin with Heidegger.

HEIDEGGER AND THE PERFORMANCE OF LANGUAGE

My aims in discussing Heidegger are fairly limited. First, I shall be talking about the later Heidegger rather than the Heidegger of *Being and Time*. And second, a comparison with Derrida will always be on the horizon. So there will be sins of omission, selection, and simplification.

Heidegger's later philosophy can be interpreted as a rejection of the subject as origin or ground of meaning. This perspective facilitates a grasp of the transition from the early to the later Heidegger. What is wrong with metaphysical subjectivity is that it has been an inadequate understanding of its own activity. It takes itself to be a source, when in fact its very status as a being is grounded in a relation to Being. The forgetfulness of Being manifests itself as a forgetfulness of the conditions of one's own being. The consequence for one's (mis)understanding of language[5] is that one takes one's possession of language for granted and treats self-expression or communication as its essence. In doing so one forgets that language is a condition for one's having anything to express, indeed for one's being a subject at all. Language is not a means but rather constitutes the world of means and ends. More strongly than most philosophers, Heidegger conveys the autonomous power of language.

Heidegger distinguishes, broadly speaking, two modes of language. The first could be called technical/logical and the second, poetic/hermeneutical. These should be thought of both as actual ways people use language and as capable of being expanded into theories about the basic nature of language. In this very duality a point for a new departure will be found later.

In its technical/logical use, language operates primarily to communicate what has already been understood. Words function as mere counters; the linguistic appropriation of the world is a thing of the past; the language space is closed. Language is a transparent medium to be filtered clean should it ever cloud up. It is the language of (at least) normal science, of some philosophy and of much everyday conversation. In poetic language, on the other hand, one listens to what language itself already says, one works with rather than uses language, one follows rather than forces language, waits rather than leads. It is not a passive "use" of language—the poet is even creative—but it is a use aware of the conditions of its own activity.

For Heidegger, the technical use of language is naive, covers over its own truth, and when he talks about "*language*" it is about that language that makes the poetic possible. Many of the remarks I have just made are captured in the following quotations from "Language" on which we will focus some attention.

> Language is—language, speech. Language speaks. If we let ourselves fall into the abyss denoted by this sentence, we do not go tumbling into emptiness. We fall upward to a height. Its loftiness opens up a depth. The two span a realm in which we would like to become at home, so as to find a residence, a dwelling place for the life of man.

> To reflect on language means—to reach the speaking of language in such a way that this speaking takes place as that which grants an abode for the being of mortals.[6]

Heidegger writes like this rather than in a more straightforward way to avoid the circuit of metaphysical repetition. One cannot distance oneself from the history of metaphysics just by exercising care over the ordinary use of ordinary or theoretical words, or by a robust confidence in the subject-predicate proposition. For these words and the privilege of certain sentence forms belong to that history. They mark the level of the ontic with the unconscious products of long-forgotten determinations of Being. At this level there are only beings, and Being never gets a word in. When describing Heidegger's views of language, I was unable to say that he sees man either as active or passive in relation to language. This active/passive opposition already presupposes the independent describability of man and language that Heidegger is questioning. It is by a transformed style that he attempts to transcend such limitations. It is now necessary to return that passage.

Heidegger began the essay with "man speaks." Now he writes the words "language speaks." The substitution of "language" for "man" is part of an exercise in undermining the status of man as a metaphysical subject, in deepening the sense of the dependence of the speaker on language. When someone speaks, is it not language that is heard? When he continues, "if we let ourselves fall . . . ," Heidegger is suggesting a response to this sentence that is not merely one of understanding it, but of allowing ourselves to be affected by it, or moved by it. He promises that the risk of falling or loss of self is illusory. Rather we will "fall upward." He is suggesting what he elsewhere calls a new experience with language. Or again, "It is not a matter of stating a new view of language. What is important is learning to live in the speaking of language." This sentiment is repeated in the next two sentences. The space opened up by the height to which we in fact rise, and the depth to which we feared we would fall, "span[s] a realm in which we would like to become at home, so as to find a residence, a dwelling place for the life of man."

Here as elsewhere it is important to see Heidegger's exploitation of ontic language—the language of houses, dwellings, abysses, and so forth—to metaphorically capture the ontological. These particular metaphors capture the decentering of man that he is emphasizing. We are *in* language.

And again, in the last sentence, which begins "to reflect on language means—to reach the speaking of language." In a grammatical transformation can be seen Heidegger's insistence on the need for a new relationship to language rather than discovering some new truth *about* it.

The way in which a style can carry a "message" could be called *exemplification*. When I discussed earlier other philosophers (from Plato to Wittgenstein) for whom style was of more than peripheral importance, the possibility was left open of an exemplification that could not be put in any other way. This I will call a *necessary exemplification*. Heidegger's general style fits this description, but in an extraordinary way. For Heidegger of course would *reject the very term exemplification* and its associated logic. For that presupposes a meaning that can be plugged into and detached from a language at one's convenience—precisely the view he rejects. Such an exemplification that rejects the logic of exemplification (that is, detachability) I call "performance." Heidegger's insistence that what is at stake is a new relation to language rather than a new view is entirely consistent with this account of his textual practice. The point of such "performance" is to speak so as to show what it is to speak.

An account of his style that cast its net over a broad selection of his writings would also remind us of several things about Heidegger.

1. Heidegger introduces new words in an innovative way, exploiting the German language to the full, and even beyond, in his imaginative etymology. Oddly enough, Heidegger always seems wholly serious about his etymology even when it seems to this reader like play. (In this respect, among others, he differs from Derrida.)

2. Heidegger yokes old words into new service, especially metaphorical. Those most favored I call pre-reflective relational words. They fall into three main categories: (a) those modeled on the body and its space, such as Zuhanden, Vorhanden, standing, reaching, remaining ... (b) those that relate me to an outer space, so to speak, such as house, world, space, openness, dwelling ... and (c) those that relate men and Being—

disclosure, concealment, withdrawing, granting, clearing. What he calls thinking unlocks for philosophy a whole barrel of previously unemployed terms. This is roughly coincident with what Hofstadter calls Heidegger's primitivism.

3. Heidegger promotes ontological language at the expense of the epistemological by what I would call Germanic nominalization. Sentences like "The greater the concealment with which what is to come maintains its reserve in the foretelling saying, the purer is the arrival"[7] stick in one's mind.

4. The provisionality of Heidegger's later philosophy—most of it is as "on the way" as *On the Way to Language*[8]—is registered in the many occasional formats he employs such as the lecture, the address, the dialogue, the conversation.

5. Heidegger's practice[9] of crossing out Being (B̶e̶i̶n̶g̶) is an attempt to both name and not name Being, for if one thought one had succeeded in catching it in a name, it would be lost again.

6. He uses what has been called philosophical tautology (care cares, temporality temporalizes, nothing noths . . .). Schöfer sees this as an attempt to name without subjecting the thing named to the dispersion of normal syntax. He closes with a remark that gratifyingly confirms the approach here. By this device of philosophical tautology, he says, "the performance of the thing and the performance of language coincide 'here.'"[10]

I have discussed six different elements of the style that Heidegger has forged for himself. Does it constitute a solution to the problem begun with—that of avoiding metaphysics? Within its own terms it is an extraordinary achievement. Metaphysics is characterized by forgetfulness of Being; that condition is finally only to be cured by a new relation to Being; that can be explained with performative consistency only by a new way of "using" language, a new way of writing. And that is the project he takes on. The achievement lies as much in the way he follows his problems through as in any one of the means he adopts to solve them. As I accept the claim that there are things that can be done with words that cannot be said—though Heidegger does not rely, as Austin does, on preexisting conventions for his performative language—Heidegger can be granted some sort of success.

However, his solution is not without its own problems.

It is easy to misinterpret the later Heidegger and just as easy not to know if you have or not. That a difficult style is not uncommon among great philosophers does not give any of them immunity from misunderstanding, and wanting to be understood—if only, as in Nietzsche's case, by another age—is a communicative a priori.

As linguistically innovative as Heidegger is, these innovations are understood only from the standpoint of an existing language to which they are bound, a language they attempt to escape. And yet the figurative nature of Heidegger's language sets severe limitations—however necessary it is—to the privileged status it seeks. For it is continually necessary to pass through the world of images—of woods, homes, hearths, and clearings—to get to his thought. And how can that not rub off on the ontological level he aims at?

Even if it is beyond all doubt philosophical, it is only doubtfully still philosophy. It is still highly relevant to philosophy because it is thoroughly determined by the whole history of philosophy and represents both a drawing of the boundary around what remains philosophy and a questioning of the continuing possibility of that discipline.

Lastly, and most critically, Heidegger seems to be aiming at an ideal coincidence between what skeptics would still call the act and the content of language. And if I am right, it is in the performative use of language that, if anywhere, he achieves this. But this is an ideal, one that the sheer materiality of language can never allow to be achieved. Heidegger is projecting on to his linguistic performance an imaginary unity, in a desire as old as philosophy.

There is some reason to suspect that Derrida might have tried to avoid some of these objections. He begins with much the same problem as guided Heidegger (distancing oneself from, if not overcoming metaphysics) and he shares some of my criticisms of Heidegger. So it is to Derrida that I now turn.

DERRIDA AND THE LANGUAGE OF STRATEGY

It will shortly be clear what Derrida's practical response is to the metaphysical predicament posed by the closedness of the conceptual and thematic repertoire of the history of metaphysics. It is first worth explaining his response to Heidegger, or at least to the style dominant in Heidegger, for he accepts that even in Heidegger's text there are "ruptures and changes in ground." Derrida's main reservation can be put simply: Heidegger has not been radical enough in his solution. Having diagnosed the privilege of presence—or the atempt to make Being present—as the hallmark of metaphysics, Heidegger seems content to give a kind of phenomenological filling out of presence, to rethink it rather than radically displace it. The very thematics of Being is of a "transcendental signified," a resource that escapes the play of language, that escapes difference. Heidegger, in short, reinstates the value of presence, and in doing so suffers the limitations of his strategy of immanent deconstruction.

For Heidegger, Being cannot be said, cannot say itself, except in *ontic metaphor*.[11] But this gives a particular interpretation to the sense of Being. In thinking through presence, Heidegger "only metaphorises, by a profound necessity which cannot be escaped by a single decision, the language it deconstructs." But the use of metaphor for Heidegger is entirely devoted to the project of bringing men nearer, closer into the *presence* of Being. This appears in Heidegger's exploitation of the language of light—brilliance, illumination, clearing, lighting, and so forth. But more important still, it appears in the privilege accorded to language, and in particular spoken language, with talk of listening, hearing, the voice, speaking, speech.

Heidegger's metaphorical style—the poetic style of many of his later writings—draws its power from the ability of metaphor to make present. For Derrida language is not so easily bent to such ends. It should not be thought that all Derrida's judgments on Heidegger are negative. This is not even largely true.[12]

The limitations of Heidegger's text lie in the area of what Derrida calls *"strategy."* At a number of points, Derrida makes thematic what could be called the logic of reappropriation or system recuperation. The sheer interconnectedness and diffusion throughout language of metaphysically burdened concepts and themes make it very likely that attempted escape moves will have already been covered, will themselves be just moves in the game. One does not, for example, escape from the sphere of influence of a concept just by negating it. What is required is an analysis of the possible strategies for success.

On at least two occasions[13] Derrida offered two possible strategies by which to handle the problem of distancing oneself from metaphysics. Each time he contrasts (a) *an immanent deconstruction*, an internal critique that would exploit the resources of metaphysical language against itself, demonstrating contradictions and so forth with (b) an abrupt change of ground, in which one would take a step *outside* philosophy. To begin taking such a step one would question "systematically and rigorously the history of these concepts"— a kind of deconstructive genealogy. He takes the first strategy—immanent deconstruction—to describe the approach of Lévi-Strauss and of Heidegger; and the second approach, certainly when characterized as an abrupt change of ground, to capture the typical choice of current French philosophy.

As has been shown, Derrida believes that the first strategy risks merely confirming the prison one is trying to escape from The second strategy runs a different risk—sterility—one ends up not being able to say anything interesting. But it is a risk he feels he has to take. He finally concludes that it is necessary to somehow interweave these two different strategies, producing complex and multiple texts.

It seemed at first that the way to contrast Heidegger and Derrida was to focus on Heidegger's style and Derrida's strategy. But in fact the difference is more subtle. It lies in the fact that Derrida's highly self-conscious styles are governed by considerations of strategy, whereas Heidegger simply does not have such a *field of options* "ready to hand." One way of answering those who complain about Derrida's trickiness is to explain this puppeting.

Derrida's references to strategy should not just be thought of as opening gestures; they are to be found throughout his texts.[14] In "Différance," for example, an essay in which he introduces the title term as a kind of cipher in the text, he repeatedly emphasizes its merely *strategic* use.[15] What he is insisting on by reference to strategy is that he is not trying to introduce a new theological constant or metaphysical mooring.[16] If it is right to focus on these references to strategy, what is being offered is a text that no longer has a straightforward surface, which constitutionally resists being read as a meta-physical text by virtue of the multiplicity of determinations and indetermina-cies to which it has already been subjected. What must be done now is to spell out some of the ground rules of Derridean writing to see just how such a text is constructed. Reference will be made to many different texts despite the danger of ignoring their proper boundaries.

Derrida owes much to Heidegger in his understanding of metaphysics. Metaphysics is a theoretical writing organized around a privileged point—a presence.[17] Derrida's basic criticism of this privilege goes something like this. Presence, even in its literal temporal sense, is never simple, but structured by a relation to what is not present, what is other or absent. This is nothing other than the basic structure of the sign, which from the beginning, if it can be put like that, involves a reference to other signs. If all meaning has the structure of the sign, then all candidates for the privilege of presence are constitutionally in debt to something outside of, other than themselves. Derrida's elements, which exhibit in their own name this deferred, derived nature, are "traces." The trace structure, and its pervasiveness, is the lever by which the privilege of presence is deconstructed.

Derrida's earliest, best-known, and perhaps most successful object of scrutiny was the logocentric or phonocentric tradition, that which privileges the voice, or the inner voice, as having some special hotline to meaning. Derrida's deconstruction of this tradition—especially as it appears in Husserl and Saussure—is father to the term *écriture* (writing, and perhaps "scrip-ture"). *Writing* serves to determine the entire field of signs or traces (includ-ing, one might add, *speech*), where this field is all the time understood as language loosened from any such privilege as a relation to meaning, or to the soul, or to the voice, and so forth. Writing, in this persuasive definition, is characteristically understood as a play of differences (in a way that takes up and radicalizes the Saussurean diacritical account of meaning).

If the metaphysical text can be described as the centered text, Derrida's is pervasively decentered. The complexity of Derrida's texts results from the attempt both to keep one foot in philosophy and yet to systematically avoid creating in his texts those structures by which philosophical effects have traditionally been created. Reference has already been made to his overriding concern for the correct strategy. I will now try to explain in more detail what his textual tactics are, distinguishing them into five categories for convenience: graphic, lexical, structural, methodological, and self-reflexive. It is by the parallel control of each of these levels that Derrida achieves his effects.

Under *graphic* tactics are included all the standard devices of scare-quotes, brackets, italics, and so on that can be deployed singly or together to indicate many different degrees of using or mentioning of terms whose full implications one might not want to embrace. Derrida also appropriates the crossing-out device Heidegger employs in *The Question of Being*, although with Derrida there is not, as there is for Heidegger, the whiff of presence around the corner.

The *lexical* level is much richer. I would like to particularly point to appropriations and modifications of words that give rise to a characteristically Derridean army of terms, some of which one hesitates to call words, let alone concepts. In the term *différance* Derrida creates a hybrid of which Hegel would have been proud, except for the fact that Derrida is adamant in retaining a proprietary control over its lack of any full meaning. Terms like *supplément* and *trace* resist ordinary incorporation by being "indécidable"—not licensing the normal inferences. A trace is not a trace *of* anything, for example. Elsewhere—with *brisure, hymen, pharmakon*—what Derrida fastens on is the whole field of associations (sometimes contradictory) that they have. The point of all these innovations must not be missed. They are intended as a kind of fifth column undermining the tendencies toward theoretical centering and congealment.

The level I call *structural* covers the surface structuring of his texts. And I am thinking of those paginal arrangements that reach their creative peak in *Glas* with insets into insets in parallel texts on the same page. In this multiple writing the very linearity of writing is questioned, and the place of the author is made problematic. In the language of intentionality, it is as if the interrelationships among the parallel texts constitute an intentional nexus that severs the primacy of the author-text relationship.

The *methodological* level is perhaps a misnomer. There is much here that escapes the category of methodology—the many subtle and detailed ways he handles the texts he parasitizes. But pretending for a moment that Derrida had a method—a pretense he once himself made, even if he later regretted the simplifying lever it gave to others—that method would be deconstruction. And this involves him in a particular textual practice, one that offers niches for the appearance of the lexical innovations already described.

If the metaphysical structure of a text is understood to betray itself in a founding opposition, in which one of the terms of the opposition is weighted over the other (think of the speech/writing relationship in Saussure), then Derrida has a definite strategy for deconstructing this so as to try to prevent it from simply closing ranks after the first assault.[18] He first of all inverts the opposition, crediting the lower term with the privilege previously accorded the first, and second, he injects a new term or an old one reworked that will permanently disrupt the structure into which he has intervened. Thus *writing* in the broad sense is what comes to disrupt the opposition between speech and writing (in the narrow sense). Deconstruction, it will be seen, is *not* merely a case of "leaving everything as it is," although the early Derrida seems to have thought of it more like this.[19] It is a procedure that aims to bring about a changed reading of a text.

Finally, *self-reflection*, or perhaps better, self-commentary. This includes all those remarks in which Derrida explicitly explains the problems of writing his kind of texts, the need for strategy, the risks of sterility, the debts he has to other thinkers, and so forth. What is so important about them is that they seem to themselves occupy a privileged position in his texts, of being meant *literally, seriously*, even *urgently*, and yet they are a part of, and necessary parts of, texts that question the very possibility of such a privilege of the serious and the literal. It is as if by going up a level one could escape the limitations of the level beneath. But there has been no escape from language, so that cannot be. This criticism will be expanded shortly.

Several avenues of criticism open up: the danger of sterility, his text centeredness, and the status of his textual reflexivity.

However much Derrida's writing has inspired literary criticism, it seems to many philosophers if not to be actually sterile, to have at least washed its hands of a whole range of classical philosophical problems. And while his own textual dissemination shows no sign of flagging, the risk of sterilizing the reader, too anxious of error to lift his frozen pen, is very real.

The criticism of text centeredness can be made at many different levels. The basic objection is that Derrida treats the central extratextual conditions on which textuality is alone possible as mere extensions of textuality. And yet the human world of action and perception is not a text, but a context. Derrida cannot handle the idea of context.[20]

Finally there is the problem of his textual reflexivity. While for Derrida the language of intention cannot be the site of a privilege, he *has* to privilege his own self-reflective remarks to ensure that his readers understand his texts properly. When trying to explain why, in Derrida's eyes, Nietzsche is innocent of Heidegger's "last metaphysician" charge, Spivak suggests, "Perhaps this entire argument hangs on who knew how much of what he was doing."[21] And she refers to an exasperated Derrida replying to Houdebine in a *Positions* interview that he "knew what he was doing" at an important gap he appeared to have left in a text.

I said of Heidegger that he aimed at an ideal coincidence of "style" and "content" and that the ideality of that coincidence was of the same order as the metaphysical dream of presence, that pure performance was a desire doomed to unfulfillment. A parallel critique could be offered of Derrida. The very requirement to expand on his stylistic repertoire in the interests of a strategy that in the absence of an end found itself showered with means[22] involved the creation of a space of textual intentionality that has to take itself to be privileged in order to assure the reader of a correct focus on the rest of the text. And yet by this very privilege the Derridean strategy undermines itself, the solvent begins to dissolve its bottle.

Derrida characterizes metaphysics as philosophy of presence. A deconstruction of such a philosophy reveals a structure of textual privilege licensing classical philosophical moves. However, while Derrida is careful not to adopt any of his terms on a permanent basis—his innovations for the most part come and go—the *common space* occupied by these terms and the deconstructive intentions they are designed to serve do claim an analogous privileged status in relation to the texts in which they are embedded. Derrida is not producing a *separate* meta-language, but he is producing a bifurcated text with an internal controlling order established only by the appeal to authorial intentions. And in this appeal there is a fundamental obstacle to the Derridean project.

On many occasions Derrida points out the impossibility of doing without metaphysical ploys to deconstruct metaphysics. In this respect he differs from Heidegger in being prepared to get his hands dirty. The inevitability of metaphysics is meant not to encourage it, but to focus attention on textual strategy and economy rather than on schemes for its final elimination. In these respects, the Derridean project is importantly modest. But this cannot serve as an escape clause to justify Derrida's resurrection of the ultimate importance of his own *intentions* at that vital hermeneutical center—a strategically informed understanding of his texts.

A CAUTIONARY TAIL

I have great admiration both for the effort and the results of Heidegger's transformation of his style, even when it takes him outside philosophy altogether. It is, however, an *experiment* in philosophical writing that should not be merely imitated but further developed and explored. Heidegger breaks through not to a world of answers but to new problems and questions. The new paths opened up by appropriating one's style in an exemplifying manner are not paved with guarantees of success. In the conditions of its use, in its semantic and pragmatic presuppositions, in the arbitrariness of meaning of so many words, in the contingent historical circumstances that shape it, a language has what can best be described as a *material* aspect. From this materiality language derives both its depth and richness and its resistance to serving any higher end such as philosophy. If a new style can mark a transformed articulation of Being, it does not protect language from its own materiality—indeed nothing could.

Heidegger's turn to style is both productive and stimulating even if it is unable to provide any ultimate solution. Much the same could be said of Derrida. But from the limitations of the Derridean project valuable insights can be drawn about the interpretation of philosophical texts.

My critical treatment of Derrida (and Heidegger) mostly took the form of an immanent critique. In this respect my criticisms followed the same line as the expository premise—that both philosophers could be seen as turning to "style" as ways of avoiding being caught in their own metaphysical searchlights. In both cases the critique took the form of claiming that they were so caught. To remain at the level of such criticism is not, however, satisfactory. One is not obliged to specify the weak spots in the position being criticized, nor to make a positive attempt to correct matters. I would finally like to indicate the direction this might take.

The weak point in Derrida's philosophical position lies in the fact that while it has certain rules for *suspecting* metaphysics—the list of structures of presence constitutes a kind of field guide—it has only ad hoc procedures for lifting this suspicion. Thus, despite the fact that the will-to-power seems to function as a *center*, an ungrounded ground, Derrida finds in Nietzsche's style(s) reason to have him released from custody. For the most part I agree with Derrida here. But if one tries to capture the grounds for this exemption of Nietzsche, in the form of a principle, one is left wondering whether many others already tried and convicted ought to be released. Nietzsche wears his style on his sleeve but might not the entire history of philosophy have been enacted by masked men?

The general question to ask of a philosopher is—What attitude does he take up to the signifier? The importance of analyzing the "rhetorical" aspect of philosophy, its "style," lies in the degree to which it betrays this attitude. But the example of Derrida's defense of Nietzsche shows us that the attitude to the signifier can manifest itself at a number of different levels of the text, and that there may even be, at least locally, a hierarchy of such levels.

For example, Nietzsche not only talks of the will-to-power, he psychologizes, he materializes, he even tries his hand at philosophical hit man. There is no doubt that both at the lexical level and at the lower level of philosophical argument, he works the old language and deftly handles the old conceptual oppositions. "How the 'Real World' at Last Became a Myth"[23] is a classic piece of antiidealism. But even at this stage Nietzsche must still be regarded as a metaphysician from Heidegger's viewpoint. He escapes because it soon becomes perfectly clear that Nietzsche's texts are elaborate *constructions*. He is not commited to the frames of reference he temporarily inhabits. The method is slash, burn, and move on, a nomadology. He has no fixed address.

If Nietzsche's attitude to the signifier could be described as *playful*, it is a feature that can be directly attributed to his texts without further evidence of a contextual or psychological nature. The way Derrida includes a theory of language within his own texts, his accounts of the strategies one needs to follow to avoid the errors of others, his warnings of the dangers of misinterpretation—all this self-documentation answers to the same demand, that a "successful" text should contain within it the principle of its own interpretation.

However, as a statement of a necessary condition for a "successful" text, this demand is a mistake, and one with unfortunate consequences. There is just no good reason apart from an idiosyncratic methodological hygiene to suppose that a text can always be accurately judged on its appropriation of language, its attitude to the signifier, its metaphysical commitment, in the absence of an investigation of its conditions of production.

A revolutionary pamphlet may invoke a range of crude conceptual oppositions. It aims to incite. It may make all sorts of questionable assumptions. But they may be shared by its readers. Such indexical, situated properties of a revolutionary pamphlet are shared by all *occasional* texts. While not all texts are written for particular occasions, they have some pragmatic, situated qualities. The *importance* of this dimension is that it is the condition for what is unsaid in the text, for what does not *need* to be said because it is self-evident or taken for granted without ever being thematized by those to whom it is directed. If this is so, one cannot read what is unsaid in a text without knowing what was taken for granted in its production.

On this view, the fact that texts have structures, centers, formal properties is an *internal* reflection of the fact that texts are produced under specific conditions, that they address certain problems, that they make certain assumptions, some of a factual nature, others about the very nature of facts. The finitude of the text—of which structures of presence are symptoms—is a reflection of the historical specificity and particularity of its Being. The *formal* analysis or dislocation or manipulation of a text brackets out concern with the context of its production and reception. And this unnecessarily restricts the value of the method.

In this limitation, the Derridean texts display their own finitude. They intervened at a time of blindness to the structures they revealed, and they exploited to the full a new liberty of style. But it is no less a mistake to treat the limitations of a particular method as limits on what can and needs to be done. Much of what is thought of as metaphysical disappears as such when the context of its appearance is supplied, for what seems to be a formal property is then seen to be a reflection of a need, an interest, a desire . . .

POSTSCRIPT to
The Question of Strategy

Tout dans le tracé de la différance est stratégique et adventureux.
                                                              Derrida, *Marges*

In the last three chapters, I have persistently (and with luck, consistently) put in question the legitimacy of Derrida's deconstructive strategy. I have argued that any positive formulations it offers are wedded to transcendental modes of thought, and that "erasure" is no protection. Surprisingly, however, all this is said in an appreciative attitude, for whether or not "legitimate," it is successful in exposing, in its own terms, the metaphysical motifs in (philosophical) texts. What it cannot do, and there is some force in the position that Richard Rorty takes,[1] is to offer anything like another positive account of anything, let alone the subject of our concern—time. But the fact that deconstruction cannot do it does not mean it cannot be done in a way responsive to its insights.

Of course, this attempt to limit the scope of deconstruction can be opposed. I have already adduced various arguments for my position, and defended it against some of the obvious replies. But the reply that it fundamentally misunderstands Derrida's moves and that it is a naive reading could still be made. In this postscript I shall try to make my critical position that little bit clearer and stronger.

Let me first rehearse some of the doubts I have voiced:

1. That neither difference nor *différance* can be thought except in relation to identity and presence. The displacement of the foundational status of the latter by the former cannot be sustained. The hypertrophic development of either pair would be one-sided and even—if that were not such a contested term—undialectical. The everyday mutual interdependence of

---

* This is a condensed and amended extract from my "Différance and the Problem of Strategy" in *Derrida and Différance*, ed. David Wood and Robert Bernasconi (Evanston: Northwestern University Press, 1988).

these pairs is the unacknowledged point of departure for any thinking involving them.

2. Deconstruction is essentially a kind of *formalism* because it interprets as symptoms of a metaphysical syndrome (fissures in a text, structures of supplementarity, positing of a transcendental signified) what are actually the internal reflections of the outer historical conditions of a text's production.[2]

3. "Presence" cannot be made into the "effect" of "*différance*" because the only language in which this makes anything like sense is the language of transcendental causation. If one uses this language under erasure, its force is illusory. If one uses it straightforwardly, one is guilty of mere (intra-metaphysical) inversion.

4. The success of Derrida's strategy seems bound up with recognition of his authorial intentions, and his generous guidance in this respect. But this gives an extratextual source (the author) the very kind of metaphysical privilege he is at pains to purge.

To each of these points Derridean replies can be made, and indeed have been made, even if these replies do not settle matters. The first and last claims will be singled out for further treatment, for it is they that are most directly concerned with the question of strategy. I shall take the last first.

Despite his warnings not to do this, Derrida must be interpreted as offering, at least formally speaking, transcendental arguments. I say "formally," for although he does not posit transcendental entities, he is offering us "conditions for the possibility of . . ." and not just logical conditions, but a "productive activity" that brings about effects. But is this reading not somewhat naive? Well, Derrida does not *always* caution us against understanding such claims in a traditional transcendental way. Great stretches of *Positions* go unprotected by such precautions. The reply to this, as Spivak says, is that there is always an "*invisible* erasure." Why? It is clear that on those occasions on which he does warn against a metaphysical reading, the force of his remarks is not restricted to those occasions, but is quite general. But if *différance* does not literally *produce effects* (because the language of production and cause/effect is appropriate only to relations to a generative ground, that is, a "presence"), then his remarks do not have the force intended. When Saussure says of language that "there are no positive terms, there are only differences," one has a sense of what he means that need not involve making difference into a principle with a power to bring about effects.

Derrida's general strategy is surely this: to infiltrate *différance* into the syntax of foundationalist and generative thinking, with a view to depriving it of its attraction. (One might compare the release of sterile male mosquitoes as an antimalarial measure.) But once it is clear that this *is* the strategy, it is possible to ask whether this substitutive infiltration is acceptable. Derrida may say that of course it is not acceptable—that it is a transgression. The question then is—what is it to *go along with* Derrida?[3]

Let me go around again, and this time with a closer focus on the paper "Différance."[4] Derrida's texts, it may be said, always imply an "invisible erasure"—he is using metaphysical concepts in a *restricted* way. That is, he would deny or refuse their full involvement in all the moves of a metaphysical discourse.[5] (Now, one could object here along the lines of my objection to the intentionalism of his insistent authorial control. Is it not an enormous claim to be able so to restrict the play of these terms that they do not start to do metaphysical work? Surely that would involve control over the reader's response? But I will leave this criticism undeveloped.) If, as I suspect, it is actually important to the ultimate shape of his thesis (what he wants to *do* with the term "*différance*," for example) that he uses such terms *out of* erasure, one should perhaps be more careful in automatically being so charitable. There are, for example, explicit contradictions between different texts on the subject of the status of *différance*. In *Positions* (1972)[6] and elsewhere, he writes repeatedly of the *concept* of *différance*, while in "Différance" (1967) it is neither a word nor a concept. This is simply to show that consistency cannot be assumed in Derrida's writing.[7] Nor should "invisible erasure" be used as a portable barrier to criticism.

So, to repeat my claim—it is that Derrida either uses transcendental forms of arguments in explaining the term "*différance*," in which case he undermines his whole project, or he does not, in which case the force of all he says about *différance* (and its intelligibility) evaporates.

Why not attribute to Derrida here a theory of transcendental textuality? Why is what he is doing not a translation from experience into writing of Husserl's account of constitution? Instead of supposing that there is this ulterior work going on in a text—of production and effacement—why not suppose that such absences and lacks and gaps are *created* by the act of transformative reflection on a text? Is not Derrida, in other words, projecting on to the texts he deconstructs a work of generation and repression that appears retrospectively only by contrast with the second deconstructive text? Surely it is *there* that all the *work* occurs, where all the action is?

To support this suggestion, an example. In his essay on Saussure, "Linguistique et grammatologie," Derrida brilliantly exposes the almost hysterical expulsion of writing from the field of linguistics proper, and the language and tone that Saussure employs makes it clear that something like *repression* is going on.[8] Writing threatens the natural life of language, and so is a monstrosity. But it is interesting that in this essay, in which Derrida is really very successful in exposing "repression," he makes constant use of the language of psychology, and "speech acts"—not of textual structure. He writes of Saussure's wishes, his "irritations," his tone, of him not wanting to "give in," of not wanting to be "too complacent." It is Saussure (not his text) who analyzes, criticizes, confronts, "says," defines, takes up, and so forth. Now I have no wish to drive a wedge between Saussure and his text, but it is true that the plausibility of treating this as a work of repression rests very heavily on the language of authorial desires, acts, and intentions (albeit unconscious) and not on an autonomous textual activity.

Elsewhere, I would claim, Derrida is making use of the language of transcendental causality, locating such work in texts, by an unjustifiable analogical extension. Perhaps deconstruction does not discover anything, but transforms texts and then allows a comparison by contrast between old and new.

It might be thought that these references to the generativity or productivity of *différance* are inadequate to establish that it is playing a transcendental role—and not because of the "invisible erasure" under which the terms are being operated (assuming for the moment that the "erasure" umbrella is always there, and that the authorial control that such "erasure" implies is not itself suspect) but simply because I am dealing with mere phrases and not with the detailed and subtle account Derrida gives of the structure of *différance* and its relation to the trace, the logic of supplementarity, and so on. This is perfectly reasonable point, and I will try to meet the challenge it throws down.

In its analytically dual aspect—of difference and deferment—and in the various contributions made to its "assemblage," *différance* is a condensation of a theory of the impossibility not of everyday presences in the "empirical" sense, but of a certain philosophical/metaphysical *value* of presence. *Meaning* is never completely fulfilled, in other words. One important consequence is that there can be no *archē*, no first point, no foundation, and no epistemological ground. For any putative *origin* has its fullness (and therefore its capacity to originate) constitutionally or essentially delayed. Once it is accepted that the possibility of philosophical discourse rests on the originating and grounding value of presence attributed to certain concepts, and to certain recourses and moves (for example, to experience, to conscience, or to truth), then there is the possibility of a reading of a philosophical text that unmasks not just the difference and deferment involved in every "presence" but the process of effacing or forgetting that difference. Thus

> the "matinal trace" of difference is lost in an irretrievable invisibility, and yet even its loss is covered, preserved, and retarded. This happens in a text, in the form of presence.[9]

I do have enormous doubts about this sort of claim. But would these doubts not be allayed if one allowed (for?) Derrida's *strategy* of writing? Am I not deliberately closing myself to what he says he is doing? For now, two replies: First, the fact that Derrida anticipates (and in *Positions*, scorns) the transcendental reading of his work, and says many times that this is wrong, is not a conclusive reason for avoiding it. One might indeed interpret these cautionary remarks as anxious premonitions of his just fate, or as themselves "merely" strategic, designed to divert the scent. Second, suppose I am *refusing* to play along. (Am I a bad reader?) Can there really be a strategy of *writing* that is not in principle compromised by the residual interpretive freedom of the reader?

Derrida could perhaps deploy here the distinction he draws at the end of "Structure, Sign and Play" between "an affirmation that plays without security" (of which he approves) and a "sure play limited to the substitution of given and existing, present pieces."[10] The reader who *refuses* to play along is the reader who plays safe, who will not take risks.

I am reminded here (and not only here, in fact) of Heidegger's "What Is Metaphysics?" and the point in that lecture at which he discovers reason's inability to deal with "Nothing."[11] These pages make us uneasy; we seem to be cast adrift. Fortunately, with Heidegger, as the shores of Reason and Logic recede, the island of Experience becomes dimly visible on the horizon and our anxiety is over. Over, that is, until we realize that it is the experience of anxiety, or Angst, that gives us independent access to Nothing, which, note, is also said *not* to be a (formal) concept.[12]

But while Heidegger ultimately redeems the danger and the risk by offering us access to Nothing through experience, Derrida's *aim* is loss of security. Heidegger's remarks about Nothing are usually questions, always tentative. Derrida's quasi-transcendental claims about *différance* are not at all tentative and are meant to be believed in some sense or another. If not, what force can they have?

The suggestion is not that risk and danger are valuable only when ultimately rewarded. Certainly Nietzsche's advice to "live dangerously" held out no such promise. But if Derrida were to reply to one who refused to play along that he/she was (just) playing safe, the obvious reply is that where ice is wafer-thin, it is not dangerous to skate, it is folly. And it would seem equal folly to talk about *différance* "producing effects" as a way of eliminating all talk of transcendental causation.

When I discussed earlier the claim that Derrida's method was "one-sided and undialectical," the question was left prematurely unresolved, and I should like now to return to that question, and in particular to see what light it throws on the question of strategy. For, as he says, "in marking out différance everything is a matter of strategy and risk."[13]

Derrida claims in "Différance" that the term *différance* has profound affinities with Hegelian language, but nonetheless works a displacement with it—one both infinitesimal and radical. In "From Restricted to General Economy,"[14] and in *Positions*,[15] it is clear that the radical aspect of the displacement is essentially Nietzschean (and Bataillean) in origin. Dialectics is understood as always a *reappropriating* method serving ultimately to restore identity. *Différance*, on the other hand, aims to break out of this system, to renounce identity and *meaning*.

Now, if one accepts that dialectic should be understood in this way— that one could not have a dialectic freed from its restitutive telos, one that charted the interminable struggle of opposites—then clearly *différance* cannot be faulted for being undialectical without missing the entire transgressive function it is designed to serve. And yet when Derrida is discussing Bataille's response to the master/slave dialectic—that of *laughter*—the response that "alone exceeds dialectics,"[16] when he claims that "*différance* would give us to think a writing . . . that absolutely upsets all dialectics . . . exceeding everything that the history of metaphysics has comprehended . . ."[17] and that "différance holds us in relation with what exceeds . . . the alternative of presence and absence"[18]—one has somehow to give a sense to "excède" that is *not* dialectical. In the sense of dialectical that means teleological/restorative of meaning, it is clear why. But in the sense of "not derived from and essentially dependent on its derivation from the oppositions between presence and absence or identity and difference or, indeed, the Hegelian dialectic," it is not quite so clear.

Derrida talks of the "displacement" of the Hegelian system—again a term which itself displaces any simpler filiation such as influence or development. But is it not quite as clear that this displacement is guided all along (and remains so guided) by that which it displaces? Surely if one were to spell out all the subsidiary operations that Derrida engages in (reversal, insertion of undecidables, double reading, displacement, and so forth, a method would be found in which a teleological dialectics has itself been transformed dialectically. There is no progressive idealization, and there is no "static" telos.

But instead: "An affirmative writing," "joyous affirmation," "the innocence of becoming," "the adventure of the trace." Not "absolute knowledge," not "spirit coming to know itself," not "the realm of transcendental subjectivity," but surely something equally idealized, and something, interestingly enough, embodying values strikingly close to authenticity and freedom. To be sure, the concept of self, and what is proper to it, have been put aside, but is there not a repetition of key metaphysical motifs at the very *end* of Derrida's project?

I do not *object* to this in principle. What is being questioned is Derrida's self-understanding, his understanding of the possibility of a discourse *other than* that of "metaphysics." It may be said that Derrida has already admitted this. It is not a confession but an important methodological claim he makes when he writes that "there is no sense in doing without the concepts of metaphysics in order to shake metaphysics."[19] And for all his complicity with these concepts, this is a *necessary means* to an end—that of "shaking metaphysics" and of exceeding it. In other words, Derrida believes in making, at least with one foot, "the step beyond"—beyond "metaphysics," "beyond man and humanism," beyond presence, beyond security, beyond the language of Being. The complexities all lie in the strategy for bringing it about. But there is no doubting, surely, the philosophical recuperability of the values informing his "goal." (And it surely is a goal; there is no reason to restrict "goals" to static states of affairs.) In brief, my admiration for his achievement does not depend on believing in his own assessment of its absolute radicality with regard to the circle of Western thought.

Derrida has transformed the way we think about, and read (or perhaps write), philosophy, he has transformed our understanding of the relationship between the inside and the outside of philosophy, but his strategic dependence on such metaphysical values as "authorial intention" and on formally transcendental arguments[20] essentially limit his achievement. There is no philosophical analogy to the chemical catalyst that facilitates the reading but remains unchanged, or the fictional number in mathematics that can be introduced and then later withdrawn from a proof.

But this limit is not a negative one. Rather, his lesson, or the lesson to be drawn from him, is *not merely* that, as he says, there is no sense in doing without metaphysical concepts in trying to overcome metaphysics, but there is no prospect whatever of *eliminating* metaphysical concepts and strategies. Rather the project of overcoming metaphysics (as Merleau-Ponty said of the phenomenological reduction) must be repeated indefinitely.

Derrida says of Heidegger that one of his real virtues lies in his *intrametaphysical* moves.[21] The same could be said of Derrida. And I will finally explain what shape this takes, consciously aware of the way in which this explanation takes for granted a particular metaphysical opposition. More important for us, in *"Ousia* and *Gramme,"* Derrida, talking about Aristotle, says that what is truly metaphysical is not the particular *question* he evades (about the being of time) but the question evaded, the covering up, the passing on, the failure to reflect.[22] Conversely, what *exceeds* metaphysics in Derrida is his writing *as and insofar as it opens up* the space of *alternative theoretical possibilities* and *as* it bears witness to the scope of its own transformative possibilities. And these occur even if the outcome may seem to be a *new theory*, or another philosophy. Philosophy on the move is the only possible transgression of metaphysics. There is no Other Place to go.

One can only hope that the break with the theoretical sterility of deconstruction (when considered from the point of view of "positive" description) adumbrated at the very least in the next chapter will fall within this description.

4 *Time and Interpretation**

The view that philosophy has a special subject matter is not a fashionable one today, unless perhaps one treats the many versions of "the linguistic turn" as suggesting such a status for language. And even then a focus on language could not define philosophy, as linguistics, communication theory, and semiology share the same focus. If language were to be treated as the subject matter of philosophy, one would have to specify in addition a particular approach to language in order to account for the specificity of philosophy. Such an approach is usually contrasted with instrumental, classificatory, or objectifying alternatives. The force of such a self-understanding of philosophy can be gauged by the number of alternative and opposed philosophical approaches that share it (such as hermeneutics, structuralism, and analytical philosophy).

However, a case can be made out for bestowing on *time* the privilege currently accorded to language. This might be thought to be an obviously regressive move. Would not privileging time be an anachronistic revival of the spirit of the nineteenth century—of Kant and Hegel—or, worse, of more recent Bergsonian vitalism? Heidegger's work would seem to allay this fear. For even if he is clearly responsive to that tradition, that responsiveness takes the form of an appreciative re-thinking rather than a repackaging of those same ideas. Moreover, the development of Heidegger's thought, perhaps rather too tidily seen as stretching from *Being and Time* (1927) to *Time and Being* (1962) (for it should not be treated as any sort of simple inversion), suggests a number of different dimensions in which the privileged position of time in philosophy can be articulated (ontic, existential, historical, ontological, and so forth). Some of the complexities of Heidegger's concern with time have already surfaced; further discussion will be reserved until later in the chapter. Meanwhile, guidance will be provided by what it opens up—the possibility of a hermeneutics of temporal structure, the scope of which will give support to the claimed privilege of time as the subject matter of philosophy. And in the metaphysical modesty of its descriptive and interpre-

* The present form of this chapter owes a great deal to the perceptive and helpful comments of Robert Bernasconi and Hugh J. Silverman.

tive procedures there will be further confirmations of its distance from idealist and vitalist ambitions. Some of these broader issues are raised again at the end.

There is no doubting the importance of time, temporal structures, and relations in understanding the identity conditions of different beings (events, things, processes). The temporal dimension of such beings is crucial to their being the beings they are. This remark applies to cultural phenomena, experience, events, and actions. They can each be shown not merely to be "in" time, but to be temporally constituted in some essential respect. Music, for example, does not just passively take place in time; it involves the organization of time in rhythm and melody. To the extent that the temporal organization of beings is what enables us to understand and relate to them as the beings that they are, an appreciation of the various types of temporal structure they can manifest will play a vital role in both the theory and the practice of such relating. This points to the hermeneutic importance of a theoretical grasp of the manifold structures of time.

As the art or science of interpretation, hermeneutics may take as its object literary, philosophical, or other texts, or it may deal with historical events or even the events of everyday existence. The aim of interpretation is basically to make sense of its object, or, if you prefer, to understand it. (A deconstructive hermeneutics, if not a contradiction in terms, would aim at teasing out a multiplicity of strands of significance and indeed of modes of significance in a text.) To see philosophy as a hermeneutical discipline is to see argument as only one of its procedures and to see careful, reflective description and interpretation of phenomena as quite as important, and the kind of patience and care that this requires would be one of the hallmarks of philosophical responsibility.

One might suppose then that if philosophy is conceived of as a hermeneutical activity and attention is directed toward the temporal structure of the objects of our interpretation, it ought to be possible to proceed without any particular expectations or intellectual baggage. One should, on this view, just be prepared to receive those temporal structures that happen to spring up in our path. But it would not too grossly compromise the values underlying such an innocence if one were also to take along a few models of how time can be organized. I offer later an analytical account of what such forms might look like.[1] (Four types of temporal structure are distinguished: reflective, generative, participatory, and active.) Here a rather different move is attempted—to show that useful specific models of temporal organization and structure can be obtained by a restriction of the scope of the most general accounts of time that philosophy has already thrown up. This move has a double thrust. I will show that such derived models have an interpretive potential by which they can supplement our stock of anticipated structures. And I make the further claim that it is the proper fate of such general accounts to be restrictively transformed into accounts with a local application. Five general ways of thinking about time can be distinguished: cosmic time, dialectical time, phenomenological time, existential time, and the time of the sign. All of these ways of thinking about time can be and at some point have been supposed to provide an account of Time as such and in general. My transformative treatment of them clearly subverts the original philosophical role they were designed to play. The point is that incompatible general theories can be rendered compatible if their scope is restricted. To say that everything is made of water is one thing; to restrict the claim to the oceans and the clouds is another. On each occasion special attention is paid to extracting and disarming the explicit or tacit principle by which the generalization to *Time as such* is brought about.

COSMIC TIME

Cosmic time can be represented as a sequence of moments characterized by singularity, homogeneity, transitivity, universality, and directionality (or asymmetry). As a general theory of Time, its obvious difficulty is that it cannot accommodate the past/present/future structure of temporality. Its moments are all related to one another by earlier than/later than relations. If somehow one were to inject a viewpoint, a privileged point of orientation, into this account, its purely serial order would fall apart. It is incompatible with and cannot recognize intentionality.

Cosmic time fills out what McTaggart calls the "B-series," but whereas he wants to claim that it is logically dependent on the "A-series" (the past/present/future model that intentionality needs, at least to start with), the simpler claim is that it cannot handle the relationships described by the "A-series" but that it can be understood (and have value) as a representation (in itself harmless) of the ordering relations presupposed by certain everyday and theoretical practices. The structure of the calendar, for example, is a form of the "B-series," albeit imperfect,[2] and it is both a representation of and a determining condition of the practice of assigning dates to events. Clocks mechanically divide the basic unit of the calendar—the day. For computational economy, the calendar and the clock both make use of cycles and nesting of orders of unit, but what they represent is a purely serial temporal order. Hence they play not just a practical role in everyday dating and structuring the synchronization of social life (timetables, mealtimes, appointments), but also a theoretical role in those disciplines for which a sequential, intentionally neutralized temporality is a prerequisite, like physics. If measurement of motion is the demand placed on a model of time by physics, it is not surprising that the resulting structure is that of "a homogeneously ordered series of points, a scale, a parameter," as Heidegger put it in an early essay.[3] Given that the success of physics is based on its mathematization of nature, and that the metrication of time is central to this project, it is hardly surprising that the structure of cosmic time, as I have called it, reflects the features of the series of natural numbers and indeed of any ordered series in the strict sense.[4]

There are, in short, a number of areas in which treating the structure of time as simple seriality should be seen as reflecting real needs or functions. The complex temporal organization of daily life requires it. But it is a mistake to take the requirements of a certain sort of civilized life as proof of some deeper truth about Time itself. The prestige of physics may tempt us to suppose that what it requires of a concept of time transcends in significance the function that the concept serves in the theory. But to succumb to this temptation is to have tacitly converted physics into metaphysics.

DIALECTICAL TIME

For Hegel, world history is Spirit unveiled in an outer, temporal form. For Marx, that history is the history of class struggle. In both cases, the shape of that history can only be adequately expressed when seen as a dialectical development rather than a mere chronological series of events. Such simple chronology is concerned only with events and with external relations of succession among them. From the dialectical point of view, however, what the surface sequentiality of events reveals when interrogated is a deeper pattern of qualitative transformations, of development through conflict, the emergence and resolution of contradictions, and so on. Essential to such an account of Time is the ability to identify and relate discrete underlying processes and forces whose interaction can be made intelligible in rational and teleological terms.

History has found many difficulties with dialectical thinking. The shape of dialectical progressions seems too formalistically represented by the thesis-antithesis-synthesis model, and yet in the absence of such a model it can too easily dissolve into woolly and unsystematizable references to struggle, the overcoming of oppositions, and so forth. More important, there has been considerable skepticism about the ontological assumptions underlying dialectical thinking. The move from Hegel to Marx, from an idealist to a materialist dialectic, is a move that in displacing Spirit in favor of social relations maintains the assumption that dialectical thought is guaranteed applicability by the nature of the subject matter in each case. Materialists claim that idealism can offer this guarantee only by inventing a subject matter—self-conscious Spirit—of which it would be true. But its own claim—that the field of social relations is actually governed by those dialectical principles spuriously attributed to Spirit—is no less metaphysical a claim. This has led some to question the very idea of a materialist dialectic.

The relation between the idealist dialectic and the materialist dialectic can be described in this way: the idealist dialectic offers a universality of scope and a more or less abstract form. The materialist dialectic, on the other hand, involves a fairly drastic restriction of scope (leaving aside such ventures as a dialectics of nature) while it is less able or willing to offer an abstract "logic" for its method. My interest here is in how the dialectical process can be seen as structuring time, and the aim is to show that while such a metaphysical ambition is misconceived, something like dialectical time can still be employed as a hermeneutical model in a more restricted way. It might be thought that the materialist version of the dialectic would happily fit with this program. However, while not in any way denying the value of dialectical thinking in trying to understand (and indeed in influencing) the course of history, there is no reason to restrict the scope of dialectics to the social/historical field. Dialectical thinking is legitimate wherever patterns of development and transformation occur that can intelligibly be said to have some sort of "logic," and that are the consequence of some human involvement. The reference to such involvement is a recognition of the need to set some limits to the scope of dialectics, but it does not, within those limits, offer any guarantee as to the applicability of dialectical thought.

The expression "dialectical thought" is employed deliberately. And something must be said about the status of such thinking. First, it can be employed to give shape and intelligibility to a sequence of connected events that have already occurred. At one level this shape may be nothing more than an aesthetically pleasing articulation. But one may in addition wish to make the claim that a particular dialectical sequence is the product of some underlying generative principle (for example, reflection, struggle, contradiction). Clearly the hermeneutic value of such thinking would vary with the kind of intelligibility claimed. Dialectical thinking can also be used predictively, but only in a hypothetical manner. If experience suggests that the situation one has encountered or finds oneself in will develop dialectically, that it is subject to particular local constraints, then it may be possible to predict its subsequent development, or the range of possibilities open. Such a prediction (as with all predictions) may not be confirmed by events. And even if it is, in fact, it is always possible that the success was adventitious, that the sequence actually had no "inner logic," but only seemed to do so. What we call the "logic" or the "necessity" of a dialectical development is in fact a product of the events or thoughts or theories working themselves out within a "closed system," and whether such a state of affairs obtains in a particular case is a matter of fact not of logic. There clearly are sequences of events in the world that have a dialectical shape to them. In addition to world history, there are arguments, emotionally traumatic periods of one's life, local political swings, and works of literature, to name a few. Whenever it helps to think of such sequences dialectically (at the very least using the language of opposition, conflict, struggle, contradiction, resolution, reflection, realization, development, and so forth), then some sense is being given to the idea of dialectical time, a time in which the principle is not quantitative succession but qualitative transformation.

PHENOMENOLOGICAL TIME

The classic source of our understanding of phenomenological time is Husserl's *Phenomenology of Internal Time-Consciousness*, a book rich in detailed analysis, and one to which I have already devoted considerable space (see pp. 37–110). And it is on this book that I am focusing the discussion of the hermeneutical value (and metaphysical limitations) of phenomenological time.

Phenomenology does not see itself as just one approach among others, but as privileged in an important respect. It restricts itself to dealing with what we have a right to deal with—the "things themselves"—namely, the data of consciousness. And whatever might be said in other cases, the initial reductive move—the exclusion of "objective time"—is one that leaves us still with access to a genuinely temporal phenomenon, "the immanent flux of the flow of consciousness." (As Husserl puts it, "The evidence that consciousness of a tonal process, a melody, exhibits a succession even as I hear it is such as to make every doubt of denial appear senseless." [p. 23])

The first section, "The Exclusion of Objective Time," seems to announce a modest scope for the study. Its exclusions of "objective time" postpone any account of its relationship to time-consciousness.[5] Between the time the main body of the book had been presented as lectures (1904–5) and their publication (1928), Husserl had shown an increasing willingness to bridge the gap between the analysis of the structure of consciousness and any possible account of the real, objective world. The theory of constitution offered in *Ideas* (1913) plays an important role. And this move seems already to have been begun in the lectures on time being considered here. Section 31 deals with objective temporal points and section 32 with "the constitution of the one objective time." The exclusion of the question of the relation to "objective time" has broken down, and for good reason. Phenomenology could never have had any interest unless its descriptions of the structures of consciousness had a value that went beyond their being an accurate account of subjective phenomena. That value lay in what was always assumed to be the epistemological and ultimately ontological significance of consciousness. The reductions allow an initial freedom from the guidance of established unities, but what they are finally guided toward is an account of the contribution of consciousness to the constitution of such unities. One of the key moves is the recognition that terms like "real," "objective," "transcendent," "outside," and "world," derive what meaning they possess from consciousness.

These remarks should be understood in the light of my general project here, which is to demonstrate the metaphysical inadequacy, but hermeneutic utility of each of these "models" of time. If one thought the exclusion of considerations of "objective time" made this argument redundant, it should now be clear that it is not.

Husserl's main concerns in this book have already been well rehearsed. How, for instance, to explain that while our perceptual experience seems to occur always at a particular time, each time being part of a temporal flux, I can come to grasp the objects of my experience as temporal unities? How is unity across time possible (experiencing a melody, itself temporally extended)? The key to this question is his distinguishing and interrelating three different "arms" of temporal intentionality: primal impression, protention, and retention. Retention, for instance, is treated as an intentional modification of impressional consciousness, which gives the two an internal linkage not guaranteed, say, to conscious recollection. Husserl is also, however, interested in recollection and memory, as well as perception and imagination. In each case what the phenomenological account of temporability allows one to grasp is the intentional structure of a certain kind of consciousness. And it is only if time could be restricted to time-consciousness that a phenomenological account could ever successfully claim universality. There is no reason to doubt its value given its own methodological premises (which an existential approach would in fact challenge). But to anyone who would be tempted to think that phenomenology has a monopoly on the description of the temporal, I will mention just three modes of "intrusion" into consciousness so conceived, which would undermine an account of Time that generalized from the time of time-consciousness.

THE TEMPORALITY OF THE BODY

The subjectivity of a temporally conscious being is an embodied subjectivity. One consequence of this fact is that even if there are clearly isolable sequences of experience such as phenomenological temporality describes, they are still the experiences of a being already occupied by and with a multiplicity of somatic temporalities. For example, when listening to music, one may get distracted, or get tired and fall asleep, or even die. These fracturings of the isolable calm of a particular series of conscious acts are the predictable consequences of the periodicities that inhabit the embodied subject.

THE UNCONSCIOUS

The unconscious can be thought of dynamically (as a set of hidden mechanisms that generate certain kinds of behavior) or structurally (as a transverse structure that can be read through consciousness). Either way the possibility arises of, if not actually interference with consciousness, at least establishing different orders of interpretation. Under the heading of the unconscious, the phenomenon of repression could also be mentioned. Even if a theory of consciousness can be developed to account for forgetting, surely repression is a harder nut to crack.

WORLDY TIME

Those concerned with internal time-consciousness cannot ignore that it has no monopoly on the forms of temporal organization. In the experience of surprise, for example, there is the recognition of the autonomous temporal orderings of worldly things. Time-consciousness has blind corners, failures, and gaps in protention. There are, of course, surprises only for a protending consciousness, but success, let alone failure, here already demonstrates a "beyond" to time-consciousness that is itself temporal.

The generalization of phenomenological time proceeds, as has been suggested, by arguing for the dependence of all meaning on the constituting activity of the subject, insisting that any other temporal order would have to be meaningfully graspable and that such constituting activity would only be made possible by a phenomenological temporality.

But this position tends toward an unacceptable idealism. The references to sleep, death, repression, and susprise provide sufficient counterexamples to such a position. With a glance toward Levinas, one could add here: fecundity.

EXISTENTIAL TIME

An existential account of temporality is essentially participatory. It treats subjects both as embodied—and thus from the beginning "in the world"—and as mortal. It is participatory in the sense that it claims that we are temporal in our very being and that the most basic temporal patterns affecting us are not those that organize the persisting objects around us, but those that involve our actions and our self-understanding as finite beings.

"Participatory" is perhaps too positive a term. "Nondetached" might be more apt. It is not the temporality of a subject whose worldliness is in doubt, but of a person whose Being is in question, and for whom the temporal dimension of its Being is a key issue. On this account of time, the possibility of its generalization is assured by its association with the transcendental character of our Being. Man is transcendental because his Being is horizonal. And the horizon in question is one of ekstatic temporality: the triadic structure of anticipation, making present, and Being-as-having-been. Because we understand ourselves in terms of possibilities of Being, and the future is the fountain of possibility, it is the future that is emphasized in existential temporality.

One specific way in which existential temporality is given a privileged status vis-à-vis other forms of the temporal is by explaining those other forms in terms of it. For example, in *Being and Time* Heidegger says of ordinary time that it consists

> precisely of the fact that it is a pure sequence of "nows" without beginning and without end in which the ecstatical character of primordial temporality has been levelled off.
>
> ... the "time" which is accessible to Dasein's common sense is not primordial but arises rather from authentic temporality. (H329)

And in claiming, as he does elsewhere, that infinite time (cosmic time) is derivative (by derestriction) from finite existential time, he further privileges the latter.

It seems possible at least at first to separate the regional application of existential time from the transcendental arguments in which it gets involved. That there should be a kind of temporality specific to finite, ontologically self-conscious, embodied, worldly beings is not surprising, and it is enormously important in, say, interpreting human action. This is not of course to say that there are no problems for an existential temporality. Think of Barthes's skeptical remarks about biography,[6] or Sartre's discussion of the retrospective nature of adventures.[7]

But if existential time can serve as the ground for cosmic time, why could it not be thought to ground all other times? This would seem to be Heidegger's position in *Being and Time*, for example. If it were successful, the interpretation of temporal structure would have found a universal basis. Its success would require satisfactory solutions to two basic difficulties.

First, existential temporality, even in its most direct application (human self-interpretation and projectivity), takes for granted as an ontological premise the value of unity. The whole thrust of Heidegger's existential analysis of Dasein is to undermine metaphysical guarantees of personal unity such as "substance," the "ego," and the "self." His account of the theological sources of the idea of substance underpinning Descartes's dualism is masterly. And yet the conceptual apparatus by which Heidegger achieves this—and in particular, the pervasive distinction between the authentic and the inauthentic, and the idea of resoluteness—ultimately serve to redefine and reestablish notions like personal identity and responsibility, and the unity of a life through time.[8] The authentic/inauthentic distinction rests on the possibility in principle of deciding what is "my own" and what is not. If it does not suppose that personal wholeness and integrity is something given or guaranteed, it does suppose it to be achievable.

Opposed to such an assumption, I have in mind the vision of radically fragmented man offered by Barthes, both in the preface to *Sade/Fourier/Loyola* and his own substitute for an autobiography, *Barthes on Barthes*:

> What I get from Fourier's life is his liking for mirlitons (little Parisian spice cakes), his belated sympathy for lesbians, his death among the flowerpots . . . How I would love it if my life, through the pains of some friendly and detached biographer, were to reduce itself to a few details, a few preferences, a few inflections. (p. 9)

It would not be difficult to see in Barthes's position that of Hume, carried gracefully to its logical conclusion. Barthes's hedonism, if one can call it that, dissolves the distinction between the authentic and the inauthentic and seems to expose its ethico-ontological status. One might admit that something like systematic self-intelligibility is something people seek, without accepting either that they necessarily or actually achieve it, or that it is the only or supreme value. And yet existential temporality does make such a claim. One should not rule out the possibility of an existential temporality that embraced without evaluation a multiplicity of "selves" or tracks within a single person, but the concern in both Heidegger and Sartre with quasi-moral notions like authenticity, responsibility, resoluteness, and commitment makes it hard for such pluridimensionality to get a grip.

Of course, one might still want to claim that difficulties in assessing the status of existential temporality in its direct application to the question of the unity of an individual life need not vitiate its role in providing a foundation for other kinds of temporality. One would still be able to think of cosmic time as a representation of the structure of the human practices of dating and measuring change, even without answers to whether such practices could or could not be existentially integrated. Even so, there is a second difficulty.

Existential time could be extended to cover all other modes of time only by supplying the ground or foundation for such times. Such an extension thus presupposes the legitimacy of the notion of such a ground or foundation. Significantly enough, Heidegger himself came to distance himself from just his transcendental form of thought.

In *Being and Time*, Heidegger's real (and declared) aim was to reopen access to the question of Being, but the vehicle he used was an analytic of human being, insofar as the latter "keeps itself ekstatically open to Being." The existential analytic is the gateway to the question of Being itself. The history of philosophy has always treated Being as, in some form or other, presence. So the opening up of the question of Being must take place within the transcendental horizon of time in order to explore the status of that "presence." In *Being and Time* the existential focus of the question of Being means that human being (Dasein) is understood within the horizon of its temporality. The development of existentialism is seen as a misunderstanding of the status of that existential analytic. And when Heidegger reasserts his real aim—thinking the question of Being—the link between time and the existential analytic gets severed, to such a point that in *Time and Being* (1962) the relation of his discussion to "the human being of mortals is consciously excluded."[9]

At the same time as the specifically existential interpretation of temporality disappears, and in consequence the capacity of "existential time" to render other kinds of time intelligible, Heidegger also seems to have assigned the concepts of foundation and ground to the very metaphysical tradition from which he is attempting to take up a certain distance.[10] He comes to see the whole model of transcendental causality as mistaken in principle. What takes its place is a new approach to both Time and Being, in which we are offered instead of a ground an account of that "presence" by which Time gets involved in Being, in terms of presencing, opening, giving, bestowing . . . in which, if one were to speculate on the place of man in this scheme, something like "active receptivity" might best describe it.

A full account of the treatment of time in the later Heidegger cannot be undertaken here, but it would not be unfair to draw the following conclusions from the discussion so far that: (a) *Being and Time* would have to be defended against his mature assessment if one were to pursue the project of treating existential temporality as a ground for other modes of time, and (b) it is not at all clear whether there emerges from Heidegger's writing an alternative understanding of Time that could stand alongside the various views being discussed. But there is no reason to deny the importance of some form of existential temporality in providing a framework for understanding human action and self-reflection. And there is no reason to rule out the possibility of a deeper understanding of the appearance of other types of time by reference to practices and attitudes the temporal structure of which can only be adequately understood existentially. This is, in effect, to endorse the descriptive if not the ontological value of much of the detail of *Being and Time*.

THE TIME OF THE SIGN

Derrida's critiques of Saussure and Lévi-Strauss[11] are among other things directed against a purely synchronic, structural account of the sign. From the point of view of a structuralist method, there is enormous value in its relative unconcern with the isolable meaning of signs, or with their referential aspects. Instead it focuses upon differential (especially oppositional) relations between signs, The representation of such relations in the form of grids or networks of opposition gives the impression—sometimes justified—that time is being altogether excluded from consideration, that *difference* is a static relation. On this view, time appears in language in the shape of tense, temporal terminology, the temporality of reading and writing, the internal development of a text (narrative, for example), and so on. But the elementary unit—the sign—would be free of temporal determinations.

Derrida's strategic introduction of the terms "trace" and *"différance"* signals a departure from the view, and although the whole idea of time proper is itself put in question,[12] the sense of deferral built into the term *"différance"* introduces an essential temporality into the sign, and hence into the whole empire of signs and signification. Derrida is in effect treating both the sense and the reference of a sign as having the structure of desire. It is part of the classical conception of the sign that it stands in for, represents what is absent. Derrida is claiming that this "absence" is necessary—not the absence of something that at some other time or place could be made present. Indeed, Derrida offers a critique of the very possibility of such a presence.[13] Meaning and reference always defer completeness. Textuality, in this light, can be seen as the movement of an impossible desire for plenitude, presence. Thinking of the signs as a "trace" involves a similar relation to an imaginary time. The "trace," he claims, is not a trace *of* anything (while the sign can be thought of as a representation [re-presentation] of what has been present). But while the trace is not a trace *of* anything (which is like saying that signs have an essential autonomy with respect to what they signify), there is, as it were, an imaginary relation to an origin that can alone make sense of using the term "trace" at all. If the term "imaginary" sounds awkwardly psychological, perhaps "virtual" would do. Signification, and hence language, would then essentially involve an imaginary or virtual temporality.

But before proceeding to discuss this notion, a number of difficulties arising from basing an account of the "time of the sign" on Derrida's treatment ought to be mentioned. The reason for this approach is that it is from the point of view of entry that the possibility of the generalization of the "time of the sign" can be contemplated. And yet, in the course of this same generalizing move, paradoxically, the very idea of Time is threatened. The generalizing move rests on two premises: that signification knows no limits—that there is little or nothing that is not caught up in it (including, for example, experience and perception); and that such terms as "*différance*" and "trace" capture the form of signification in general. But if this is how the generality of Derrida's position is produced, it is not obviously a generalization of a form of time or temporality. Indeed, he sees it as deconstructing the inherently metaphysical concept of time itself. (On this view talk about imaginary or virtual time could be permitted only on the strict understanding that these were not *kinds* of time at all, but at bottom, mark the absence of the temporal.) But this would not help demonstrate the limitations of scope of each of our models because it fails to establish that it is a way of thinking about time at all. The terms "*différance*" and "trace" seem to be used as part of a negative transcendental argument to deny the possibility of any concept of time dependent upon the idea of the present—that is, any concept of time at all.

However, whatever status is ultimately given to this argument, it is clear that Derrida cannot want it described in this way—as a (negative) transcendental argument. It would, for example, make "*différance*" into a ground and thus condemn it to the status of a new metaphysical concept. Indeed, having identified the concept of time as such with the metaphysical tradition, it is as we have discussed at length, surprising, although gratifying, to see him referring to "pluri-dimensionality" and "delinearized temporality" in another paper.[14]

It is precisely such notions of pluridimensional time that the field of signification—particularly textuality—opens up. A discussion of the temporality, say, of a narrative text, would have to bring out not only the real chronology of events, but also the chronology of their presentation, the structures of internal repetition of words, situations, themes, the mapping onto the text of different types and modes of time (imaginary, symbolic, biographical, historical), and so on. There is an important sense in which the operational value of Derrida's deconstruction of the concept of the sign lies in its liberation of *textuality* from the interpretive constraints imposed by traditional concepts of meaning and of time. Even if one were to continue to suppose that real, objective time was linear, Derrida's critique of the sign as representation would undermine any attempt to restrict textual temporality to a linear form. The text is a privileged site for the liberation of time.[15]

This has particularly important consequences for interpretation. For the discovery of multiple strands of meaning within a text—part of what would be involved in a deconstructive hermeneutics—is heavily dependent on the discrimination of a variety of types of temporality. But strictly speaking, this status has been gained for a manifold of textual temporality by narrowing down rather than generalizing its scope. Can one really treat the world as a text? Has the age of hydrosemantics arrived?[16]

My original aim was to demonstrate how a number of different highly general ways of thinking of time can be successfully transformed into local hermeneutical "models" by relinquishing their claim to a universal scope. This whole procedure is conceived of not as a salvage operation, not one that adopts the strategy Nietzsche attributes to the worm,[17] but more precisely as a restoration of these models to the site of their more intuitive application. If the shape of such an enterprise has become clearer, I can now turn to consider some of the difficult question it raises.

Has this discussion not been operating under a mantle of innocence as to the real compatibility of each of these "models"? Has not reconciliation been brought about simply by silencing rival claims, and by ignoring all the important questions? Has not a healthy dialogue between theories been exchanged for a mere patchwork of their shriveled remains?

There is no denying the desire for some further account of what if any the principle of unity of these various models might be. One might reply that it is the desire for a ground, or a foundation, an *archē*, and that such a desire rested on the mistaken belief in the possibility of such a first point. It is just this impossibility that the deconstruction of presence (discussed in the last section) demonstrates. But it might seem that the result of accepting this diagnosis would be equally unpalatable. What more is it than a barren empiricism that simply notes the variety of modes of temporalization and says, "How interesting!" or a set of operational models, or descriptive techniques that might be packaged ready for use by history, literary theory, the phenomenology of music, and so forth? And is not the value of such a result severely limited? Can the production of such models be the aim of philosophy?

Heidegger might have put the "barren empiricism" charge in another form. He might have said that the concern throughout is with beings and not with Being. He might cite in evidence the fact that I seem to want to keep the analytical detail of *Being and Time* while leaving aside the question of Being, the horizon within which that detail appears. The fact that I retreat from any serious engagement with *Time and Being*, in which the whole existential perspective is so clearly dropped, is further proof.

Of course, the objection is not merely that talk about Being is absent from this account. The history of metaphysics is full of philosophers who have talked about Being without understanding "the ontological difference" (between Being and beings). And in principle it should be possible to respond to Being without naming it. (Indeed, as Heidegger discovers, naming it brings its own problems.) The objection would have to be that there is not, in this discussion, a responding to Being as such, but only to beings.

My reply to this must be seen as tentative. First, I would insist on the permanent, or better continuing, importance of detailed description of ontic structures. In an important sense there can be no discussion of Being without careful consideration of the realm of beings. It would be ironic if reawakening the question of Being were to encourage yawning about beings.[18] And yet just such a possibility of disassociation seems dangerously suggested at the end of *Time and Being*:

> The task of our thinking has been to trace Being to its own from Appropriation—by way of looking through true time without regard to the relation of Being to beings.

> To think Being without beings means: to think Being without metaphysics. (p.24)

But doubts about Heidegger's formulation here do not constitute a reply to the objection that I am are forgetting Being. I am drawn to make two apparently very different responses to this. The first response, put in a strong form, is that Being has indeed been abandoned, and rightly so. Nietzsche was right to think of it as "the last cloudy streak of evaporating reality" or as "an empty fiction."[19] And Derrida's own suspicion that it represents an unspent yearning for presence is fully justified. Is there confirmation in Heidegger himself of the impossibility of any coherent theoretical treatment of Being? Is one not always left with the *question* of Being, rather than answers? Put a little more positively, perhaps Being is *nothing else* but "beings" grasped, understood, related to *in a certain way*? And does not the recognition of the sparkling play of different types and orders of temporalization of beings offer just such a new horizon for relating to things? This horizon is not just wonder, but wonder informed by a sense of the variety of the forms of emergence, change, openness, and so forth which collectively *constitute* the horizon of temporality.

The second response would be this: If something remains in the later Heidegger of the recognition that the question of Being can be reopened only by the location of the dominant value of presence within the wider horizon of temporality, might it not be that the reopening of that horizon just *is* the reawakening of the question of Being, and the thought that there is some further task to be undertaken is an illusion? The account offered here has demonstrated a *polyhorizonality* of time.

But am I not then committed to the perspective of *Being and Time*, the very perspective I have just criticized? And would that not involve just the sort of privileging of one temporal "model" against which I have been arguing?

There is a sense in which the original definition of the problem makes this inevitable. If one is concerned with interpretation, with hermeneutics, not with epistemology or metaphysics, it is not giving hermeneutics a metaphysical grounding to recognize that it is a human activity, one that draws on existential "categories" in pursuing its interpretive ends. Accepting this as an internal consequence of the way the problem of Time and Interpretation has been posed does not force the withdrawal of our critical remarks about existential temporality as formulated in *Being and Time*. The most attractive corrective to that account would be one that attempted to blend an account of the multidimensionality of the text with a version of the account given in *Being and Time* from which the concept of authenticity has been dropped. The claim would not be that life is a text, but that it is textured. And that man is a tissue of times.

5 Some Temporal Structures of Language: Prolegomena to a Future Theory of Time

Quite by chance, or so it seemed, two particular books recently lay side by side on my desk. One was Plato's *Republic* and the other, the multiple authored *I Ching*. I was enormously struck by this coincidence, as they seemed to represent a disagreement of great import on which I had already taken sides. For Plato, time and change lead down the slippery slope to chaos, and intelligibility begins with the elimination of the temporal. The *I Ching*, the book of changes, teaches us to recognize intelligible patterns of change, and that time is a condition for these patterns and not a threat to them.

The recognition that change can be intelligibly structured results, I would argue, from the absence of an enormous burden that Western philosophy has had to carry—the task of discovering *necessary* truths. To the extent that time is the source of changing circumstances, it poses a threat to necessity. And it is no accident that in one of the most self-conscious and dedicated attempts to rescue time from the intellectual wilderness—I refer to the writing of Hegel—it can return only in the form of necessity. The more it is recognized that there is a middle ground of intelligibility between necessary truth and bare contingent facts about the world, the more plausible becomes the project of providing some sort of account of temporal structure and the less this phrase will provoke howls of mental anguish.

This project has its limits. What is offered here is a framework for a theory of intelligible temporal structure, and many traditional problems about time will not be directly touched upon at all. First, some points of orientation and clarification. I assume that whatever is true of time is first true of the temporal, that the adjective "temporal" qualifies in particular relations, and that complexes of relations may be called structures. What is required is an analytical vocabulary by which such structures can be illuminatingly discussed. I am not the first to suppose that language itself (or at least reflection on language) can supply such a vocabulary.

What is claimed is that the visible and reflectively discoverable structures of language evidence a wide range of general temporal structures; that in language the structures of time are writ large. Language, it is claimed, is an exemplary and a privileged temporally ordered phenomenon. If my account here can sustain this claim, one welcome consequence would be that it would not be necessary to *choose* between the alternatives: time *or* language suggested earlier.[1] I do not suppose, as it might sometimes seem, that without structure there is no time. No doubt time can appear in such primitive forms as simple duration, as desire, as flux. What is wrong is using these phenomena as paradigms, for they are neither typical, nor perhaps fundamental. Language is an *exemplary* phenomenon. The temporally informing features to be discussed are not restricted to language, and a number of examples will illuminate this.

The use of language as a site for excavation—and language here means speech, writing, and interior discourse—has important advantages over a study of the structures of time-consciousness. First, the structures of signification—intentional structures—that language makes possible are infinitely more complex than would be available to a being without a language. The "objectified intentionalities" of a natural language not only massively expand the possibilities of consciousness, but are more readily accessible for analysis. To the extent that temporal structure is intelligible it is ultimately, and in the broadest sense, an intentional phenomenon. So a study of temporal structure that did not take into account the possibilities that language opens up would be importantly defective. Second, focusing on language makes it possible to consider temporal structures without the doctrine of "one man one time" hanging over us. If what is discovered is useful for understanding experiential time, it has no a priori assumptions about unity, identity, or continuity to hinder it, although these may of course creep in unnoticed. And third, the use of language as a mine of temporal structures provides a breathing space of ontological neutrality, and a longer one than any analysis of consciousness can possibly hope to have.

Having established the site for excavation, I will now explain my particular interest in one sort of temporal structuring, the less-favored sort.

Language structures time in two ways which could be called explicit and implicit. The explicit may involve (a) (in languages with which I am familiar) such modifications of the verb as tense, aspect, and mood, (b) the use of temporal indices such as now, then, and once, and (c) the use of various languages of time and date—both everyday and specialist—from the calendar to the measuring systems of physics.

It is in these areas that much of the work on the relationship between language and time is done. In my opinion, however, this explicit treatment of temporality by language is subject to certain a priori constraints that I shall suspend, or bracket. These constraints are associated with the thesis of the a priori unidimensionality, unidirectionality, and continuity of time. This will be called, for short, the thesis of the a priori unity of time. This thesis has an application to the objective world—that (at least from a given position) all events can be uniquely ordered in a single temporal series; and it is applicable to subjective time in the view that for each person there is a single stream of consciousness in which each experience can be uniquely ordered. Obviously there are various ways of handling the relationship between these two applications of the thesis, but these are not here my concern. I argue that the unity-of-time thesis is an unnecessary limitation of the investigation of temporal structures and particularly on those we call intentional. Accordingly, I propose to put to one side the unity of time as an a priori assumption. For this reason, and because much work has already been done in this area, the focus will not be on the ways in which language *explicitly* structures time insofar as these are linked to some form of the unity of time thesis.

But before proceeding to my main topic—the implicit temporalizing of language—there are two important respects in which the explicit temporalizing of language seems to already point beyond the unity-of-time thesis.

First, one considers a discourse in which tense or mood or aspect (or all three) are employed—the unity-of-time thesis will appear in the form of rules for assessing the *consistency* of these uses. In particular one thinks of tense logic. In my view, however, logic is unsuited to legislate for temporal relations in general. It is itself dependent on further explanations of such key temporal notions as "at the same time" (a point Geach makes in another context[2]) and is only ever introduced and comprehended via our ordinary natural language. However, tense or mood (and aspect somewhat less) each allow the creative expansion of what we call ecstatic virtuality, or intentionality. They allow the subject to specify the most complex existential orientations, and indeed to conceive of them in the first place. And in doing so, one-dimensionality is implicitly expanded into a multi-dimensional texture, which opens up a crack in the unity-of-time thesis.

The other respect in which explicit temporalizing can lead beyond this thesis is found in the possibility of multiple histories sharing the same dating system. Here I am thinking of Foucault's insistence that history be considered plurally[3]—that there are histories of toy soldier manufacturing, book binding, and soap as well as those of the kings and queens of England. So while dating systems in principle allow any event to be uniquely located in a single temporal series (the history of the world), there remains the possibility of establishing a number of more limited special series restricted by content and in which actual serial order is not always of overriding importance. In studying the lifework of a particular painter, it will only sometimes matter that one knows which painting he painted first—when there is a significant change of style, say.

There are two respects in which the explicit structuring of time by language can overcome the a priori constraints already mentioned. Within these constraints much valuable work has already been done. My concern, however, is with the ways in which language structures time *implicitly*, and for the clues this offers for more general insight into temporal structure. I begin looking at the implicit structuring of time considering temporal order.

Consider first of all seriality. Metrical time is merely a succession of instants, drawn out in a line, but language offers us more complex forms of seriality. I distinguish just four:

REPETITION

In which the same element (a sound, or letter, or word, for example) is repeated over and over again.

SIMPLE PROGRESSION

In which different words succeed one another without establishing any larger units.

SYNTACTIC CONNECTION

In which words that succeed one another do form a new unity (for example, a phrase, a sentence) and in which the relations of succession may be much less important than, say, *agreement* (in case, number, person, and so forth) with a more distant element. One way of looking at such connections is to treat their successivity as only the surface consequence of generation from a deeper structure by some rule.

TEXTUALITY

Which is defined by the *articulation* of these syntactic units and which is most important to complex temporality.

If one considers next the fact that language offers a number of different significant units, something like a nested structure of articulation becomes visible. Again, in a lengthy text, one may find a plot, a collection of chapters, paragraphs, sentences, words, and letters. At each level, a shift in attention occurs that changes the intentional focus and opens up a new horizon of meaning. And at each level a distinct series is discovered. Ordinary empirical studies reveal what could be called different orders of magnification of temporal structure—from the vibration of crystals of history. The example of language suggests that at each of these different levels of magnification, distinct series can be found.

At each level of unit a new horizon of function appears. Words and sentences have distinct kinds of unity. The sorts of series just mentioned are combinatory or syntagmatic. But there is another axis of seriality, as it were, that opens up a virtual horizonality based on selectional or paradigmatic chains. Each time a particular word is used, a chain of the repetitions of its token equivalents is activated and extended. And this chain can have different sorts of significance. The etymological roots of the word may, for example, be unconsciously or explicitly alluded to. A word may take on a special local significance through textual repetition. (Think of *Dasein* or *différance*). So one can distinguish within the cosmic list of a word's repetition both the regional series to which it belongs and the relationship between that series and the wider one.

What has this to do with time? The recurrence of the same is a primitive temporal structure. It too has its own horizonality. (The consequences of developing this insight can be momentous, as became clear when discussing Nietzsche.)

Yet another advantage of taking linguistic productions as exemplary for understanding temporal structure arises from the plurality of texts, each of which to a greater or lesser degree allows distinct temporally ordered series to be identified, and "times" within these series. Reference is made to the "early" chapters of a book, the opening lines of a poem or an argument. Where there is some ground for intertextual grouping, this intentional seriality can be extended to a whole opus, for example. Consider the early and later Heidegger, Kant's precritical writings, and so on, even up to such large-scale series as the phenomenological movement or the logocentric tradition.

In the world of texts and discursive sequences, there is no *one* time. Multidimensionality is the rule. This is not to say that by some magical process texts escape the possibility of being objectively dated. Rather the relationship between a text and physical time is likely to be less significant than between one text and another (or indeed between a text and the nature of its historical context). Foucault is right to emphasize histor*ies* rather than history. And yet multidimensionality does not require the sacrifice of the intentional—of meaning, horizons and so forth—rather its expansion.

My use of the text as a model for temporal complexity has so far been to expand and pluralize time understood as serially ordered—linear time. Before leaving this valuable but limited perspective, attention must be drawn to another consequence of multidimensionality. The overall continuity of a multidimensional text permits *discontinuities* of particular strands, both in the sense that a particular series may begin or end, but also that it may be interrupted. The overall structure of a text allows such discontinuities to be located and identified. The pluralization of time has as its consequence that discontinuity becomes conceptually admissible. This will be discussed more fully when dealing with narrative.

So, the text can be used as a model for understanding a historical break. I do *not* dismiss the idea of continuity or its importance. Rather, I argue that consciousness, experience, or epistemological history *may* be characterized by discontinuities. To admit discontinuity does not require one to abandon all interest in temporal order. It only suggests that one avoid a mistakenly a priori understanding of the structure of time which would find radical discontinuity hard to handle.

So far, with the concepts canvassed—multidimensionality, internal horizonality, nested articulation structure, and discontinuity—I have tried to illustrate the way in which linguistically structured temporality transcends simple seriality, drawing heavily on the structure of texts. But these particular concepts disclose only some of the most visible temporally constitutive intentional aspects of language. In order to try to anticipate in principle some of the diversity of other such structures, I will now look at three of the most important principles responsible for their appearance, which, with certain reservations, may be called determination, reflexivity, and "presence."

First, *determination*. Put most simply, a word, a phrase, or a sentence can be determined by a rule. But this opens up a number of different senses of "determined" and of "rule." There are lexical selectional rules, and syntactic combinatorial rules. These could be called rules of *construction*. It is already clear that a single utterance can be determined by many different rules simultaneously. Moreover, there are other ways in which utterances are determined which allow them to be multiply determined. In particular, one thinks of all those conditions of appropriateness that a linguist would treat under pragmatics.

This talk of *determination*, however, treats a phenomenon from only one point of view. The rules involved would appear to function as limitations, constraints, conditions. They present themselves as demands that one must either meet or risk falling into babble or foolishness. But the other face of this same phenomenon is rather different. A single utterance can *satisfy* a multiplicity of conditions so as to be "just the right thing to say," the "mot just." And the very same rules, constraints, and conditions can serve this creative role as well as the conforming role that the other model suggests. To understand and appropriate these conditions of sense and force, and to bring them to bear, consciously or otherwise, on one's speech or writing has as its limit what could be called *discursive saturation*, the closest thing to a redefinition of authentic discourse.

There are two important respects in which multiple determination or multiple satisfaction are temporally significant. First, if one allows that some of the rules governing the surface structure of utterances are rules that relate that surface order to a deep structure—a generative model—then an important sense has been given to the idea of surface order being *derivative*. It is not actually important for present purposes that agreement is reached on a generative model for syntax. For it clearly applies to linguistic production at other levels (such as writing a book from an outline) and to quite different fields, such as acting from a blueprint, as when one follows a recipe or a map or a flow diagram. In these cases the temporal order of one's actions or decisions is derived from the order of another schema or blueprint. It may be wrong to treat world history or my monadic autobiography as the conscious or unconscious unrolling of some such deeper order, but the fact that a model can be misapplied does not invalidate it, it only reminds one of its limits. The other respect in which multidetermination or satisfaction is temporally constitutive is more general. It exemplifies temporal focusing or condensation. An act in general, and a linguistic act in particular, with a single physical description can be part of a number of different intentional series. Painting a fig leaf in the right place can satisfy the censors as well as improve the color balance. A sentence can inflame and inform at the same time. Or consider the note at which two pianists moving along a piano in different directions cross over. That single note will have a different meaning depending on whether it is seen as part of an ascending or descending scale.

The point is this: language offers us examples of a very general phenomenon, which is the concretely multifaceted nature of its instants, its moments.

After multiple determination or satisfaction, the next specifically intentional feature of language to be singled out for its temporal importance is reflexivity. The fact that language can be about itself, or that one can talk or write about other sentences, or even the very ones being uttered, means that the temporal structures constitutive of texts or discourses must be thought of as doubled back, or folded over on themselves. In such a reference back, the actual temporal gap—when there is one—is only there to be ignored. Reflexivity establishes loops of immediacy amidst what remains of linearity. But it does more. Reflexivity could be restricted to the clearest case in which S' refers to S either by name or in quotes. But the same principle of reflexivity is at work in anaphora, and in clause modification. It is no accident that Hegel associated the possibility of the dialectic with what he called the *speculative proposition*, and, as I have argued elsewhere,[4] the progress of the

Spirit should be seen as at last modeled on, if not at times dependent on, a reflexivity that belongs primarily to the Hegelian text itself rather than to some independent object to which it might refer. Dialectics, in short, would be in large part an explanation of the phenomenon of textual reflexivity. This is not a reductionistic thesis. Complexly intentional time is best revealed by the structures of language, but it need not be reduced to them.

The last particularly intentional feature of language is "presence." I have discussed this already but it is worth repeating here. A certain confusion has proliferated around this term. According to Derrida, presence is a feature improperly attributed to terms or utterances because of the trace "structure" of any sign. And if there is *différance* wherever there is signification, even the purity of that presence called self-presence is threatened, to the extent that presence in its temporal aspect is infected by its relationship to past and future. "Presence" is an illusion of false immediacy, as Hegel might have put it. Or in Derridean parlance, language is a play of differences and deferments. But need one choose between *différance* or presence? In our view, the deconstruction of presence is effective against taking presence—in one or other of its forms—as a metaphysical foundation. And if phenomenology is committed to the value of presence then (insofar as it is a metaphysics) its plausibility would be threatened. It is worth considering, however, that (a) the development of phenomenology since Husserl has been consistently away from the metaphysical temptations to which he arguably fell victim, and (b) phenomenology is not committed in principle to *presence* in some absolute sense. Phenomenology can survive without supposing, for example, that meanings can be perfectly fulfilled.

So one need not believe that a deconstruction of presence is an automatic deconstruction of phenomenology. Although it rightly breeds discontent with a narrowly intentional account of language, as Merleau-Ponty was beginning to see in his last work.[5]

And this critique of presence is quite compatible with presence being an essential phenomenon of language. It is quite true that words do belong to series, and do have meaning by virtue of unstable differences of oppositions to which they are related. But for all that, people use words to refer to actual and possible worlds. They take some words to be apt and others awkward, and they freight their discourse with meaning. A phenomenology of discourse cannot do without the concept of presence. Even if, as has been argued earlier,[6] one cannot straightforwardly accept that *différance* is what makes presence possible, there is undoubtedly a complex relation of mutual interdependence. If presence cannot serve as a foundation, it is equally no illusion.

Presence as understood here is the primary *phenomenon* of language. It is made possible by the relationship that the term or utterance in question has to the various differential series to which it always belongs and to the intentions and references it may bear. Presence is difference focused by a linguistic subject's desire. The product of this focusing is the ideality that Nietzsche rightly attributed to language. Making present draws in pasts and futures. Presence in language is the phenomenon of which difference is a condition and ideality the abstraction.

So far I have dealt with what could be called *structural features* of language. Most, if not all, the claims made could have been illustrated by a written text. Textuality has been explored as complex temporality. It has been argued that certain structural features of texts exhibit the general properties of this complex temporality. Three important problems arise out of this treatment and need to be disposed of before going on to look at the most obvious standard form of textual temporality—that of narrative.

1. I seem to have licensed references to *intentionality without a subject*. Texts have appeared in a kind of uncommitted limbo between being objects of analysis and being enlivened by reading or writing. References to speech, on the other hand, have not been sensitive to the specific temporality of the act of speaking.

2. I have still not explained how it is that time can be thought of as *structured* without ignoring its essential spontaneity and creativity. Does not any talk of temporal structure inevitably spatialize time? Is it not a mistake to apply the static products of reflection to the prereflective level? Am I not confusing Being with its representation?

3. What is the intended scope of this analysis? Is it only subjective time, or existential time, or what?

First, the problem of intentionality without a subject. There are two justification of this.

First, that the properties of texts and utterances so far discussed are neutral as to being written or read, being uttered or listened to. It is unnecessary to specify the position of the linguistic subject on each occasion, even if an adequate account of the ontological status of a text must acknowledge the need for its animation by linguistic subjects. But there is a more important reason for suspending the question of the status of the subject, and the precise sense of intentionality employed in the analysis of texts. Such a strategy allows the possibility to emerge that the concept of the subject, and of intentionality, is considerably transformed when language is taken as its field of operation rather than consciousness as traditionally understood. If the subject is treated from the outset as the subject of consciousness, not only would one be unable to understand a text except by problematically invoking an *unconscious*, but more important here, one would still be rattling the chains of unidimensional temporality.

This struggle has an important historical dimension. In my view, the only way of avoiding psychologism, in the battle against which phenomenology was born, without reaching for a transcendental phenomenology, is to recognize the determination of our lives and experiences by structures of signification, structures that are most complexly developed in language. While language most thoroughly exploits signification, it is not claimed that there is no signification without language, rather that simpler forms, such as association by similarity or proximity, do not provide us with an adequate paradigm. This is why Hume's associationism won't do. Transcendental phenomenology is understood as a reflection of the inability of a unidimensional temporality—like Hume's—to handle on its own the complex structure of experience, and particularly to guarantee continuity and the identity of the subject. This flight into ideality loses its appeal if instead the temporal structure of our everyday experience and action are rethought along the lines suggested here.

The second problem is that one might be thought to be an accessory to the age-old crime of killing time by spatializing it. Is that not what is implied by talking of structures? The importance of this charge is that it is an opportunity to make historically overdue clarifications about the representation of time. Some philosophers seem to have believed that the attempt to represent time spatially was either in itself a fatal error or led with some kind of necessity to making such an error. In whichever form it is taken, the belief is not only mistaken, but theoretically stultifying.

My view, on the contrary, is this: that there has never been anything wrong with the spatial or other representation of time *as such*. The error can always be found in the *interpretation* of the representation or of time's representability. Put another way, the error has always been in failing to realize that a representation of any sort requires an interpretation. I take it that a version of our claim here is commonly accepted—that, for instance, even the most representational painting, the most descriptive statement, the most realistic novel utilizes conventions, and that not all of the features of a representation can be assumed to be features of what is represented.

But if this has been generally agreed, the consequence for representations of time does not seem to have been drawn: that one can describe temporal structures—those that I have located in language, for example— and diagrams can even be provided to display these structures more vividly—without killing time.

Let me illustrate this point. A spatial representation even of what appears to be a simple linear unidimensional temporal sequence can be used to capture at least four different types of temporal structure, which could be called reflective, generative, participatory, and active (see the diagrams, "Four Types of Temporal Structure"). These types of temporal structure differ in the sorts of relations that hold among successive elements. In the *reflective* case, the principle by which the succession is unified is one reflectively projected on the events after the fact—as one might look back on the outcome of a series of accidental occurrences. In the *generative* case, the serial ordering is understood to be a product of some underlying formative principle—such as a genetic code or a transformation of a deep structure. In the *participatory* case, the order is one that a subject is essentially involved in unrolling. Think of listening to a joke, watching a play, or reading a book. Each successive step is marked by anticipations confirmed or denied, and others born. Finally, in the *active* type of seriality, the sequence of events reflects a plan of action in which some order at least is already determined.

Although something like a diagrammatology is required to adequately construct and interpret them, each of these types can be spatially represented, and this involves neither their reduction to a simple type nor the loss of their authentic temporality.

All of these, as it happens, are modulations of simple temporal seriality. I do not claim that they exhaust the field of temporal intentionality. Indeed, they are consistent with the unity of time thesis, whose limits I reject. It is a confirmation of our orientation to language that each of these types of interpretation of temporal order has a clear linguistic embodiment. I have already directly discussed *reflexivity*, and the idea of a succession indebted to a rule that generates it was covered by the more complex principle of multiple determination.

The two remaining ways of interpreting representations of time—as participatory or active temporality—both have their place in language, indeed a place that constitutes one of the most fertile resources of temporal structure. They lead us back to those structures particularly characteristic of communicative interaction.

In the active production of contextually appropriate grammatical sentences, both speaker and listener will modulate the way they *follow* utterances with various kinds of anticipation and retention. *What* is retained and anticipated will reflect linguistic probabilities resting on both formal and contextual conditions. More important, what is actually spoken, and thus whether a particular anticipation was correct, will reflect the very participatory temporal structure that utterances have. Thus, to take an extreme but clear case, the punch line of a joke is both something actively waited for and successful only if it is so anticipated. Part of the art of telling a joke (or writing a thriller) lies in the orchestration of clues—both leading and misleading—to the existence and nature of the climactic structure of the discourse.

The importance of a joke or, say, a Hitchcock film, for this project is that it exploits the active and participatory temporal structuring of language as this structuring gets elaborated by communicative interaction.

The temporal complexities of communicative intersubjectivity cannot

FOUR TYPES OF TEMPORAL STRUCTURE
(a diagrammatic representation and interpretation)

REFLECTIVE

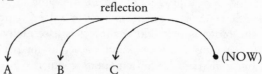

The connections recognized between A, B, and C are simply a product of a later reflective grasping of their serial ordering.

GENERATIVE

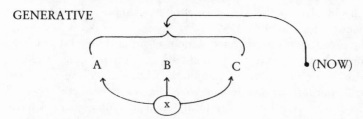

The connections recognized between A, B, and C are deemed a consequence of some underlying formative principle (x).

PARTICIPATORY

Lists of
words
guessed at
each stage

IN ------>	IND ------>	INDI
INNOCENT	INDECENT	INDIRECT
INNOVATE	INDUSTRY	INDIFFERENT
INWARD	INDIA	INDIA
INHALE	INDOLENT	INDICATIVE
ETC.	ETC.	ETC.

INDIA ------>	INDIAN ------>	INDIANA ------>	INDIANA
INDIA	INDIAN	INDIANA	INDIANA
INDIAN	INDIANA	INDIANAPOLIS	
INDIANA	INDIANAPOLIS		

Imagine watching a skywriter spelling out a word and guessing it at each stage.

ACTIVE

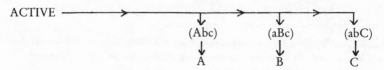

The chosen structure ABC guides the successive unrolling of three stages of action.

here be explored in any depth, but some of the key factors involved may be mentioned. The central condition is the fact of another linguistic subject with his or her own temporally structured life. Communication requires some sort of interarticulation of these temporalities. Whether one is dealing with a monologue or dialogue, the intelligibility and the effectiveness of discourse will in various ways depend on the speaker's correctly judging the progressive structure of his audience's participations—the phenomenon known to theoretical linguists as "garden-pathing." This is equally and more vividly true of the writer/reader relationship, or the relationship between a director of a film and his projected audience. Communication, in short, involves anticipations of anticipations, and ongoing modifications of one's discourse that accommodate these anticipations. The precise nature of these adjustments of cotemporality will vary with the type of communication involved. Making music together is often formally constrained by a score;[7] the communicative aspects of climbing a mountain together will involve a mixture of verbal exchange, silent gesture, and the interpretation of involuntary bodily movements. The feeling of copresence, of shared exhilarated exhaustion on the summit is a *product* of all these previous exchanges.

Earlier on three problems were raised with our treatment of time structure, and I have so far dealt with only two: intentionality without a subject and the alleged dangers of spatializing time. The third is the question of the status and scope of our investigation, which I shall now try to answer.

So far we have been guided by two ideas. The first is that structures of time are intentional structures, and that intentional structures are structures of signification. As language offers the most visible development of such structures, it is to language that I have brought the spade.

One consequence of the view that all structures of time are structures of signification is that these conclusions are not restricted in principle to time-consciousness or to existential time. Wherever change is patterned in ways that can be repeatedly recognized or accommodated to, or intelligibly grasped, a structure of signification is involved. I retain a certain sympathy for Heidegger's view that intelligibility and signification are rooted in our worldly involvement, even if discourse subsequently develops a certain autonomy. So the scope of this theory is intended to be quite general.

The second guiding idea has been this: if it is true, as Derrida claims, that our ordinary concept of time is metaphysical, this is seen, perhaps with a certain deliberate naivety, as a condition to be changed. It is quite unacceptable that temporal language should be squeezed out of philosophy just because of the temptations it puts before us.

Indeed, as Derrida knows, much of the evidence points in the opposite direction. Rather than supposing that time is a metaphysical concept, it is more fruitful to consider the temporal determinations and commitments of metaphysics, such as the desire for the end of time. The same goes for a wide range of philosophical concepts, not least that of *theory*.

Something of this problem will affect what follows—a consideration of one of the classic ways in which texts exhibit temporality—that of narrative. The importance of narrative structures to literary studies goes without saying. But much broader claims have been made for its significance. For Ricoeur, it is the key to understanding existential time and for mediating between individual and historical existence.[8] For Gallie, it is the defining feature of historical discourse.[9] For Jameson, it is "the central function or *instance* of the human mind" around which we can "restructure the problematics of ideology, of the unconscious and of desire, or representation, of history and of cultural psoductions."[10] For Lyotard, it is the irreducible form of intelligibility in a sea of multiply differential discourses ("petits récits").[11] For Barthes, it completely surrounds and permeates us, it transcends genre, and it is "international, transhistorical, transcultural: it is simply there, like life itself."[12] The interest in narrative structure is as old as Aristotle, and can plausibly be said to have "taken a quantum leap in the modern era."[13]

The question of narrative is very often and understandably the focus of interest for the study of the very intersection between temporality and textuality begun here. The reasons for not taking this approach would be a story in itself. It would have been more difficult to resist the conveniences of an increasingly established frame of reference. But more important, many of the structures pointed to function at a more primitive level than that of narrative. (Unless, of course, the sense of narrative is extended to include any and all linguistically mediated temporal intelligibility.)

In the face of such a wealth of narrative theory, one or two critical questions will be selected to which this approach makes a positive contribution.

The first question is one raised by Barthes in his seminal paper "Introduction to the Structural Analysis of Narrative": "Is there an atemporal logic lying behind the temporality of narrative?" On the one hand, there is Propp, "whose analytical study of the folktale is totally committed to the idea of the irreducibility of the chronological order: he sees time as reality and . . . is convinced of the necessity of rooting the tale in temporality."[14] On the other hand, there is the modern consensus that takes the opposite view:

> . . . all contemporary researchers (Lévi-Strauss, Greimas, Bremond, Todorov) . . . could subscribe to Lévi-Strauss's proposition that "the order of chronological succession is absorbed in an atemporal structure." Analysis today tends to "dechronologize" the narrative continuum and to "relogicize" it . . . or rather . . . to [give] a structural description of the chronological illusion—it is for narrative logic to

account for narrative time . . . temporality is only a structural category of narrative (discourse), just as in language (langue) temporality only exists in the form of a system; from the point of view of narrative, what we call time does not exist, or at least only . . . as an element of a semiotic system. Time belongs not to discourse strictly speaking but to the referent; both narrative and language know only a semiotic time, "true time" being only a "realist" referential illusion, as Propp's commentary shows. It is as such that structural analysis must deal with it.[15]

I shall focus our attention on two segments: ". . . from the point of view of narrative, what we call time does not exist, or at least only [exists functionally,] as an element of a semiotic system." and "It is as such that structural analysis must deal with it."

This account raises some most difficult issues, two of which I shall briefly discuss: the question of a fundamental atemporal structure, and the general question of the dependence of temporality on narrative discourse.

First, for Lévi-Strauss, "the order of chronological succession is absorbed in an atemporal matrix." As a *description of the method* of structural analysis it is quite accurate. But does it reveal anything about narrative itself, or only about the limitations of structuralist method? This question can be hidden by the analytic success of a structuralist illumination of, say, myth. But it should not be lost for ever.

Consider the following. A matrix structure qua spatial representation is *necessarily* atemporal. At this level so are representations even of the most obviously temporal phenomena—such as musical scores. But this gives us two senses of "atemporal matrix structure." In the first sense, it is the representation itself which (qua spatial) is "atemporal"; in the second, what is represented is represented *as* atemporal.

This latter, however, can again be divided. For what is represented *may actually be* atemporal (for example, a map of the London Underground, conceived of as labyrinth of tunnels) or it may, for whatever reason, just be represented *as* atemporal.

Clearly narratives *are not* atemporal. The mythical narratives Lévi-Strauss studies are *recounted* over and over again, and they have orders of events, orders of action that are essential to the intelligibility of that *telling*.

What Lévi-Strauss's myth analysis claims (or presupposes) is that this order is not *essential* and that the *binary oppositions* at work are. But perhaps it is this atemporality that is the illusion.

What *is true* is that "binary logic" does not involve time or temporality. But it does not follow that *instantiations* of binary oppositions are free from temporality. And this is the case at two different levels. Take Lévi-Strauss's analysis of the Oedipus myth. Central to his analytical matrix is the opposition: born from the earth/born from man. Clearly, (a) the individual *components* are as temporal as one could want, and (b) for the myth *to be a myth*, these components have to be articulated in *some* narrative sequence, however complex.

Freud (psychoanalysis was one of Lévi-Strauss's acknowledged "mistresses") can spawn a similar confusion when he declares that the unconscious is atemporal. What *is* true is that it is no respecter of *simple linear* order. But that does not make it *atemporal*; it rather suggests a new complexity to temporal organization. If the unconscious *were* atemporal, then no significance could be attached to repetition. And the same point can be made against Lévi-Strauss. It may be possible to identify versions of "the same myth" in which the structurally significant elements are distributed in a plurality of narrative orders. But that only shows that a *particular* narrative sequence is not important, not that narrative order is not, on each occasion, vital. And the residual and ineliminable temporality of the *elements* which are then organized in binary oppositions (birth, marriage, death—all figure in the Oedipus matrix) surely clinches the argument. Lévi-Strauss is talking of the atemporality of his *representations* of the structure of narratives/myths. But, as I have argued, it tells us nothing about the temporality/atemporality of *what is represented*. Moreover, one could ask, What is it to "read," understand, grasp, one of Lévi-Strauss's matrices? If that were an essentially temporal process, what would that say for the claimed atemporality of these structures? Finally, it is worth noting the clear parallel Lévi-Strauss sees between the structure of myth and that of *music*. (His *Mythologiques* are organized in a "musical" fashion.[16]) This only intensifies the problem. Clearly, music can be talked about structurally, spatially (Haydn's *Creation* has been called "a cathedral of sound"), and so on. But temporal articulation is quite simply indispensable for the appreciation of music (even for those who "read" scores and "hear" the sound). And that appreciation is never just the construction of an edifice complete only at the end, but an ongoing temporal experience, a play among structure, process, and event.

There are enormous gains to be made by looking at the various "logics" of narrative. Barthes discusses (a) Bremond's attempt "to reconstitute the syntax of human behaviour utilized in narrative, to retrace the course of the 'choices' which inevitably face the individual character at every point in the story . . . an energetic logic," (b) (linguistic) Lévi-Strauss, Jakobson, and later Greimas, showing how "paradigmatic oppositions . . . are 'extended' along the lines of the narrative," and (c) (analysis of action) Todorov's attempts to "determine the rules by which narrative combines, varies, and transforms a certain number of basic predicates."[17]

In each case, the atemporality is that of the representation alone, not what is represented. The time that *is* excluded, perhaps, is that radical sense of time as the opening on to the radically other that we draw out of Nietzsche (see Part One). But this is excluded not by an atemporal logic, but by a temporally articulated narrative.

Consider Barthes's famous account of "the structuralist activity."[18] It is a veritable self-subversion of any "structuralist metaphysics." Central to structuralism, he avers, is "the notion of the synchronic . . . which accredits a certain immobilization of time." What is meant by "a certain"? Is this an expression of *reservation* or of ontological limitation of such "immobilization"? Barthes *seems* to be distancing himself when he adds, in parentheses, "(although in Saussure, this is a preeminently *operational* concept)." Is it *more* than that for Barthes? For him, the structuralist activity aims at producing an intelligible simulacrum of an object, by "dissection and articulation." This simulacrum is atemporal, and in this project, Barthes surely reaffirms the Platonic connection between the timeless and the intelligible. But without Platonic metaphysics. For he never loses sight of the fact that these simulacra are human *creations*, which suggests that for Barthes, too, the synchronic perspective may be purely "operational." Indeed, the language of his account is shamelessly temporal, even teleological. Not only does he *insist* on the structuralist *activity*, he describes it as "the controlled succession of a certain number of mental operations,"[19] operations that have as their goal "to reconstruct an 'object.'"

Anyone who supposes that the detemporalization of narrative is anything more than a methodological ruse should reflect on this essay.

I have also insisted on the inherent temporality of those narratives ("myths") analyzed by Lévi-Strauss. How can one deal with his description of myths as "machines for the suppression of time"?[20] My response would be this: it rests on the assumption that myths are reducible to a binary logic, that "logic" is timeless, and that their *purpose* is to induce this timelessness. The first has already had doubt cast on it, and the last is wholly unproven. There is an explanation for Lévi-Strauss's remark, however. Lévi-Strauss suggests the possibility that his own *Mythologiques* might itself be taken as a myth.[21] Perhaps indeed it might not be the common myths of primitive people that are "machines for the suppression of time," but rather the Lévi-Straussian structuralist "myth" of a logic of myth that "suppresses time" by its atemporal analysis of myths. Structuralism, after all, does have that as its avowed aim; primitive people do not.

The second question to be asked of Barthes is this: Precisely what relationship is being claimed between temporality and narrative? As I understand it, Barthes is claiming that extratextual "time" is a referential illusion—"temporality" is a product of narrative, and not vice versa.

In relation to the first question, it has already been argued that language in the ordinary sense is exemplary for revealing those temporal structures that transcend simple linearity. But the stronger claim, that temporality is essentially dependent on narrative textuality, would depend on our being able to show that the temporality of "consciousness" and of "existence" was already organized in narrative fashion.

My position is slightly different. I claim that narrative, time-consciousness, and existential time each *share* certain primitive features—many of the features analyzed earlier in this chapter—but this does not mean that full narrative intelligibility links all the phases of my experience; and Ricoeur is closer to the mark when he suggests in a most illuminating way "that narrative activity, in history and in fiction, provides a privileged access to the way we articulate our experience of time."[22] Further justification of this theoretical preference will not here be offered, but I do share Ricoeur's genuine but *limited* appreciation for those authors engaged in the structural analysis of narrative for whom the aim is the reduction of narrative to atemporal structures. Referring to Barthes and Greimas, he writes that "the result is that the narrative component as such is identified with the level of manifestation, whereas only achronological codes would rule the level of constitution."[23] Ricoeur suggests that this leads to the overlooking of "the temporal complexity of the narrative matrix," and suggests focusing (again) on "plot,"[24] not in any naive way, but seen as the rich "matrix" of narrativity. My appreciation of the structuralist approach is quite genuine—Barthes, Greimas, Genette, and many others have made an enormously valuable conceptual contribution, but it is a mistake to confuse the analytical power of structuralism with philosophical completeness. What is required, one might suggest, is a *structural hermeneutics*. At the very least the limitations of a plain structuralist approach make it very difficult to pose *my* question: How is the relationship between narrativity and existence to be understood?

I will try to approach this question through what might be thought to be the exemplary case of *biographical* narrative. There are two views of the relationship between existential time and narrativity, one being that *narrative* (for example, biography and autobiography) involves an aesthetically motivated selectivity essentially distinct from life as lived. This difference can be understood positively (narrative *articulates* one's existence *as* an intelligible story) or negatively (narrative is a retrospective, rationalizing distortion of life as actually lived). The ending of Sartre's *La Nausée*, in which narrative intelligibility is offered as an aesthetic solution to the problem of the meaning of one's existence, seems to leave this question open, undecided.[25] Earlier on in that same book, Sartre had taken a more critical attitude to the gap between a retrospective construal of a series of events as an "adventure" and the actual experience of those events at the time.

"Adventure" is treated as a fictional *distortion*. In Sartre's partial auto-biography and his other "biographies" (of Flaubert and Genet) his position seems more positive.[26] But one has to ask what basis there could be for a positive treatment.

In Hugh Silverman's "The Time of Autobiography,"[27] which, curiously enough, begins by looking at Lévi-Strauss's own autobiographical *Tristes Tropiques*[28] (in which, surely, there is little evidence of an atemporal logic underlying the narrative), he makes a number of important distinctions between the different modes of temporality in "autobiographical textuality": "historical time," "chronological time," "the internal textual marking of autobiographical time," and "autobiographical temporality, or the re-marking of lived time." But for the most part the paper operates with an unproblematic notion of "lived time" as a pliant material on which one works when "writing one's own life." At one small point, however, he makes the suggestion that narrativity might appear at the level of existence itself. And surely it *is* important to consider the role this might play.

There is clearly a literal (literary) sense of "narrative" which applies specifically to written texts. But equally, in a great diversity of ways, our everyday experience is full of partial, and sometimes complete, "narrative" reflections, projections, memories, imaginings, and so on. Lived experience *cannot be purified of these sequential constructions*, understood both at the level of (unreflective) protention, say, and more reflective and explicit projection and planning. It may well be that there is some limited principle of *dependence* here—that the explicit may often involve the articulation of the inexplicit. But we need not suppose that things always work that way. One's ongoing reflection may also work on actual or possible past or future sequences never prereflectively strung together.

Of course, the ongoing reflection of (auto-)biographical narrative will itself include pieces of writing (in the ordinary, pre-Derridean sense)—diaries, letters, poems, lists, perhaps even articles and books. And they cannot be thought of just as "partial narratives" in comparison to a final summing up. For they actually *affect* the way one continues to organize one's experience. Lived experience is dialectically related to reflection on it, *in* it, whether written down or not.

But is it proper to talk of reflections on/in lived experience as taking a narrative form? Do they always do so, or just sometimes, or what? What is meant by "narrative" here? I am going to assume that the question of author and audience can be put to one side. If autobiography is narrative, then my mental musings about my last year in Paris *could* also be, even if only I *happen* to be aware of them. But what form must any such musings take to be considered to be a "narrative"?

Let me be clear: my theoretical interest in "narrative structure" for its own sake is limited. The project of a general theory of narrative must be deferred. Our question is whether "narrative" can teach us anything about the ways in which the standard model of time as a single linear sequence of moments can be enriched, displaced, transformed, deconstructed, and so on. My answer to the question of what intra-existential reflection must be like to count as "narrative" will be content with the limited aim of showing that it can display some of the most commonly recognized features of narrative. And the claim will be that these features offer us a model by which discontinuity, nonlinearity, and pluridimensionality can each be thought of as dimensions of both existential and textual temporality. The wealth of possible *differences* between existential and narrative time will not here be discussed.

One can distinguish the following *levels of time narrative*: (1) the time of the reader (and the reading), (2) the time of the narrator, (3) the time of the plot, (4) the time of actions, (5) the (real) time of events, (6) the time(s) of the characters, and (7) the time of the narrative discourse. In making these distinctions I have already given some of the game away. It would be hard to establish more than one of these and not find oneself with the possibility of pluridimensional time. And the fact that they do not evenly cross-correlate will allow discontinuity. The interrelations among such times will allow such departures from simple linearity as circularity, splitting/rejoining, and so on. I shall briefly comment on each of these categories before drawing some conclusions about their relevance for understanding existential time.

THE TIME OF THE READER

Unread, the book is dead. (Without an ear there is no speech.) And yet the reader is no passive receiver of strings of words. Not only is the reader already lodged in a multiplicity of time tracks (writing a book, moving, struggling to defeat an illness, digesting a meal, waiting for a phone call, wondering about salvation, imagining unlikely possibilities, remembering his/her childhood) but these can impinge on the practical business of getting through the book, on the understanding of what the book is about, of what the characters are doing, of its literary merits, of its construction and organization and so forth. If there is no narrative except in relation to a reader, then the reader's range of temporal insertions is a vital element. All this is true even of the "ideal reader." Should he be a first century Greek, should he be able to read the book at one sitting . . .? Reading is also an activity involving a very complex ongoing activity of retention, comprehension, protension, and imagination. The inner complexity of the book may set a certain threshold of synthetic power on the reader's part, bound up with his (or her) capacity to "organize time."

THE TIME OF THE NARRATOR

The same description could be repeated for the activity of writing. But in principle the *product* is independent of how continuous, how long, how interrupted the writing was, so this aspect will be ignored. What cannot be ignored is the time of the "implied author," the author as textually evident. Consider the difference among a retrospective narration, a stream of consciousness narration, and a dream. (Whether the text is written in the first or third person is in principle a logically distinct question.) One can ask, Does the temporal position of the narrator remain fixed? For it can begin with "memories" and work up to present description, or play with its own temporal position in more complex ways.[29] The time of the narrator also allows for complex interplay with the other levels of narrative time, especially those of the plot and the characters. Barthes gives the wonderful example of a story by Agatha Christie ("The Sittaford Mystery") in which the narrator, who is one of the characters, turns out to be the murderer the story seeks to reveal.[30] Here the epistemological interaction (*essentially* bound up with time, and exceptionally so in a mystery or suspense story) between narrator and characters takes the form of the explosion of a conceit at the moment of disclosure.

THE TIME OF THE PLOT

This category is as old as Aristotle (see his *Poetics*) but substantially reworked by Ricoeur, as I have suggested. The need to rework it results partly from the way modern narrative has produced variations on the simple plot—multiple plots, ambiguous plots, plots without resolution, as well as the more obviously subordinated subplots. Unless we extend narrative "downward" to include a simple recitation of events, or "upward" to include any piece of prose, something like the notion of plot will remain essential. It does not require the structure of stage setting, development, resolution of action. Nor does it require that the narrative be able to be reduced to a "sentence" (see Barthes[31] and Genette[32]). But it does require some thread of developmental intelligibility to which the other *orders* of the narrative are more or less subordinate. (A postmodernist text could of course make a plot out of refusal of one of these other elements to remain subordinate to the plot. What makes the British "Goon Show" scripts merely modernist is that the continuous rebellions against the plot by the characters [for example, Bluebottle's "I don't like this game"], and even on occasion the narrator, Greenslade, are always quashed, or at least reappropriated. A postmodern reading might argue that the "plot" was only a device on which to hang these acts of textual transgression.) If the plot cannot be summarized in a single sentence, it can in principle be briefly explained. Its narrative unrolling takes the form of a series of *actions*.

THE TIME OF ACTIONS

By "action" is understood a stage of narrative description that contributes to the development of the plot. The *agents* of such actions may be characters, but, equally, natural forces (such as a flood); and actions may be "physical' or "mental." What is critical is that the *order* of actions can be quite different from the *order* of *events*. A narrative may begin with the death of the hero and set itself the (retrospective) task of tracing how that happened.

THE TIME OF EVENTS

"Event" here includes "actions" and other events that make no contribution (or no essential contribution) to the plot (or to the plot complex). The usual way of describing the order of events is to talk of their "real" chronological order. It is assumed that however subtle the internal temporality of a text, it nonetheless projects (and rests on) a "real time" beyond itself. A flashback requires a "real past" to flash back to. Clearly, however, it is something of a literary convention that all events can be located in a single sequence. It is a convenient backdrop against which one can construct a temporal play, but it is equally a convention that can itself be played with. A flashback may be a memory, or a "touched-up" memory, or even a fantasy, and the question remain unresolved at the end.[33] Science fiction can operate with uncoordinated time frames in which questions of relative priority are again undecidable. Geographical, institutional, and personal "time bubbles" can thwart the establishing of a single sequence of events. However, the convention of "real time," "the *actual* sequence of events," is clearly very common and, if only as another fiction, essential even for these ways of playing with it to make any sense, and in particular for the time of action to have any autonomy.

THE TIME(S) OF THE CHARACTERS

Whether or not the narrator is actually a character in the narrative, characters have their own temporal involvements and intentionalities. Everything claimed initially about the reader's multitracking, and memory and projection, equally applies to the characters. Of course one must also distinguish (a) what one is *told* by the narrator, directly or indirectly (for example, via other characters) about their own temporal insertion, (b) what one can gather, and (c) what one imagines or supposes without much evidence. The point is that the characters have their own individual time horizons, some of which will be shared among themselves, some not, and these will differ from those of narrator and reader. Again, the most obvious differences are ones of knowledge and perspective; some characters will know more, some less than the "implied" narrator and the actual or ideal reader. In such differences lie the possibilities of irony, comedy, tragedy, and so forth. One might add that, exceptionally, characters can of course have their own sophisticated views of time itself, which the text can endorse or refute.[34]

THE TIME OF THE NARRATIVE DISCOURSE

This is the temporality of the "thick surface" of the narrative conceived of as a string of words. At the *most* superficial, it is a simple succession of words. But it will equally be broken into clauses, sentences, paragraphs, and chapters and chapter headings. Slightly "thicker" still, we find various forms of repetition of words and names. (See the earlier account p. 338.) And there are variations of tense, mood, direct and indirect speech, and so on. The thickness of this layer goes as far down as it is possible to go without touching "meaning." And wherever what is discovered exhibits a temporal organization (including paradigmatic oppositions "extended along the line of the narrative,"[35] it will be added to this account of the temporality of narrative discourse. I am being radically unjust to this category, for a large proportion of the structural analyses of narrative would be contained within it. But my narrative has its own plot requirements.

If temporality is as all-pervasive as it sometimes seems, these seven levels of narrative temporality cannot claim completeness. But they are sufficient for present purposes. And first I shall demonstrate that they permit the divergencies from our ordinary account of time. Then next they allow us to exhibit certain features of "existential time" as well.

In themselves, these seven different levels of narrative temporality make narrative time *multidimensional*. There is just no *one* answer to the question of what "the temporal sequence of the narrative" consists in. And these various times will not just run parallel but will modulate their independence/interference with one another. One of the *radical* consequences of this is that the *unity* of narrative time can also be put in question. Or rather, one can propose a kind of unity not dependent on linear continuity. For the interweaving of seven different temporal dimensions can generate a kind of continuity without there being a "single thread" running through it. I have in mind here the model of a piece of string composed of overlapping strands not one of which runs throughout the length. It might be said that it is precisely the role of the plot in establishing such a unity that distinguishes narrative from other forms of prose. That can be admitted. But the two points made earlier deserve repeating: (a) that "plot" can itself be the subject of a disconcerting play, generating, for example, multiple plots and ambiguous and unresolved endings; and (b) that one can produce a "narrative" in which the plot is only an arbitrary device for bringing about a more complex literary effect.

Finally, this wealth of levels allows the possibility of temporal splitting (*plots* can split or rejoin — or not) of *circular* time, cycles of repetition, dispersal, irresolution, and so forth. The key to all this is that one or more temporal dimensions can fragment or break, or disperse as long as the narrative line can be continued, for the moment, along another dimension. (A bacchanalian revel in which all [but one] are drunk!)

Yes, it may be said, such is the freedom of language (which even so runs the risk of unintelligibility). But surely that shows just how different it is from ordinary experience.

At this point one could return to the question of whether (auto-) biography necessarily "distorts" life, and look at specific examples of this in the light of our current claims. But this will be left for another occasion.

I have *endorsed* the Heideggerean understanding of the tri-ekstatic finitude of human temporality and his disturbing of the linearity of time, but cast doubt on his residual commitment to temporal unity as it is reinscribed in the gathering and focusing of anticipatory resoluteness. This resoluteness is fundamentally a demand that the multidimensionality be fused together, that the Babel of voices be silenced, that the different levels of our (temporal) worldly involvement be assigned a strict order of subordination. In plurality, Heidegger finds only "turbulence," "ambiguity," and confusion. And yet his account of everydayness is an acknowledgment of the fact of this multiplicity.

It would be too trite to compare the temporality of the narrator to that of "conscience," of the characters to that of the multiplicity of our inner voices, of the plot(s) complex to our "fundamental project," and of the narrative discourse to the sequence of our lived thoughts and words. It could be done, but I shall not do it here. I will simply look at the difference between the temporality of events and of actions, for this surely captures a key duality in the temporality of our lived experience. It is overwhelmingly clear, even on one particular time track (say, writing a book), that the sequence of one's experiences needs bear no relation to the sequence in which they can best (or at all) be rendered intelligible. One may discover how it all fits together only at the end. (Compare picking up jigsaw puzzle pieces as opposed to successfully putting them down. Or listening to a long German sentence with the verb right at the end.)

The multiplicity of only contingently coordinated timetracks is surely not restricted to avant-garde literature, but a perfectly common experience.

One need only think of the way we weave together a life of reverie, of dream, of different kinds of perceptual experience (natural and artistically constructed; temporally highly structured and not), a whole variety of *activities*, some nested within others, many quite independent of one another, and the way the various more or less sealed or communicating *worlds* we live in have their *own* inner temporal development (home, school, job, national politics). And so on. It may be that one does not need narrative, or the most general structures of temporal textuality with which I began, to be able to focus theoretically on this multidimensionality; the advantage of considering textuality is that it is more easily objectified, and, in consequence, easier to develop a set of analytical categories for understanding it. The claim is, of course, that the correspondence goes beyond mere analogy, and that "lived" experience is already impregnated with those very reflective structures elaborated in textuality in general and narrative in particular.

I said I would not consider the differences between them. However, there is an obvious question it would be hard to ignore: Is there not some parallel between human finitude and that of a text, of which biography would represent a convergence? There is though a fundamental difference, that between the *ethical* and the *aesthetic*. I *am* my finitude; that of texts is a quite different matter.

After all this, however, is there not a sense in which this whole discussion of temporal intelligibility rests on a complacency about the values of unity, sense, self-understanding, which it endorses by the work of theoretical elaboration? To put it crudely, might not the narrative impulse (including perhaps the theory of narrative) be a flight not just from a contingent disorder, but from a fundamental obstacle to the kind of closure of meaning that it systematically induces? The "fundamental" obstacle would be dispersion, death, madness, error, forgetfulness, and so forth as they appear within language itself, in the form of its irremediable fragility. If the philosophical ideal would render language "transparent"—referentially and semantically— the same can be said for narrative—that its "fictions" can satisfy the desire for wholeness, unity, and intelligibility only through the mirage of language.[36] Narrative achieves aesthetically a unity that would wither outside the framework of linguistic "representation."

I have three responses to this argument, which have a more general application to the double strategy of the book as a whole.

First, one can surely "admit" something like the claim that madness lurks beneath the skin of order, and that narrative provides just such a "skin," without supposing that this would in any way detract from the *importance* of narrative.

Second, if narrative *is* as universal as is claimed, one ought to investigate the *wealth* of its resources for generating sequential intelligibility.

And third, there is surely a sense in which "the function of narrativity" performs a kind of *reversal* on attempts to restrict its scope. If we say that madness lurks beneath the fictions of order, are we not telling another story, and necessarily so? The moral of Derrida's demonstration that Foucault's project of writing a history of madness is dominated by the ideal of Reason[37] is that at some level there is no alternative. It is not just that narrative intelligibility has an infinite power to subvert the discovery of what exceeds its bounds. Even *such* a story of necessary subversion reinstates narrativity.[38] This is surely a sobering thought. What it suggests is necessary complicity between the excessive and the limits exceeded, madness and the law.[39]

6   *The Philosophy of the Future*

So far the attempts at philosophical reform have differed more or less from the old philosophy only in form, but not in substance. The indispensable condition of a really new philosophy, that is, an independent philosophy corresponding to the needs of mankind and of the future, is however that it will differentiate itself in its essence from the old philosophy.[1]

The author continues:

The new philosophy makes man—with the inclusion of nature as the foundation of man—the unique, universal and highest object of philosophy.[2]

The candidates for inclusion under the heading of what has been called "postmodern philosophy"[3] are so varied that they probably do not illuminatingly share any common features, but they could be negatively unified by their distance from such a program as this. The author knows what the needs of mankind are, what philosophy requires; he has no doubts about the concept of man, about man's roots in nature, or about the distinction between form and substance. (What would he have said to the suggestion that it is perhaps a change of style that we need?)

These remarks are taken from Feuerbach's *Principles of the Philosophy of the Future* (1843) and the title itself calls for comment. Feuerbach was confident about what "the future" needed, that philosophy could supply it, and that the very idea of the future was unproblematic. We, who now exist in Feuerbach's future, are less sure. What is called poststructuralism seems to articulate a withering away, or perhaps better, a mutation in the concept of the future. But is it the future as such, or is it only a particular conception of the future (teleological, utopian, and so forth) that is put in question? If the very horizon of futurity is suspect, are practical and intellectual defeatism and inertia appropriate responses? Is the power to provide intellectual and practical ideals still a sound way of ranking competing discourses or should

361

possessing such a feature be seen as the legacy of a metaphysical desire?

The first alternative, which gives a positive value to a vision of a future, was spelled out recently by Cornel West, an American critic of deconstruction, who wrote that "post-modern American philosophers . . . have failed to project a new world view, a counter movement, a 'new gospel' of the future."[4] He accuses them of offering no way past the nihilism they usher in. Instead, he goes on:

> Their viewpoints leave American philosophy hanging in limbo as a philosophically critical, yet culturally lifeless rhetoric mirroring a culture (or civilization) permeated by the scientific ethos, regulated by racist, patriarchal, capitalist norms, and pervaded by the debris of decay.[5]

Clearly West can have no serious dispute with rhetoric per se. But a gentler version of his views must surely be quite widespread. If they seem harsh, it is worth remembering that we are living in Nietzsche's future too, and that in his preface to *The Will to Power* he claimed to be describing

> . . . the history of the next two centuries. I describe what is coming, what can no longer come differently: the advent of nihilism.[6]

How does Nietzsche know?

> . . . necessity itself is at work here. The future speaks even now in a hundred signs; this destiny announces itself everywhere.[7]

West's argument is that Nietzsche offers the possibility of a stage beyond nihilism, a postnihilism, but that postmodern American philosophy is stuck at the stage of nihilism itself.

In the winter of 1887, Nietzsche wrote of nihilism as a necessary stage:

> We must experience nihilism before we can find out what value these "values" really have—we require sometime new values.[8]

It is a stage that he sees himself as having completed only at that point. Until then—which would cover most of his writing—he had been a thoroughgoing nihilist. He describes himself as

> the first perfect nihilist of Europe, who . . . has now lived through the whole of nihilism, to the end, leaving it behind, outside himself.[9]

Clearly the overcoming of nihilism is an operation vulnerable to mirage and illusion. But there is a beyond for Nietzsche—affirmation, or what he sometimes calls active or affirmative nihilism.

Accepting for a moment Nietzsche's characterization of "modernity" as "overabundant development of intermediary forms; atrophy of types; traditions break off,[10] schools [appear] . . . the willing of end *and* means has been weakened," then arguably postmodern American philosophy has not even left modernity behind. Postmodernity, on this view, is a treat in store, is yet to come. It would not so much lack a future, or any projection of the future, as *be in the future*. And the present, in West's words, would just be "lifeless rhetoric" and the smell of decay. How fair is this judgment?

Clearly, at this point it matters whom one chooses to look at, but suppose one takes the case of Derrida, who has had such an impact on the scene West describes. One can easily conclude that there is some justice in the verdict, not by a politically motivated analysis of Derrida's writing, but by reflecting on a remark he once made in (published) discussion.

I don't see why I should renounce or why anyone should renounce the radicality of a critical work under the pretext that it risks the sterilization of science, humanity, progress, the origin of meaning, etc. I believe that the risk of sterility has always been the price of lucidity.[11]

Can one conclude then that Derrida, for one, took the risk, and lost? These remarks need more careful attention. Contrary to what these sentences might suggest, Derrida is not actually wagering anything he values. The values of "humanity," "progress," and "the origin of meaning" are precisely the sort that, following Nietzsche and, with reservations, Heidegger, he puts in question. To sterilize them, to render them no longer *productive* would not be an unwanted by-product that deconstruction has to risk but is in fact, if one can say this, a central aim. It remains true, then, that for those subjects whose hopes and dreams are organized by reference to these values, sterility might indeed be the result of their being deconstructed. And then West's challenge to postmodern philosophy to produce Nietzsche's new values returns in force.

I have not yet made good or made precise my claim that postmodernism allows the idea of the future to wither away. But to set the scene for my discussion of that theme, let me offer, perhaps prematurely, two opposing views of the philosophical status of the future, at a fairly general level.

On the one hand, there are those for whom some sort of projection of a future would be a condition for the intelligibility of the present. And ranged against them are those who would treat such references to the future with a certain caution. I shall begin with the first option which is that no theoretical account of any present (whether it be the present epoch or lived experience) is complete, adequate, or even (perhaps) intelligible without an account of the future. An account of the future does not mean a prediction, but rather a projection. I would distinguish two different ways in which this positive account of the importance of the future can be articulated, which I shall call hermeneutic and ethical.

THE HERMENEUTICAL IMPORTANCE OF THE FUTURE

This view can be traced back to Heidegger's concern with the role of the future as a *horizon* of intelligibility for action, for meaning, for truth, and so on. Understanding is never a mere grasping of the present, or the presence of something, but always occurs within a triple ekstatic horizonality—of past, present, and future. For Heidegger, of course, this complex projectivity comes in different forms (authentic/inauthentic), but the underlying horizonal structure remains. This is meant to be true at many different levels. The understanding of a sentence as it is spoken is an achievement in which a mobile synthesis of retention, current awareness, and anticipation operates at fairly short range. But equally, one's understanding of the events in the morning paper is necessarily predicated on anticipating their consequences, how things might turn out. To multiply examples here would be otiose, for each aspect of everyday life—listening to music, watching a film, reading books, eating a meal, writing a letter, going for a walk—exhibits this same structure of horizonality. And even if Heidegger was later to drop the privileged position he had given to the future in *Being and Time*, there is no doubting the hermeneutic importance of the asymmetry between past and future; the future is the horizon of possibility in a *strong* sense. And it is a horizon that is essentially *finite*,[12] limited for each individual by death. I will take this up again later.

Although it stretches the word "hermeneutic," it is worth adding here that on a dialectical view of history, an analogous general point can be made, in a rather different way. To understand any present, one has to grasp the contradictions at work within it, by which its capacity for (self-) transformation is marked. Understanding a state of affairs is to grasp its possibilities of change. Here, it might be thought, the future is only logically or abstractly required; but in another sense, of course, this transformed state *is* the future. On Hegel's version of this view, an easy transition opens up from the general hermeneutic role of the future, to the specifically *reflexive* importance of the future to philosophy itself. Two different versions of this can easily be distinguished. The first envisages the future as fulfilling the ideals of philosophy in a way continuous with the past; the second announces the possibility (or necessity) of a radical break either within philosophy or with philosophy itself. For convenience, I shall discuss the role of the future in philosophy's own fulfillment within the framework of the *ethical* importance of the future. The projection of the future required by the various forms of break, closure, fulfillment, or end of philosophy cannot be dealt with just yet. For such issues mark the point at which the whole question of the future becomes problematic. So I will turn to the *ethical importance* of the future.

THE ETHICAL IMPORTANCE OF THE FUTURE

This thesis could be stated as follows. No account of an ideal state of human affairs, no account of basic human values, and indeed no prescriptions as to how things ought to be, or what ought to be done, are complete without a projection of the future as the condition of their realization. The practical move from the real to the ideal requires the positing of a future. Thomas McCarthy, in words that echo Feuerbach, describes Critical Theory in just these terms, as

> a theory of the contemporary epoch that is guided by an interest in the future, that is, by an interest in the realisation of a truly rational society in which men make their own history with will and consciousness.[13]

And such projections are clearly not confined to Critical Theory. Phenomenology was from the very beginning inspired by the prospect of bringing about the realization of such ideals. For Husserl, for example, philosophy

> has claimed to be the science which satisfies the loftiest theoretical needs and renders possible from an ethico-religious standpoint a life regulated by pure rational norms.[14]

Philosophy undertakes to satisfy

> . . . humanity's imperishable demand for pure and absolute knowledge (and what is inseparably one with that, its demand for pure and absolute valuing and willing).[15]

Philosophy will ". . . teach us how to carry on the eternal work of humanity."[16] His claim, of course, is that it has failed miserably, and that only phenomenology can *actually* satisfy this demand. I shall modify this claim a little later.

One could multiply examples of this theme. Kant's essay, "Perpetual Peace," would be another obvious source.[17] At one level, of course, there is nothing problematic about this at all. What is not yet finished needs more time to be completed. But what if "the future" were nothing but the horizonal complement of a *desire* that *in principle* could not be fulfilled? The problematic referentiality of the future is notorious. Is it not extraordinary how easily one slips between indexical, substantive, and "qualitative" senses of this term? Some such slippage might actually be part of the mechanism by which the future plays this ethical role. But before turning to some of the caution about the future that just such considerations can engender, let me rehearse in a positive form this claim about the ethical value of the future. *If* life is judged to be improvable, then our hopes and plans demand a dimension of futurity as a condition for their realization. *If* philosophy is thought of as having a telos as yet unrealized, then the future will be projected in such a fashion as to allow for this. If the improvement of life can come only through, or with, the cooperation of philosophy, then these two futures fuse. And, on the whole, it is this strong version that I have been looking at.

There is little doubt that at whatever level of explicitness and intensity, such views are widely held even today. They are, however, increasingly difficult to defend. That may only be symptomatic, but the proliferation of reasons for treating references to the future with caution is worth closer scrutiny, and constitutes the second, alternative general philosophical projection of the future.

There are a number of grounds for such caution.

THE FUTURE AS UNPREDICTABLE

First, and without my being able to offer an adequate account of its philosophical significance, there is renewed vigor to plain old skepticism about our knowledge of the future. This has been brought about, paradoxically, by the massive expansion in human technological control. I say "paradoxically" because this would *seem* to suggest an increase in predictability. And yet the necessary incompleteness of this control has wildly unpredictable consequences, not least of which is the *possibility* of total self-destruction—at which point issues of technology and theology could be said to merge.

THE FUTURE AS RIPE FOR IDEOLOGICAL APPROPRIATION

The second ground for caution can be put succinctly. There is a clear ideological potential in normative references to the future. In the absence of a *predestined* path, interpretations of its proper course have a field day. The hand that charts the future rules the world.

However momentous, these first two grounds for caution do not seem *conceptually* problematic. But there are others—I shall select three—that surely are, and each of them belongs to poststructuralism.

THE FUTURE AS MYTH OF FULFILLMENT

The most common theoretical role played by reference to the future is as a supplement to the deficiencies of the actual present. The future then will bring a time, a future present that will not have these deficiencies. Heaven, utopia, the war to end all wars, absolute knowledge, "a truly rational society" are all examples. The critique of the future, so understood, is continuous with the critique of the value of presence, of fullness of meaning, of freedom from "*différance*," from a relation to the outside . . .

THE FUTURE AS UNINTELLIGIBLE DISCONTINUITY

If words like "present" and "future" are understood to refer to something like current and future paradigms or general frameworks of discourse, rather than merely to events in chronological time, then any difficulty one might have of thinking outside our current framework of discourse will set an immediate limitation on our ability to project the future. Here "the future" would be another way of referring to what is (absolutely) other. That, I take it, is the sense of Derrida's claim that "the future can be anticipated only in the form of absolute danger."[18] What is the absolute danger? I take it that it is the breakdown of all intelligibility—madness: a radical discontinuity.[19]

"THE FUTURE" AS A CONCEPT WITHIN A DISCREDITED LOGOCENTRISM

If it is accepted that references to the future are inseparable from the general apparatus of historical discourse, and if it can be shown that the concepts involved in historical discourse (development, progress, evolution, influence, cause and effect, context, and so forth) belong to metaphysics, to logocentrism, a consistent attempt to mark out a distance from logocentrism would require what I called a certain caution about the future. This argument would rest on the *complicity* of references to the future with traditional historical discourse. Somewhat differently, it could also be argued that a deconstruction of the future as telos, as end, as fulfillment, is possible by a straightforward transposition of the kinds of analysis Derrida gives of Husserl's discussion of the idea of "origin," of the transmission of meaning, of reactivation, in his introduction to *The Origin of Geometry*.

Husserl had realized that it is only through writing that geometry is guaranteed historical continuity and hence the possibility of a reactivation of its origin. Yet this, at the same time, is the source of the very danger that threatens it. For writing is ruled by the absence of any absolute origin, an original absence. Might not the same paradox affect the future—that it is not so much what writing demands for its fulfillment (such as the ideal cashing out of its meaning in experience) but that it is the always withheld impossibility of that fulfillment. If the future were such an impossibility of achievement, if such distance were *inherent* in the philosophical functioning of the future (consider the Spanish *mañana*), then one might come to think of the future as the objective counterpart to the way Lacan understands desire. And the link between writing, and indeed language in general, and the future would be supplied by treating language as the primary site for the investment of that (metaphysical) desire.

Not only are appeals to the future necessarily expressed in language, but it is only *as* so expressed that projective clarity and focus can be retained. Is it an *accident* that the response to a philosophical crisis is so often the proposal of a new method—in which one further layer of self-monitoring and control is established for writing? A successful transposition of the deconstruction of origin to that of the future as telos would argue that such projections are retained and refined only in writing. And if writing is thought of as involving a perpetual deferment of presence, then teleological fulfillment would rest on an impossibility of fulfillment, which cannot be a satisfactory state of affairs.

So, whether one thinks of the future as involved *fairly generally* in a traditional historical discourse viewed with logocentric suspicion, or as subject to a deconstruction symmetrical to that of an *archē*—either way it would come out as an essentially logocentric concept.

I have now outlined both the view that the future is an indispensable philosophical category—for reasons both hermeneutical and ethical—and the view that such references to the future should be treated with great caution, both for traditional reasons—its unpredictability and vulnerability to ideological abuse—and for more powerful and complex reasons linked broadly speaking with poststructuralism. It is these last reasons that must now be considered.

On this view, which will now be elaborated, the typical philosophical role of the future is always or typically to project the value of presence on what is not present, to reduce the unknown to the known, the other to the same.[20] I turn now to Derrida.

His critical interest in the theoretical value of references to the future rests without doubt on his linking it with teleological and eschatological views of history, with linear temporality, and with the metaphysics of presence. I shall outline the argument more precisely, draw out a crucial topological schema that reinforces the argument, ask whether there is any nonteleological, nonmetaphysical way of thinking about the future that might emerge from the deconstruction of History, and so forth, and then look at the way some of these themes are drawn together in one of Derrida's most interesting recent papers.

First, the argument.

When the future has been mentioned in the context of poststructuralist caution, I have often myself exercised a certain caution. I tried not to talk of the future per se, but of the theoretical value of references to the future. Perhaps *textual* would have been a better word than theoretical. This is important. For Derrida, it is not enough to know whether a text uses this or that term, but how it is being used. What is metaphysical about a term is the work it does, or its textual deployment. There is a difference between a mere prediction of a future course of events and the use of such a prediction to legitimate a course of action. But there is a further level, and that is the use of the future to satisfy an ontological desire—the desire that history both past and future be seen to exhibit a continuous unity, which he calls the history of meaning—of its production and fulfillment. Models for such an account might, for example, be Hegel's philosophy of history and Husserl's account of the *Origin of Geometry*. History is the unbroken transmission and development of meaning. Contingency, plurality, death, breaks, circles, regressions are all to be appropriated within a wider continuity. Progress, truth, wisdom, freedom, and so forth are all names for what history is or can be made to generate. Derrida's general interpretive claim is that philosophy has always understood history in this way, and consequently has always interpreted the future in this light. His explanation is that history so understood is the recuperation of *différance*, of what has always already exceeded and threatened presence, as well as making it possible, under the value of presence. Origin and end are equally examples of this value of presence. They always function as textual legitimations of one sort or another. Once one accepts that history so understood exemplifies the structure of presence, Derrida then argues that the value of presence is not fundamental after all, but "constituted" by the very differences it seeks to appropriate. The dependence of a sign's identity on its difference from other signs is given the strongest possible reading—a critique of any identity that might be deemed original. Where Hegel had criticized Schelling for the unmediated nature of his account of Absolute Self-Identity, for Derrida it is precisely this principle of mediation, by which time and history are reduced to serving the principle of identity, that is criticized.

Derrida's position has much in common here with Lyotard's critique of "grands récits,"[21] which in a historical context would take the shape of all-embracing narrative structures, drawing every event in history into a single story. Derrida's caution, then, is toward the subjection of the future to metaphysical ends. What is important is that he sees this subjection as almost universal. It is not *just* the mistake of isolated metaphysical excess or a lapse of judgment. Moreover, as will be seen, any attempt to break out of it is

threatened by reappropriation. Before coming to that, however, I would like
to fill in some of the complexities of Derrida's position.

I begin with the reflexive problem lodged within the idea of transgress-
ing "history"—that one does not know how to describe the move itself. In his
"Exergue" to *Of Grammatology*, Derrida is writing about science's emerging
unease with phonocentric writing.[22] This has always been there,

> but today something lets it appear as such, allows it a kind of takeover
> without our being able to translate this novelty into clear-cut notions of
> mutation, explicitation, accumulation, revolution or tradition. These
> values belong no doubt to the system whose dislocation is today pre-
> sented as such, they describe the styles of an historical movement which
> was meaningful—like the concept of history itself—only within a lo-
> gocentric epoch.[23]

This paradoxical position calls for what Derrida describes as a double
strategy, which will shortly be outlined.

Derrida's second ground for caution about the future lies, I believe, in
his fascination with invaginated topologies. Formally speaking, what is
important about such topologies is that their outside, or part of their outside,
is also inside, or alternatively, that at certain points the distinction between
outside and inside becomes problematic, undecidable. These structures have
all sorts of physical and mathematical exemplars of varying complexity, from
the rubber ball with a deep dimple pressed into it to Klein bottles. It is not
perhaps insignificant that all the erogenous zones of the body (and certainly
the typical points of fixation, according to Freud) are just such structures,
points at which pleasure and discipline mingle. As models, they allow
Derrida to displace the metaphysical paradigms of circle and line, which
maintain the strictness of separation of inside and outside, or unity of
direction.

But how do such models get textually articulated in a way that bears on
the future? Derrida writes that there is no possibility of doing without the
concepts of metaphysics in the attempt to "escape" from metaphysics.[24]
Theoretically speaking, these concepts, with all the dangers of recuperation
and reappropriation they bring with them, are the only tools there are. There
is no choice but to use them. The way out, the way of going beyond
metaphysics, involves a strategy of displacement *within* the metaphysical
text. The possibility of something other than metaphysics lies within, the
outside inside, the future in a mutation of the present. Here, it should be
clear, references to "the future" are to be read as references to the textual
work of disentangling itself from metaphysics that poststructuralism would
involve. This topology of invagination that puts in question the distinction
between inside and outside, what belongs and what does not, plays a more

general critical role in the form of the structures of supplementarity, but this cannot be developed here.

It is important now to realize that this search for a beyond within, which could be called immanent deconstruction, is treated by Derrida as just part of the double strategy he recommends in a number of places. The second strand attempts to stand *outside philosophy*,

> to determine from a certain exterior that is unqualifiable or unnameable by philosophy—what this history has been able to dissimulate or forbid.[25]

He continues:

> By means of this simultaneously faithful and violent circulation between the inside and the outside of philosophy—that is of the West—there is produced a certain textual work that gives great pleasure.[26]

But this second strand offers us no more assured grasp of, projection of, the future than does the first. To the first corresponds Derrida's reference to Nietzsche's suggestion that what we need is a change of style,[27] and that "if there is style, Nietzsche reminds us, it must be plural."[28] To the second corresponds Derrida's reference to

> what is proclaiming itself and which can do so, as is necessary whenever a birth is in the offing, only under the species of the non-species, in the formless, mute, infant, and terrifying form of monstrosity.[29]

Either way, the idea of the future as something representable is radically undermined. The future can be approached through the interweaving of these two strategies, a new kind of "writing."

It is worth saying in passing that Derrida leans heavily on a post- or at least *a*teleological motivational structure, which again displaces the idea of the future as the projected dimension for the achievement of one's goals.

Heidegger wrote about not wanting to "get anywhere," but first of all and for once "to get to where we are already,"[30] and Derrida, who perhaps could not quite swallow that formulation, writes of wanting to "get to a point at which he does not know where he is going."[31] This implied deconstruction of the linearity of ordinary motivational structures undoubtedly owes something to what I have elsewhere called the intensification wrought by Nietzsche's eternal recurrence,[32] and it is again no accident that when Derrida actually does get around to naming the unnamable, or at least offering for its nominal effects the term *"différance"* (which supplants the possibility of a future that would be a future present, by denying that possibility), he refers explicitly to Nietzsche. The term *différance* implies that there will be no unique name, not even the name of Being. This must be conceived without *nostalgia*, that is, it must be conceived outside the myth of the purely paternal or maternal language belonging to the lost fatherland of thought. On the contrary we must *affirm* it in the sense that Nietzsche brings affirmation into play—with a certain laughter and a certain dance.[33]

Translated into our problematic of the future, this means that rather than mourning the demise of the future as the horizon of intelligibility, we should celebrate it.

I have dealt so far with various reasons for caution about the future—the reflexive problem that when one is trying to go beyond the very categories of historical understanding it is hard to see what terms one should use; the use Derrida makes of an invaginated topology, which makes the reference to a "beyond" undecidable; and his account of the need for a double strategy of interpretation and writing, which makes utterly obscure any sense of the possibility of representing the future.

Finally, I would like to return to the question of whether Derrida has not "merely" put a particular concept of time, history, the future, and so forth into question rather than these concepts "themselves." The two distinct answers he gives to this question seem inconsistent. The first one can be found in "*Ousia* and *Gramme*," in which he denies the possibility of a postmetaphysical concept of time.

> Time belongs in all its aspects to metaphysics, and it names the domination of presence . . . In attempting to produce (an) *other* concepts, one rapidly would come to see that it is constructed out of other metaphysical or ontotheological predicates.[34]

But this is perhaps a little hasty in view of the claim we noted before, that there *are no* metaphysical concepts per se, only metaphysical moves or modes of textual articulation. It is perhaps this that, second, allows him to acknowledge in "Grammatology as a Positive Science" the possibility of a "pluri-dimensional and delinearised temporality."[35] And this must be his considered view, because when, in *Positions* (the title interview), he is asked whether he can conceive of the "possibility of a concept of history that would escape . . . [linearity] . . . as a stratified, contradictory series," he makes explicit that it is the "metaphysical concept of history that he is against." What he endorses is the view he attributes to Althusser and to Sollers that criticizes the Hegelian concept of history, the "notion of an expressive totality," and

> aims at showing that there is not one single history, a general history, but rather histories different in their type, rhythm, mode of inscription—intervallic, differentiated histories.[36]

The claim is now not that there is no alternative concept of history, but that it would be a mistake to underestimate the interconnectedness of the concept of history with the rest of the philosophical armory, nor the difficulty of displacing it. He writes:

> The metaphysical character of the concept of history is not only linked to linearity, but to an entire *system* of implications (teleology, eschatology, elevating and interiorizing accumulation of meaning, a certain type of traditionality, a certain concept of truth etc . . .). That being said, the concept of history . . . cannot be subject to a simple and instantaneous mutation, the striking of a name from the vocabulary. We must elaborate a strategy of . . . textual work . . .[37]

His strategy involves constantly reinscribing the term "history" in his own texts with the aim of transforming it.

It might seem reassuring, having been forced to give up History, to be given back lots of little histories, and by analogy, lots of little futures in place of the Future, but there are still one or two further questions. First, how possible is it to shake off metaphysics, or even to try to do so, just by going plural? Second, is there not a sense in which Derrida is offering two different levels of description, one that competes with grand teleological history at its own level, and the other that would have more obvious empirical applications? Third, how does this pluralism of histories affect my reading of Derrida's understanding of the future of poststructuralism?

First, my doubts about pluralism, I shall develop the argument here with reference to a plurality of histories. The basis of the worry, I take it, is that if each of these "histories" is stamped with logocentrism, then one has merely multiplied the heads on the monster, and not slain or even tamed it. I think that is a misreading. Two points are worth making. (a) These histories would not be, or would not need to be, histories of meaning, constituted by guiding ideas, and so forth. They could be documentations of discursive formations in Foucault's manner, partial histories of particular institutions within a particular society, histories of particular scientific, cultural, agricultural, and so on practices. Any regulative involvement of "intentions," ideas, or concepts would have only a nominal status and be readily dissolved back into the histories themselves. (b) *The very plurality* of these histories, histories that do not require, and would almost certainly resist appropriation by History, is for that very reason a threat to History. Indeed, this would be so even if each microhistory were cast in a teleological mold, for there would be nothing to guarantee their commensurability.

An acceptance that a radical plurality of futures was *in itself* a sign of a postmetaphysical stance would allow one to forge links with a rather different tradition. Consider the words of Ernst Bloch:

The concept of progress . . . requires not unilinearity but a broad, flexible and thoroughly dynamic "multiverse": the voices of history joined in perpetual and often intricate counterpoint. A unilinear model must be found obsolete if justice is to be done to the considerable amount of non-European material. It is no longer possible to work without curves in the series; without a new and complex time-manifold . . . [38]

Of course, the stress must be on the word "radical." The power of pluralism to undermine logocentrism is shown by the considerable role it plays in the work of philosophers such as Lyotard and Rorty, each of whom in his different way offers a new way forward.[39] In both their cases, the pluralism in question is radical in the sense that there is no enveloping unity by which this multiplicity can be contained, and no theoretical place for such a unity.

Second, the suggestion that Derrida is offering two different levels of description. What I mean by this is that he is prepared to talk about *the future*, as if that singular term still made some sense. It appears, as he says, as a monstrosity from the point of view of logocentrism when thought of in terms of radical otherness. The passage beyond metaphysics is a certain sort of textual activity that, in all its duality of strategy, can be straightforwardly described, as I have done. But the future as such cannot. It cannot be represented. And yet is not this talk of pluridimensional time and multiple histories giving us precisely that—representations, a promise of a complex cautious pluralism where once there was just the affirmation of the absence of meaning? If this is so, how should these two levels be ordered? I would suggest that the first be treated as opening up the possibility of the second. And if something of the subtlety of Derrida's cautions about premature exits can be maintained, this is a really promising suggestion. But perhaps this is to move too fast. There is still the next question.

Third, what has been said here about history be transferred to a consideration of the future? Is not the term "history" entanglingly ambiguous? Does not the pluralism Derrida embraces really only make sense if history is thought of as limited to the past? One could hardly have documentations of *future* discursive formations.

The answer to this question is surely that one simply admits, indeed encourages, a plurality of incommensurable and even incompatible projections of the future. But then a second question arises: Is the future perhaps being treated not as a future present but as a future past, by extending to it the idea of a plural history? Is one not, in effect, closing off the future by acknowledging the multiplicity of possible projections on to it? Is not the future potentially the source of the radically other, the unprojectable? This it surely is, but it would seem perverse, at least at some level, to expect that an account of the future should "take account" of that radical otherness, for it is precisely what cannot be taken account of. If anything, it is what the multiplicity of our projections are the vain attempt to anticipate. But even with this question tentatively resolved, there must be a certain unease about reading Derrida as inviting a thousand flowers to bloom. What a naive pluralism forgets is the caution to be attached not just to the idea of a single, universal history, but to its linearity. Plurality could leave linearity untouched. What is still required is an account of the complex subversions of linear order—of delated effects, inversions of order, structures of repetition, substitution, supplementarity, and so forth.

For philosophy, surely, the problem of the future is a reflection of a crisis of *method*. The problem can be outlined in the following general way. Philosophy is a striving after truth, and a striving, as has often been noted, marked by constant failure. Philosophy is a history of disappointment and disillusionment. Every "success" is trampled on by what succeeds it. The future is bright only to the ignorant and innocent, the children in philosophy, those who have not grasped the folly of its ways, those seduced by its promise. What is needed is a *radically new way* in which the ideals of philosophy can once and for all be realized. What is needed is a proper *method*.

With the idea that a proper *method* could achieve what the undisciplined wrangling, inspiration, and muddied thought of the past could not achieve, the future seemed brighter. With Descartes's *Discourse on Method*, philosophy was to be systematically purged of what could not be clearly and distinctly perceived, and it became a slogan of modernity. Kant too turns his back on the mock battles of metaphysics by directing philosophy along "the secure path of a science."[40] And he claims to have found "a way of guarding against all those errors which have hitherto beset reason," and that "there is not a single metaphysical problem that has not been solved."[41] Hegel insists that philosophy cease merely being the *love* of knowledge and achieve knowledge itself. As he put it:

> The systematic development of truth in scientific form can alone be the true shape in which truth exists. To help to bring philosophy nearer to the form of science—that goal where it can lay aside the name of *love* of knowledge, and be actual knowledge—that is what I have set before me.[42]

And mention must be made of Husserl, who would violently object to being harnessed with Hegel in this way—he saw Hegel as "weakening the impulse toward philosophical science"[43] in that it "lacks a critique of reason." (Strangely Kantian words, given that he had only just lamented

Kant's claim that one could not learn philosophy, only to philosophize: "What is that but an admission of philosophy's unscientific character?") Husserl aims to restore philosophy "to its historical purpose [as] the loftiest and most rigorous of all sciences, representing . . . humanity's imperishable demand for pure and absolute knowledge."[44] In each case it is (scientific) *method* that supplies the horizon of the future for philosophy. For it is only by a rigorous reaffirmation and redefinition of philosophical method that metaphysical *desire* (for knowledge, understanding, truth, clarity . . . and, fundamentally, for presence) can be guided in a direction not immediately guaranteed to disappoint. When Kant decries the mystagogues,[45] it is surely because they usurp the role of proper philosophical method in articulating the horizon of the future. Hegel's attack on Schelling is similarly motivated. Philosophical method is a conceptual labor that cannot be replaced by intuition. No royal roads. And if Husserl seems to return to intuition, no one can doubt the *work* of detailed description this involves. It is not the speechless ecstatic revelation of a mystical vision.

And reference to Husserl suggests a rather interesting possibility. If the desire to avoid philosophy's traditional failure is taken as the driving concern of Husserl's philosophy, it suggests, at least initially, the subordination of the value of *archē* to that of telos. Secure foundations are not ends in themselves; rather, they ensure that the building does not collapse when completed. I say "initially," because of course one can, symmetrically, claim that the goal for Husserl is the reestablishing of contact with the "things themselves," intuitive fulfillment, the reactivation of origin. In that case, the phase of subordination of *archē* to telos would be followed by an inversion of that relationship: ends and origins would cycle back into each other.

What this suggests is that the initial cyclic fracturing of linearity we claimed for Nietzsche could be found just as well in Husserl. And the case of phenomenology, and of Husserl in particular, is surely paradigmatic for philosophy. It could be argued that while phenomenology may have limitations, it is still the healthiest embodiment of the ethico-rational ideal, one that has not yet failed for the simple reason that it can be seen to have eventually taken the choice recommended by Lessing—the infinite striving after truth.[46]

If phenomenology cannot be completed, this could be *affirmed* in a positive way rather than lamented. Is not the ideal of perpetual activation and reactivation of the primal sources of intuition one we could embrace? It would draw in the past, deepen the present, and give sense to the future. "The greatest step our age has to make is to recognise that with philosophical intuition . . . a limitless field of work opens out. . . ."[47]

This understanding of Husserl's phenomenology is persuasively presented by Merleau-Ponty: "The most important lesson . . . the reduction teaches us is the impossibility of a complete reduction."[48] And Husserl often refers to himself as a perpetual beginner, and his own constant inventiveness even in respect of major concepts would support this self-assessment. Philosophy is an "infinite meditation," and when "faithful to its intention . . . never knows where it is going."[49]

> The unfinished nature of philosophy and the inchoative atmosphere which surrounded it are not to be taken as a sign of failure. They were inevitable because phenomenology's task was to reveal the mystery of the world and of reason.[50]

I offered earlier a limited defense of a certain slippage between philosophical discussions of the future and of the future of philosophy. So far I have listed a little on the side of the former. But the more reflexive questions about the future of philosophy—briefly alluded to when discussing Derrida's double strategy—can profitably be attended to again. For while philosophers such as Rorty and Lyotard chart both the death of philosophy and, via radical pluralism, its rebirth and reorientation, and while there is a real chastening of philosophy's commonly totalizing pretensions, radical pluralism leaves *linearity* untouched. However, with the kind of discussion of the end or closure of philosophy inaugurated by Nietzsche, and carried through by Heidegger and Derrida, this whole question of linearity is put to the test.[51]

Nietzsche offered us both synchronic and diachronic versions of the closure thesis: the synchronic in section 20 of *Beyond Good and Evil* and the diachronic in "How the Real World at Last Became a Myth" in *Twilight of the Idols*. There is a mutation of linearity in both senses of "circle" involved in these accounts—the circle as a space of limitation or enclosure, and the circle as a movement of return. But the most radical break with linearity comes (see Part 1) in his discussion of the eternal recurrence and the possibility of its *affirmation*. With this affirmation, it could be said that one "opens oneself" to the future without reservation. And such opening is no longer captured in linear terms.[52]

A full discussion of the question of the end(s) of philosophy in Heidegger and Derrida is undoubtedly called for here, but the call will have to remain unanswered. Other thinkers have already applied themselves most satisfactorily to the job.[53] Important distinctions have to be repeatedly emphasized, and further analytical clarification and distinction between different versions of this thesis would not be out of order. It would be necessary to distinguish end as exhaustion of possibilities, as extreme focusing of possibility, as termination of activity, as fulfillment, as goal, and so on.

What it is important to extract from this discussion is the double inscription of what has been called "caution" about the future, one response among many to crisis. On the one hand, the announcement of the closure of philosophy (as metaphysics, as logocentrism) declares a break within the history of thought through which no continuous thread may pass. What the future offers (a change of style, the step beyond, "writing," "thinking," and so forth) lies beyond our current projective base, and it is quite the opposite of a fulfillment or completion of the present. But in addition—and here there is a doubling back—the place of the future is an issue fundamental to determining as distinct the content of what precedes and succeeds the rupture. For traditional philosophy, the future is an *essential* category (ethically and/or hermeneutically, as described earlier). For poststructuralism, it may not be. The question is—is it merely not *essential*, is it excluded altogether, or does the future reappear in some mutation or other? What future has the future?

I have tried to develop the idea that the crisis in philosophy's future (and hence its very nature) has been seen, certainly by a string of the most important thinkers in modern philosophy, to be coextensive with a crisis in method, for which a redefinition of its status as a "science" is in each case required. I have suggested that while Husserl in one respect clearly belongs to this series, there is in his work a positive recognition of the impossibility of finality, of completing the task of philosophy. Where would this leave poststructuralism? And where, more particularly, would it leave Derrida? Does this eminently reasonable restatement of the idea of philosophy as a perennial activity not deflate the very project of deconstructing the future?

Let us be clear that Derridean deconstruction is heavily committed to a transformation of our ways of thinking about the future. His remark that "the future can only be anticipated in the form of absolute danger"—which is at the very least a radical discontinuity thesis—has already been quoted. And while it always forms part of a double strategy, and so is coupled with immanent deconstruction, Derrida constantly refers to the need to take the step beyond—beyond "metaphysics," "beyond man and humanism," beyond presence, beyond security, beyond the language of Being.

Such an alternative philosophical(?) practice can presumably claim to be radically different from everything hitherto known as philosophy. I have my doubts about this. But a question arises here that will help us clarify just what kind of transformation of our ordinary view of the future is being proposed. Leaving aside the radicality of the destination, surely the nature of the transition, and of the language in which it is announced, is very far from being radical. Indeed, it is very traditional. The name for it is apocalyptic discourse.

On the last day of the 1980 conference at Cérisy, Derrida presented a paper on Kant's "lampoon," "Von einem neuerdings erhobenen Vornehmen Ton in der Philosophie" (1796) ("On a Genteel Tone Recently Sounded in Philosophy"), in the title of which the word "Apocalyptic" replaces

"Genteel."[54] Kant had taken to task those mystagogic pseudophilosophers "who by the tone they take and the air they give themselves when saying certain things, place philosophy in danger of death, and tell philosophy, or philosophers the imminence of their end." Derrida is not only fascinated by the categories Kant uses to judge these philosophers, the fact that it is their *tone* that Kant is worried about, and the whole significance of Apocalypse as Revelation, he is clearly touched by the fact that Kant's essay can be seen to apply to him. This is not the place for a detailed analysis of Derrida's paper, but here is a brief account of part of the argument.

For Kant, philosophy's basic commitment is to "Wissenschaftlichen Lebensweisheit"—the ethico-rational life. The mystagogues borrow the word "philosophy," but not its significance. They claim "an immediate and intuitive relation with the mystery," they are organized in tiny exclusive sects, use coded language, and set themselves apart as superior beings. They prefer intuition to concepts, "gift" to work, and value genius above scholarship. Clearly they have a total disdain for Kant's commitment to rational inquiry, and indeed for its democratic implications.

Derrida begins by suggesting in analytic fashion that we treat apocalyptic discourse as essentially predictive eschatological discourse, discourse about the end or ends, or the limit (. . . of the world, of time, of philosophy). He then makes three moves. The first, specific one is to argue that Kant himself produces just such a discourse in his "marking a limit, indeed the end of a certain type of metaphysics."[55] His second move is one of generalization—to argue in various ways that eschatological discourse is unavoidable. It is clear that he means to encompass himself (and Heidegger) in these remarks:

> . . . whoever would come to refine, to tell . . . the end of the end, the end of the ends, that the end has always already begun, that we must distinguish between closure and end, that person would, whether wanting to or not, participate in the concert.[56]

A few pages later Derrida repeats this argument, even to the point of allowing its force to override the ironic distanced tone he believes is the most he has ever allowed himself when making his apocalyptic pronouncements.[57] Here it seems neither mental reservations nor a change of style will do.

The importance of *this* claim is that it supports the view that Derrida is taking up a radical attitude to the future, and it rules out the response that his references to a beyond, to the future as absolute danger, should be taken with a pinch of French salt. The argument, however, and perhaps also his "tu quoque" response to Kant, is surely a little misleading, for it would obscure the difference between the Revelation of John of Patmos and the claim that philosophy, or a certain Western tradition of philosophy, has exhausted its possibilities. Derrida, like Lacan, has little time for the distinction between use and mention, or between announcing the end and charting its limits. Clearly both are concerned with marking limits, but to identify the two would conflate apocalyptic and eschatological discourse completely. And when one talks of end as *closure*, the distinctively *predictive* element of apocalyptic discourse makes no sense, for "closure" does not have chronological consequences in the ordinary sense at all.

But if that is so, surely it would be wrong to try to use the "end of philosophy" debate as the basis of a claim that these various thinkers (Nietzsche, Heidegger, Derrida) are transforming or deconstructing[58] the future. The point, however, is that the substitution of considerations of exhaustion, possibility, repetition, closure, and so forth for those of philosophical progress *is* in itself a displacement of philosophy's understanding of its own future. It could be said that the ordinary *temporal* future has lost its dominance.

It might be replied that the *ordinary temporal future* never has dominated philosophy. Certainly, teleological projection is far more than an extension of a series of now-points beyond the present. That is quite right. What is being claimed is that teleology nonetheless represents the preservation of the values of self-identity, of fulfillment, of . . . "presence." After Nietzsche, Heidegger, and Derrida, the projective scene can be redescribed at two different levels. (1) Philosophy may well carry on as always, blind to its own limits, like a fly in a fly bottle, perhaps. (2) A radically distinct possibility (of "thinking," of "writing") begins to emerge. This possibility, it may be said, surely gives us hope (indeed, Derrida even goes so far as to endorse what he calls "Heideggerean hope" on one occasion![59]) and thus revives the "ethical" importance of the future! This possibility will reappear at the end of the chapter.

I will turn shortly to a considered interpretation of this stereoscopic, or bifocal projection, but first let us consider the third move Derrida makes in the Apocalypse essay. It is a further *generalization* of the invaginated reversal by which apocalyptic discourse is no longer seen outside, to be excluded. He asks (in other words, he suggests), "wouldn't the apocalyptic be a transcendental condition for all discourse, of all experience itself, of every mark or trace?"[60]

I have elsewhere had strong words to say about Derrida's reliance on such transcendental manoeuvres.[61] If his general antifoundationalism is accepted, it has no force at all: the "sous rature" is no portmanteau protection. But this *particular* argument seems to me to commit a simple logical fallacy. The argument essentially is that apocalyptic discourse involves an indeterminacy of authorial voice, and of its audience, and that this indeterminacy, if only we were to realize it, is the condition for the possibility of any signification whatsoever. But even if one accepts that this generalized "*écriture*" could be so characterized, it simply does not follow from the fact that the apocalyptic has these properties (among others) that it is the condition of possibility of discourse in general, which also has these properties. Structural corrrespondences have no transcendental consequences at all, and if by "the apocalyptic" Derrida means "apocalyptic discourse," he would have to be claiming that a particular discourse was a transcendental condition for discourse in general, which seems a priori implausible. If all discourse did share this property (authorial/audience indeterminacy), then nothing would privilege the way in which apocalyptic discourses embodies it. And quite apart from this, it is hard to see how one discourse can have a transcendental relationship to other discourses, unless one has redefined the relationship as some sort of mapping, or translatability.

Derrida's difficulty here is that such displacements and reversals are often productively provocative, but that very success can breed routinization and even recklessness.

That said, there is still something of interest in this last move. I believe it is all to be found in the last pages of his paper, in which he offers us a "meditation" on the word "viens" (come). With a commentary on this, the discussion will be drawn to a close.

Allow me, for the sake of economy of orientation, merely to *indicate* some of the lines of thought *condensed* in this word. First, the apocalyptic gesture could be said to involve a general invitation to the reader, to the listener, to "come," to follow, to open himself/herself to the light. Second, the word "come" (*viens*) in either English or French is immediately linked to the language of the future (*avenir* itself, and many phrases involving *venir*; in English, consider "the time to come," "coming events," and so on). Third, Derrida is clearly using this word, much as Heidegger uses the words "ereignis" and "es gibt," to mark a site at which a new kind of thinking might begin to crystallize.[62] And here, I claim, there is a clue at least to how the future might still have a role—indeed, the central role—after philosophy's closure.

As I understand Derrida, the apocalyptic "viens" (come) has the force of a primordial *event* in a Heideggerean sense, with a fundamental ethical flavor superimposed on it (in response perhaps to Levinas). It could be seen as an attempt to bring about a confluence of Heidegger and Levinas. At this point, yet another paper could begin, but I will try to make brief sense of these obscure allusions.

How can an *event* be fundamental? Every event, as such, marks a rupture with the past, has its own temporal integrity. Some events can be thought of as historically (or biographically) momentous. They open a whole world of possibilities. Such an idea could be used to understand the activity of philosophy itself—as an opening on to, the opening up of possibilities of thinking, indeed, of living.

Philosophy so understood could be called the "fundamental event." But so defined, it has the most intimate bond imaginable with the future. The common trait of those views of the future (and of philosophy's future) that deconstruction sets its sights on is their embodiment of the value of presence, usually via some anticipatory representation of What Is to Come. For fear of prejudging the issue, no positive alternative has been offered. But the account given of the "fundamental event" surely supplies one. The future "is" fundamentally an event—the opening of possibility. As such, it is, literally speaking, a continuous Apocalypse, or revelation. The future opens on to what is other.[63]

Something of the force of these remarks comes from the impersonality of the idea of event being employed here. The same can be said, for example, of Heidegger's "es gibt." One cannot ask who gives or what is given. Once again, the attempt is being made to demonstrate a space, an *opening* and indeed to provide one. But Derrida's employment of "viens" is not simply impersonal. In the apocalyptic texts in which the "come" is uttered, there is, to be sure, a movement away from the personal.[64] But equally, the word "come" has an essentially interpersonal significance that Heidegger does not to my knowledge suggest for his focal words.[65] There is little doubt that what Derrida is doing here is trying to bring Levinas into the picture for the latter's insistence that the ethical relation have primacy over the ontological.[66] For Levinas, the relation to the Other (the other person) is the scene of the opening on to the infinite. And in the discontinuity, the otherness of the Other, can be found the basis of the discontinuity of time. Moreover, in the gift of pardon the Other has the power to heal the past. . .

What Derrida is trying to do is to capture a postmetaphysical sense of the future that does not entirely abandon the ethical dimension (that was so prominent at the beginning of the chapter) but rejoins it at a deeper level.

This may be thought to be simply a fascinating topic for philosophical discussion, but it is clearly *more* than this. It is meant, I believe, to offer some sort of basic insight, if I can use that word, into how one might come to think of philosophy or of its successor. It suggests that it will rest ultimately on the invitation that we each extend to the other to open ourselves to the adventure (ad-venture) of thinking, without the prospect of completion, and without aiming at some prescribed destination.

Philosophy would be something of a *double* event: an invitation to the Other, a vulnerable opening of oneself to the Other, on the one hand, and a thinking about the open, and about opening itself, that is, about the space and time within which philosophical thinking is at all possible, on the other.[67]

An acknowledgment of the event character of philosophy so understood is of fundamental importance. For it provides both a principle of selectivity in reading the history of philosophy—texts vary in their grasp of the space of possibility in which they work—and it would tend to protect those texts against reductive readings. (One should always ask what they are *doing*, not just what they *say*.) It is this—at least in principle—that prevents the history of philosophy from being totalized prematurely. And interestingly, it suggests a way in which a deconstructive history must be supplemented by a parallel history of philosophical openings, of philosophy as an opening. Historical retrospection easily suffers from the illusions of objectification. One takes for granted the space or spaces that the texts one studies were responsible for opening up. But one could instead focus on the possibilities they foreclosed. This would be a rather different suggestion for a double strategy of reading.

In conclusion, I would like to return to the point at which it was asked rhetorically why Husserl's account of the future of philosophy, formulated as a ceaseless activity of exploration, of reactivation of primordial experience, was not an adequate answer to (post-)modern doubts about philosophy as a progressing or completable enterprise. Do not Husserl and Derrida's projects represent the same mutation of that future-projecting metaphysical desire?

Derrida is certainly more the child of phenomenology than many care to admit, and the extent of his debt to Husserl's spiritual fecundity is easily underestimated. Perhaps I could conclude by formulating the result of this discussion in the shape of a question: Is it possible that Husserl's logocentric concern with the primordial, with origins (and indeed Heidegger's increasingly precarious commitment to the question of Being), could each be treated as strategies subordinated to a conception of the activity of philosophy—as a sustained opening (up) of the space of thinking—that Derrida (only) reopens, albeit on new terrain? Where exactly is the difference? And what would come of such a thought?

But these questions I must leave for another day, another time.

NOTES

INTRODUCTION

1. Jacques Derrida, "*Ousia* et *Gramme*," in his *Marges de la philosophie* (Paris: Minuit, 1972), p. 23 ("*Ousia* and *Gramme*," in his *Margins of Philosophy*, trans. Alan Bass [Chicago: University of Chicago Press, 1982], p. 63).
2. This view of metaphysics and its determination of the history of philosophy, shared in broad outline by Derrida and Heidegger, will be discussed in a critical way later.
3. It is only for the sake of simplicity of presentation that we assume for the moment that there is only one concept of time.
4. The precise sense of "critical" here will emerge later. It is closer to the Kantian sense of "critique" than to any negative sense of criticism, but it cannot be identified with the Kantian sense.
5. See Jacques Derrida, "De la grammatologie comme science positive," in *De la grammatologie* (Paris: Minuit, 1967), p. 130 ("Of Grammatology as a Positive Science," in *Of Grammatology*, trans. Gayatri C. Spivak [Baltimore: Johns Hopkins University Press, 1976], p. 87.
6. Edmund Husserl, *Die Krisis der europäischen Wissenschaften und die transzendentale Phänomenologie* (The Hague: Nijhoff, 1962 [actually written in the mid-1930s] (*The Crisis of European Sciences and Transcendental Phenomenology*, trans. David Carr [Evanston: Northwestern University Press, 1970]).
7. Martin Heidegger, "The Task of Destroying the History of Ontology," section 6, *Being and Time*, trans. John Macquarrie and Edward Robinson (Oxford: Blackwell, 1962).

PART 1

1. See J. L. Austin, "A Plea for Excuses," in his *Philosophical Papers* (Oxford: Oxford University Press, 1961).
2. Nietzsche uses two different expressions: return (*Wiederkunft*) and recurrence (*Wiederkehr*). These two expressions could be used to mark a strict distinction between the recurrence of *events* and the return of *people* or things. But no such systematic usage is found in Nietzsche, and we have not attempted to impose one on his text. Joan Stambaugh, whose *Nietzsche's Thought of Eternal Return* (Baltimore: Johns Hopkins University Press, 1972) is one of the best things written on the subject, further points out that in his critical passages Nietzsche usually uses the expression *Wiederkunft* and hardly ever talks of *Wiederholung* (repetition) (pp. 29–31), which again suggests that exact reruns were not part of his favored version of eternal return.

3. See below, pp. 276–7.
4. See, for example, "The Intoxicated Song" ("Das trunkene Lied"), the penultimate section of the fourth and last part of *Thus Spake Zarathustra*, trans. R. J. Hollingdale (Harmondsworth: Penguin, 1961).
5. A selection of Nietzsche's unpublished writings, his *Nachlass*, was assembled under the title of *Der Wille zur Macht* (1901). Quotes in this chapter are from *The Will to Power*, trans. R. J. Hollingdale and Walter Kaufman (New York: Vintage, 1968). The section containing the most important attempt at a scientific proof is section 1066.
6. An excellent paper by Robin Small, "Three Interpretations of Eternal Recurrence," *Dialogue* 22 (1983): pp. 91–112, makes this point more systematically.
7. The title for this chapter was to have been "Nietzsche's *Deconstruction* of Time." I was persuaded that this was too loose a use of the term. But there are parallels with even the technical account Derrida gives of the general strategy of deconstruction (in his *Positions*, trans. Alan Bass [Chicago: University of Chicago Press, 1981]). I will suggest that Nietzsche reinscribes "becoming" in a way parallel to the way Derrida reinscribes "writing." I claim, too, that the concept of eternal recurrence is undecidable in terms of the framework it puts in question.
8. Nietzsche, *Will to Power*, section 1066.
9. Arthur Danto, "The Eternal Recurrence," in *Nietzsche: A Collection of Critical Essays*, ed. Robert Solomon (Garden City, N.Y.: Doubleday, 1973).
10. The letter is dated March 8, 1884, and is in Stambaugh, *Nietzsche's Thought*, p. 87.
11. It is quite true that Nietzsche was strongly and positively influencd by F. A. Lange's *Geschichte des Materialismus . . .* (1863) (see Ronald Hayman, *Nietzsche: A Critical Life* [London: Quartet, 1980]: pp. 82), but we prefer to treat such positivistic and scientific streaks as there are in Nietzsche as weapons in an antimetaphysical struggle rather than as beliefs strongly held in their own right. Causal determinism, for example, would be hard to square with his account of causation in "The Four Great Errors," in Friedrich Nietzsche, *Twilight of the Idols*, trans. R.J. Hollingdale (Harmondsworth: Penguin, 1968). Where Nietzsche's rhetorical and scientific tendencies clash, we favor the former.
12. Nietzsche, *Will to Power*, section 1062.
13. Ibid.
14. Ibid., section 1066.
15. And the significance of the horizontal axis—the serial order of time—is itself compromised by the addition of the second. A parallel to Nietzsche's construction of an account that is deconstructive in its effects, by a simple modification of seriality, can be found in the Moebius strip, beloved of Lacan, in which the absolute difference between the two sides of a ribbon is transformed into a continuity merely by a twist and a join.
16. Nietzsche, *Zarathustra*, p. 288.
17. See Nietzsche, *Twilight of the Idols*, section 9. See also Martin Heidegger, *Nietzsche*, vol. 1: *The Will to Power as Art*, trans. David Farrell Krell (New York: Harper & Row, 1979), section 14, "Rapture as Aesthetic State."
18. Ibid.
19. Friedrich Nietzsche, *The Gay Science*, trans. Walter Kaufmann (New York: Vintage, 1974).
20. The references here (*Nachlass*, chapter 14, p. 306, and chapter 12, p. 371) are taken from Stambaugh, *Nietzsche's Thought*, p. 23.
21. In this word "intensity" we should hear the work of condensation—primarily of

the ideas of *tension* and concentrated focus, a felt intensity—ideas that both inhabit and displace a psychological interpretation insofar as they suggest all sorts of difficulties with any traditionally substantive account of the "subject."

22. "The eternal is present with us in every moment; the transitoriness of time causes us no suffering,"quoted by Karl Löwith in his *From Hegel to Nietzsche*, trans. David Green (Garden City, N.Y.: Doubleday, 1967), p. 211.

23. Pierre Klossowski, *Nietzsche et le cercle vicieux* (Paris: Mercure de France, 1969).

24. For Nietzsche, Heidegger writes, and clearly with Schelling in mind, "Will is primal being. The highest product of primal being is eternity. The primal being of beings is the will, as the eternally recurrent willing of the eternal recurrence of the same. The eternal recurrence of the same is the supreme triumph of the metaphysics of the will that eternally wills its own willing." Martin Heidegger, *What Is Called Thinking?*, trans. Fred D. Wieck and J. Glenn Gray (New York: Harper & Row, 1972), Lecture 10, p. 104.

25. Nietzsche, *Thus Spake Zarathustra*, "Of Redemption," p. 161.

26. Ibid.

27. This problem is plausibly represented by Vincent Descombes, in his *Modern French Philosophy* (Cambridge: Cambridge University Press, 1980), as *the* problem inherited by the "désirants"—the French philosophers who took up Nietzsche's problems in the 1970s, especially Deleuze, Lyotard, and Klossowski. Descombes's book is more revealingly titled in French: *Le Même et l'Autre*.

28. Nietzsche, *Thus Spake Zarathustra*, "Of the Vision and the Riddle," p. 178.

29. Nietzsche, *Twilight of the Idols*, "Maxims and Arrows," p.8.

30. I am grateful, again, to Stambaugh's *Nietzsche's Thought* for this quotation from Friedrich Nietzsche, *The Wanderer and His Shadow*.

31. Vincent Descombes argues convincingly that this problem haunts French Nietzscheans and that they do not escape its grip.

32. We could find here a parallel in Heidegger's account of authenticity and his contrast between finding oneself and forever losing oneself.

33. Gilles Deleuze, *Nietzsche et la philosophie* (Paris: Presses Universitaires de France, 1962) (*Nietzsche and Philosophy*, trans. Hugh Tomlinson [London: Athlone, 1983] pp. 71–72).

34. Nietzsche, *Will to Power*, section 617.

35. Deleuze, *Nietzsche and Philosophy*, p. 72.

36. Nietzsche, *Thus Spoke Zarathustra*, p. 179.

37. The *locus classicus* of this view is probably Roman Jakobson's (with Morris Halle) *The Fundamentals of Language*, in particular "Two Aspects of Language: Metonymy and Metaphor" (The Hague: Mouton, 1956).

   Jacques Lacan puts it well: "There is in effect no signifying chain that does not have, as if attached to the punctuation of each of its units, a whole articulation of relevant contexts suspended 'vertically,' as it were, from that point." *Écrits* (1966), trans. Alan Sheridan (London: Tavistock, 1977), p. 154.

38. G. W. F. Hegel, *Phenomenology of Spirit*, trans. A. V. Miller (Oxford: Clarendon, 1977), p. 152.

39. See "The End of the Book and the Beginning of Writing" and "Linguistics and Grammatology," in Derrida, *Of Grammatology*.

40. The non- (or post-) dialectical possibilities of Nietzsche's thought rest on such a relation.

41. See n. 17.

42. I allude here to John Sallis, "Dionysus in Excess of Metaphysics," in *Exceedingly*

Nietzsche, ed. David Farrell Krell and David Wood (London/New York: Routledge, 1988).

43. See Deleuze, *Nietzsche, and Philosophy*, p. 48.
44. Ibid.
45. Compare Kierkegaard's position here, when he discusses the self-relatedness of the Self, in *The Sickness Unto Death*.
46. Heidegger, *What Is Called Thinking?*, p. 104.
47. Deleuze, *Nietzsche and Philosophy*, p. 48.
48. Ibid.
49. Ibid.
50. Ibid.
51. Derrida, "Différance," *Margins of Philosophy*, p. 17.
52. Heidegger, *Nietzsche*, vol. 1: *The Will to Power as Art*, p. 19.
53. Ibid.
54. Ibid., p. 20.
55. Heidegger, *What Is Called Thinking?*, p. 101.
56. I have David Krell's remarks, in discussion, to thank for this point.
57. Heidegger, *Being and Time*, H264.
58. Ibid.
59. There are many of these: "Way Back into the Fundamental Ground of Metaphysics," "Letter on Humanism," *Time and Being*, and so forth.

PART 2, CHAPTER 1

1. This is in large part closely matched by Husserl's "lebendige gegenwart" (living present), and was illuminated as a value central to the history of metaphysics (that is, Western philosophy) by Heidegger (variously as Praesenz, Anwesenheit, and Gegenwart), and then again by Derrida.

   The drying up of the wellsprings of philosophical thought was itself only a symptom of a culture-wide crisis for which phenomenology was to be the cure. Husserl suggests this evangelical point of view at both ends of his career. See "Philosophy as a Rigorous Science" (1911), in his *Phenomenology and the Crisis of Philosophy*, trans. Quentin Lauer (New York: Harper & Row, 1965), and his *The Crisis of European Sciences and Transcendental Phenomenology* (1936) (hereinafter *Crisis*).

2. Although Heidegger made this distinction *eigentlich/uneigentlich*) central to his *Being and Time*, it was important to Husserl too: "since authentic experience, i.e. the intuitive and ultimately adequate, provides the standard for the evaluation of experience, the phenomenology of "authentic" experience is especially required." Edmund Husserl, *Phenomenology of Internal Time-Consciousness*, trans. James. S. Churchill (Bloomington: Indiana University Press, 1964), p. 28 (hereinafter *PITC*.)

3. Husserl, *PITC*, p. 28.

4. The stabilization of meaning through language is an important theme for Husserl (see, for example, his " The Origin of Geometry," section 66, and our discussion in the next chapter), but this is a regulative rather than a creative or productive phenomenon, as it would be for one such as Nietzsche. See Edmund Husserl, "The Origin of Geometry," appendix 6 of his *Crisis*.

5. J.N. Mohanty, in *Husserl and Frege* (Bloomington: Indiana University Press,

1982), convincingly disputes Dagfinn Føllesdal's commonly accepted account of the decisiveness of Frege's review. See, for example, p. 13.

6. A version of such a reduction can be found in Yehoshua Bar-Hillel, "Husserl's Conception of a Purely Logical Grammar," *Philosophy and Phenomenological Research* 17 (1956–57): pp.362–69.

7. For an excellent account of the difference, which amounts to a defense of Husserl's solution, see Dallas Willard, "Logical Psychologism: Husserl's Way Out," *American Philosophical Quarterly* 9, no. 1 (January 1972): pp. 94–100.

8. Richard Rorty's *Philosophy and the Mirror of Nature* (Oxford: Blackwell, 1980) offers a now-classic version of such a general antifoundationalism.

9. See, for example, Ross Harrison, "The Concept of Prepredicative Experience," in *Philosophy and Phenomenological Understanding*, ed. Edo Pivcevic (New York: Columbia University Press, 1975).

10. This statement is meant to mark a position in what has become known as the "end of metaphysics" debate that has arisen from the work of Nietzsche, Heidegger, and Derrida, and was much in the air in the late 1960s. Are such interventions really of any avail? They can be confusing. Compare, for example, Kant's discussion (B21–23, *Critique of Pure Reason*) of how metaphysics *as science* is possible, where he can be found castigating, as *dogmatic* metaphysics, a use of reason that transcends the limits of its proper application. I am generalizing that ideal of adequacy, while jettisoning the ideal of philosophy (or metaphysics) as a science. My suggestion that "metaphysics" should be understood as blindness to the question of limits rather than as having recourse to the value of "presence" has the merit of releasing *experience* from the demand that it perform an impossible (foundational) task all by itself. "Presence" is indeed an illusion if seen as an absolute origin or foundation, but that experience can seem, both at the time and on reflection, to have this value, that "presence" is phenomenally real, is not an illusion but a fact. What one has to assess are the *limits* of its significance where once none were even suspected.

11. It might be thought that Husserl's emphasis on the descriptive aspect of phenomenology would make the reference to this *theory* of time inappropriate. My reply would be fourfold. (1) While he does always write of his *analysis* of time-consciousness, when he discusses Brentano's account of time, he uses the words *analysis* and *theory* interchangeably. (2) He is quite happy writing of his *theory* of noetic/noematic structures at a point at which phenomenological description is still to the fore. (See Edmund Husserl, *Ideas*, trans. W. R. Boyce Gibson [New York: Humanities Press, 1931]: section 3, chapter 4.) (3) If we were to conclude that Husserl is not offering a theory of time, it could only be because the term "theory" was inapplicable in principle to an account of time, not that Husserl had not gone that far. Certainly none of the other "interesting" ways in which one might study time that he enumerates (*PITC*, section 1, "The Exclusion of Objective Time," pp. 22–23) could *more* suitably wear the mantle "theory of time." (4) Finally, there is an important and straightforward sense in which his "analysis of time-consciousness" is, or is part of, a "theory of time" in that he wants to argue that our grasp of the "objectivity of time" is predicated on internal time-consciousness.

12. Of course, these alternatives might be quite distinct. "Intuition" could be valued as hypothesis-generating procedure, without any status as an epistemological foundation. Yet interestingly, Popper, to whom one might attribute such a view, does speak of the continuing need to "clarify" the "foundations" (of differential

and integral calculus). See Karl Popper, *Conjectures and Refutations* (London: Routledge and Kegan Paul, 1972), p. 70.

13. Precisely here Heidegger can be seen to have diverged from Husserl. For Heidegger, this is no mere formal or even methodological circle, but one with a vital ontological dimension, in which the *preliminary* considerations are seen as both *rooted* in our existence and *foreshadowing* the later reflexive elaboration. See Martin Heidegger, *Being and Time*, section 68f, and the discussion of phenomenological method in the introduction, especially section 7 (c).

14. It would perhaps be stretching a point to describe Derrida as being to Husserl what Gödel was to Hilbert. And yet . . .

15. In "Philosophy as a Rigorous Science," Husserl attributes the confusion in mathematics and physics in the early twentieth century to a failure adequately (that is, intuitively) to ground their basic concepts. What he was clearly not prepared to accept is that it was precisely the status of such intuitive foundations that was in crisis.

16. See n. 10.

17. Husserl, *Ideas*, part IV, chapter 2, section 136f.

18. After analysis into simple truths, "what I have to do is to run over them all repeatedly in my mind, until I pass so quickly from the first to the last that practically no step is left to the memory, and *I seem to view the whole all at the same time*." See René Descartes, *Rules for the Direction of the Mind*, Rule XI (in *The Philosophical Works of Descartes*, vol. 1, trans. Elizabeth S. Haldane and G.R.T. Ross [Cambridge: Cambridge University Press, 1970]).

19. This involves transforming "consciousness" from an adjectival form qualifying a particular type of worldly existence to being an independent field of investigation with only contingent worldly embodiment. Husserl's precise position on the autonomy of consciousness underwent certain shifts as his thought developed. The following three positions stand out of a more complex picture:

    1. *Logical Investigations* (and the first decade of the twentieth century)

        The shift to describing the formations of consciousness could be described as purely methodological, without commitment to the existential autonomy of consciousness.

    2. *Ideas* (and after)

        Husserl writes of the realm or sphere of transcendental subjectivity as logically prior to any positing of the existence of an "objective" world. See, for example, *Ideas*, section 46 (p. 145: "I myself or my experience in its actuality am *absolute* Reality, given through a positing that is unconditioned and simply indissoluble," and section 49 (p. 151): "the Being of consciousness, of every stream of experience generally, though it would indeed be inevitably modified by a nullifying of the thing-world, would not be affected thereby in its own proper existence."

    3. *Crisis of European Sciences and Transcendental Phenomenology* (1936)

        Husserl attempts to correct the worldlessness of the "reduced" ego in *Ideas*, introducing the idea of the lifeworld as the complex worldly involvement that all our philosophical activity takes for granted. See, for example, *Crisis*, section 44ff.

20. " . . . no direct experience can ever deceive the understanding if it restricts its attention to the object presented to it, just as it is given to it either at first-hand or by means of an image; and if it moreover refrains from judging . . . for in . . . judgements we are exposed to error." Descartes, *Rules*, p. 44.

21. A science of necessary truths, established with certainty and clarity. Kant had used the term before Husserl.
22. It might be thought that echoes, or at least parallels, to Wittgenstein's argument against the possibility of a logically private language are to be found here. But, perhaps oddly, I think Husserl would not disagree with Wittgenstein. The intuitive adequacy my words ideally have to me is a private but not idiosyncratic relationship. Such privacy is a condition of significance logically generalizable to all other language users.
23. Husserl's solution, much discussed by Derrida, is to separate two forms of such "clothing"—*indication* and *expression*. In the latter, such problems of adequacy are supposed to disappear. Derrida claims *indication* is ineliminable. See the discussion in ch.xxx.

PART 2, CHAPTER 2

1. Edmund Husserl, *PITC*.
2. G. F. W. Hegel, *The Phenomenology of Spirit*, Introduction.
3. Even though this is not the place for a discussion of the varieties of hermeneutic theory, it ought to be said that even within "hermeneutics," there are considerable differences of opinion as to its scope and its basis. Broadly speaking, there are those for whom hermeneutics is a supplementary method of "scientific" inquiry, especially adapted for the human sciences—a view derived from Wilhelm Dilthey and continued by Emilio Betti and E. D. Hirsch, and those for whom it has an ontological significance, in which the main question to be answered is always man's relation to Being. This latter we associate more with Heidegger and Gadamer. See, for example, Richard E. Palmer's *Hermeneutics* (Evanston: Northwestern University Press, 1969), chapter 4. The account given here that focuses on the inadequacy of the subject/object schema is common to both.
4. Even Husserl's own discussion of the Absolute Flux, developed especially in *PITC*, section 36ff, marks within the text itself a limit to the phenomenology of time because it suggests that time-consciousness is itself founded on a more fundamental flux to which temporal predicates are at best only analogically or imperfectly applicable. Does this mark the point at which the phenomenology of time deconstructs itself?
5. In his account of the "dreamwork," Freud of course claims that the unconscious is oblivious to both logic and time, but might it not be a particular kind of time that it ignores? After all, in talking about the interpretation of dreams he tells us that we must be on the lookout for temporal inversions (sequences that begin with their conclusion). Husserl's answer to this possibility can be deduced from Eugen Fink's brief appendix 8 to Husserl's *Crisis*, "On the Problem of the Unconscious." He claims that for Husserl, the common (including the Freudian) understanding of the unconscious rests on a naive (prephenomenological) understanding of consciousness. The concept of the unconscious can only function adequately at all *after* the phenomenological clarification of "consciousness." This seems a reasonable claim, but is it compatible with the radical (Freudian) claim that phenomenological clarification of the unconscious would only help us to see just how and why consciousness is a layer floating on top of it, like froth?
6. We gratefully acknowledge a general scholarly debt to two previous commentators in particular: John B. Brough, especially his "The Emergence of an Absolute

Consciousness in Husserl's Early Writing on Time-Consciousness," *Man and World* 5, no. 3 (August 1972), pp. 298–326, and Robert Sokolowski, *The Formation of Husserl's Concept of Constitution* (The Hague: Nijhoff, 1964), esp. chapter 3.

7. The *PITC*, the shorter collection of lectures, is in many ways to be preferred to Rudolf Boehm's definitive longer edition, *Zur Phänomenologie des inneren Zeitbewusstseins (1893–1917)* (The Hague: Nijhoff, 1966), published as vol. 10 of *Husserliana*. It already contains quite as much as we need to grasp the character of his thinking about time—indeed, there is already considerable repetition in the shorter version; it is also the manuscript with which, as editor, Heidegger was most familiar, the one that is referred to in the literature, and the one to which Derrida refers in his *Speech and Phenomena*, trans. David B. Allison (Evanston: Northwestern, 1973).

8. Sokolowski, *Formation*, p. 75, n. 3.

9. He goes some way toward this in section 33.

10. Oskar Kraus, "Toward a Phenomenognosy of Time Consciousness" (1930), in *The Philosophy of Brentano*, ed. Linda L. McAlister (London: Duckworth, 1976), pp. 224–39. Kraus edited and introduced Brentano's *Psychologie vom empirischen Standpunkt* (2d ed.) in 1924, and regarded himself as Brentano's disciple.

11. William James, in his *Principles of Psychology* (1890) (New York: Dover, 1950), endorses this view of time, attributing the phrase to E. R. Clay. See vol. 1, p. 609.

12. Friedrich Nietzsche, *Twilight of the Idols*, p. 50.

13. "If we now again take up the question of whether a retentional consciousness that is not the continuation of an impressional consciousness is thinkable, we must say that it is impossible, for every retention in itself refers back to an impression." (Husserl, *PITC*, pp. 56–7) It is just such a logical impossibility that Derrida "braves" (or perhaps "flaunts") in the term "trace," which is, in effect, a retention of what has never been present.

14. Husserl, *PITC*, section 11, p. 53.

15. By "presentification," Husserl means a reproductive (that is, nonoriginary, nonself-giving) mode of givenness.

16. "In mere phantasy there is no positing of the reproduced *now* and no coincidence of this *now* with one given in the past." (Husserl, *PITC*, section 23, p. 74)

17. Within an experimental psychological setting, for all its alien conceptuality, it has been shown that ability to recall number sequences is considerably dependent on whether the subject makes use of subsets. These enormously expand one's power, and might well take the form of spontaneous glancings.

18. My point in bringing out the dual way in which Husserl is operating here is ultimately to suggest that the foundationalist ambition of phenomenology—to provide for philosophy (and indeed for Western thought and culture, if Husserl's "Philosophy as a Rigorous Science" is to be believed) a new grounding—cannot be thought of as being totally reconstructive. Very many rational principles and procedures and schemas are taken for granted, and it is not enough to say that they can, one by one, be scrutinized, because they are not independent of one another, as such a strategy would require. Husserl takes for granted not merely existence but the value of concepts like "unity," "reliability," "identity," "genuineness," "authenticity," "certainty," the part/whole relationship, form/content, and so forth, even at the very point at which he subjects such concepts to a phenomenological elucidation. What is then in question is not necessarily Husserl's procedure, but his claim as to its status. He believes it is possible to "begin again," to put all previous philosophy in brackets, to immunize oneself

NOTES 393

against history. But it may turn out that a much more plausible view would present phenomenology as a procedure that always and only can work with and from what is already given, with strictly limited powers of radical recommencement.

19. Derrida's name for this field of operation is "presence."
20. This language is reminiscent of P. F. Strawson's *Individuals* (London: Methuen 1959). There are strong affinities between Strawson's Kantian position and that of Husserl on both time and experience. The chief difference is that for Husserl the necessary independent temporal series is constituted subjectively.
21. John Brough's "The Emergence of an Absolute Consciousness in Husserl's Early Writings on Time-Consciousness," from which my title page quotation is drawn, was referred to at the beginning of this chapter. It offers the best account of the moves by which Husserl was led to posit an Absolute Flux.
22. Brough interestingly notes Husserl's shift from "now-consciousness" to "primal impression" at this point in his discussion. This shift might perhaps be linked to what could be called Husserl's *deconstruction of the now*, as it has already been discussed.
23. It is surely curious that when this problem reappears in appendix 6 (p. 154), "self-constitution" is not proposed as a way of avoiding infinite regress. He seems to be concerned that reflection inevitably objectifies Absolute Flux, and so makes it something temporal and in need of constitution. Abstaining from actual reflection leaves the possibility unaffected, and the possibility is sufficient to keep the regress problem alive. His concluding remarks suggest he has not solved the problem.
24. One important point to realize about "retention," which allows it to play its part in Absolute Flux, is brought out very clearly in appendix 9. In particular, retention is not an act; it does not objectify. It is an intentionality. It is not an act, because that would be "an immanent unity of duration constituted in a series of retentional phases," which would be a confusion of levels.

The act of apprehension qua objectifying consciousness is not the same as primal impression, for the act of apprehension is a unity of some sort.

Retention *is* a condition for becoming aware of the original impressional consciousness, but that is itself a consciousness, not something requiring retention to bring it out. This all suggests the possibility of talking about an "unconscious," and Husserl's important rejection of this idea we shall quote in full:

> It is certainly an absurdity to speak of a content of which we are "unconscious," one of which we are conscious only later. Consciousness is necessarily consciousness in each of its phases. Just as the retentional phase was conscious of the preceding one without making it an object, so also are we conscious of the primal datum—namely in the specific form of the "now"—without its being objective ... Were this consciousness not present, no retention would be thinkable, since retention of a content of which we are not conscious is impossible. (PITC, p. 162)

These remarks have had some notable readers. Sartre's account of prereflective (self-)consciousness would seem to be modeled on them, and Derrida, who also quotes this passage (*Speech and Phenomena*, p. 63), brings it into critical focus.

It might finally be worth quoting someone else, who most likely never read these words of Husserl, though he could be read as referring to them:

To most people who have had a philosophical education the idea of anything
mental which is not also conscious is so inconceivable that it seems to them
absurd and refutable simply by logic. . . They have never studied hypnosis
or dreams." (Sigmund Freud, *The Ego and the Id*, trans. Joan Rivière
[London: Hogarth Press, 1949], p. 10)

25. These remarks might be compared with the following: "the existence of what is
transitory passes away in time, but not time itself. Time [is] . . . itself nontransitory, and abiding . . ." (Immanuel Kant, *Critique of Pure Reason*, trans. Norman
Kemp Smith [New York: Macmillan 1968]: "Schematism," B183) Kant is here
drawing parallels between *time* and *substance*, insofar as each is nontransitory. It is
perhaps such a parallel that allows Husserl to generate an account of transcendental subjectivity that will displace the idea of an absolute self-constituting flux.
Transcendental subjectivity gives a *substantialist* interpretation of the timelessness
of the form of the flux. Moreover, in appendix 6, Husserl writes of the flux as
"abiding" (*Verbleibendes*), and adds that what is abiding above all is 'the formal
structure of the flux, the form of the flux . . . an abiding form [which] however,
supports the consciousness of a continuous change."
26. See Martin Heidegger, "Time and Being" [1962], trans. Joan Stambaugh, in *On
Time and Being* (New York: Harper & Row, 1972), pp. 1–24.

PART 2, CHAPTER 3

1. Jacques Derrida, *La voix et le phénomène* (Paris: Presses Universitaires de France,
1967) *(Speech and Phenomena*, trans. David B. Allison [Evanston: Northwestern
University Press, 1973]).
2. The obvious word to use here might be thought to be *critique*. But while Derrida's
readings are clearly *critical* in a number of senses, including "historically
momentous," "discerning," and "careful," he has taken pains to dissociate
deconstruction from the negative (or indeed positive) implications of *critique*.
3. See part 4, chapter 2, p. 283ff, for a discussion of the status of *différance*.
4. See part 4, chapter 3, pp. 293–318.
5. Richard Rorty, in his *Philosophy and the Mirror of Nature*, suggests that the risk any
"edifying" philosopher runs is that of having one's work transformed into a
"systematic" form. This possibility is surely the counterpart to the possibility of a
deconstructive self-unraveling latent in any theoretical text. Perhaps such "syste-
matization" should not be so much deplored as treated as a valuable index of
decline in the "edifying" power of the original text, and one that is rooted in
possibilities *necessarily contained* in every text, however eccentrically transgressive.
6. Derrida's arguments in this book could be seen to reformulate the criticisms
made by French existentialists of the Husserlian project of phenomenological
reduction, which similarly attempts to exclude what cannot be excluded, namely
the "world." Merleau-Ponty's version, somewhat different from Sartre's, that the
reduction *is* a proper part of philosophical method but can never be completed
(see the preface to his *Phenomenology of Perception*, trans. Colin Smith [London:
Routledge and Kegan Paul, 1962), converges with Husserl's own eventual
position. The relationship between Sartre and Derrida is surely worth pursuing.
Derrida has only rarely mentioned Sartre and has seemed touchy about the
subject ("Is this an interview about Sartre?" he asked, in the *Nouvel Observateur*

"Interview with Derrida,"trans. David B. Allison and others, in *Derrida and Différance* ed. David Wood and Robert Bernasconi (Warwick: Parousia Press, 1984; republished Evanston: Northwestern University Press, 1988). An excellent start can be found in Christina Howells's "Qui Perd Gagne: Sartre and Derrida," *Journal of the British Society for Phenomenology* 13, no. 1 (1982): pp. 26–34.

Throughout this chapter, references to and quotes from Derrida, *La voix et le phénomène*, given only by page number (s), are from the English edition translated by David B. Allison (see n. 1).

7. See, for example, "The End of the Book and the Beginning of Writing," in Derrida, *Of Grammatology*.
8. See Martin Heidegger, "The Anaximander Fragment," in *Early Greek Thinking*, trans. David R. Krell (New York: Harper & Row, 1975).
9. This strategy has previously been announced in the shadows of a footnote, from which I here quote:

> In affirming that *perception does not exist* or that what is called perception is not primordial, that somehow everything "begins" by "representation" (a proposition that can only be maintained by the elimination of these last two concepts: it means that there is no "beginning" and that the "re-presentation" we were talking about is not the modification of a "re-" that has *befallen* a primordial presentation) and by reintroducing the difference involved in "signs" at the core of what is "primodial," we do not retreat from the level of transcendental phenomenology towards either an "empiricism" or a "Kantian" critique of the claim of having primordial intuition; we are here indicating the prime intention—and the ultimate scope—of the present essay. (*Speech and Phenomena*, p. 45, n. 4)

What these remarks leave open, perhaps deliberately, is whether there is any retreat *at all* from transcendental phenomenology. Derrida will continue to make many of its moves while denying certain of its vital implications.

10. Ludwig Wittgenstein, *Philosophical Investigations*, 2d ed. (Oxford: Blackwell, 1958).
11. Ferdinand de Saussure, *Cours de Linguistique Générale*, ed. T. de Mauro (Paris Payot, 1973) (*Course in General Linquistics*, trans. Wade Baskin [London: Fontana, 1974]).
12. Whether "in the last analysis" these relationships (as inverted by Derrida) should be treated as "constitutive" ones is a crucial question, dealt with later. I argue that the philosophical strength of Derrida's arguments rests on their being transcendental arguments that contest the direction of constitution. To accept the "sous rature," Derrida's denial of their being transcendental arguments, is to render their course uncompelling. Moreover, I claim that the orientation to a transcendental level (albeit now formulated in terms of presence and absence) is such as to impede the development of a positive alternative understanding of time. Instead, we are left with the emptiness of terms in which the varieties of representation (and hence of the invasion of presence by representation) are lost. See the end of this ch. and Part 4.
13. See especially Derrida's long footnote in *Speech and Phenomena*, p. 84, n. 9.
14. See the discussion on pp. 234–5.
15. For a discussion of the broader issues involved here—the interconnection between philosophical strategy and the limits of language, with its consequences for our discussion of temporality—see pp. 293–310

16. See, for example, David F. Krell's "Engorged Philosophy," in Wood and Bernasconi, *Derrida and Différance.*
17. The two chapters are " . . . That Dangerous Supplement . . ." and "From/Of the Supplement to the Source: The Theory of Writing."
18. Derrida, *Of Grammatology*, p. 144.
19. Further investigation of Rousseau reveals that the issue is not so simple. Rousseau, after all, has taken up writing because of the way in which, in speech, in the immediacy of social interaction, he constantly *misrepresents* himself. He is forced into positions in which he says things he does not mean. Far better to write, when one can give considered thought to one's words. From this point of view, writing helps us to *restore* presence; culture comes to the aid of nature.

But equally, Rousseau sees writing as a *threat*. For it can easily be seen as a distortion of what is natural, a mere technique added to natural speech, which threatens to displace it. This theme of the twofold sense of the supplement as completing an original presence, and as a substitution that threatens it, is apparently repeated over and over again in Rousseau, usually centering on the nature/culture opposition, and it underlies his account of education.
20. Compare, for example, Nietzsche's account of reversals in the cause/effect relation in *Twilight of the Idols*, "The Four Great Errors."

PART 3, CHAPTER 1

1. This corrects the Macquarrie and Robinson translation (1962), which for some reason has "unity" instead of "temporality" for *Zeitlichkeit.*
2. "Der Zeitbegriff in der Geschichtswissenschaft" ("The Concept of Time in the Science of History," trans. H. S. Taylor and H. W. Uffelmann, in *Journal of the British Society for Phenomenology* 9, no. 1, (January 1978): pp. 3–10).
3. See William Richardson, *Through Phenomenology to Thought* (The Hague: Nijhoff, 1963), pp. 28–29.
4. Martin Heidegger, *Die Grundprobleme der Phänomenologie* (1927), in Gesamtausgabe, vol. 24 (Frankfurt: Klostermann, 1975) (*The Basic Problems of Phenomenology*, trans. Albert Hofstadter [Bloomington: Indiana University Press, 1982]).
5. See Heidegger, *Being and Time*, pp. H432–433.
6. Derrida, *"Ousia* et *Gramme."* This essay is subtitled "Note to a footnote in *Being and Time."*
7. See, for example, *Zur Sache des Denkens* (Tübingen: Niemeyer, 1969) ("The End of Philosophy and the Task of Thinking" in Heidegger, *On Time and Being*).
8. Martin Heidegger, *Nietzsche* (Pfullingen: Neske, 1961) (*Nietzsche*, 4 vols. ed. David F. Krell, trans. David F. Krell and Frank Capuzzi [New York: Harper & Row, 1979, 1982, 1984]). See also my Part 1: "Nietzsche's Transvaluation of Time," above.
9. Heidegger, *Was heisst Denken? (What Is Called Thinking?).*
10. "Der Ursprung des Kunstwerkes" (1935–6/1950), in *Holzwege (Gesamtausgabe* 5) (Frankfurt: Klostermann, 1975), pp. 1–74 ("The Origin of the Work of Art" in Martin Heidegger, *Poetry, Language and Thought*, trans. Albert Hofstadter [New York: Harper & Row, 1971], pp. 17–87).
11. L. M. Vail, *Heidegger and the Ontological Difference* (University Park: Pennsylvania University Press, 1972).
12. See Heidegger, "Time and Being," p. 5ff.
13. See Martin Heidegger, *Identity and Difference* (1957), bilingual ed., trans. Joan

Stambaugh (New York: Harper & Row, 1959), p. 14ff.

14. See Martin Heidegger, *The Question of Being* (1955), bilingual ed., trans. William Kluback and Jean T. Wilde (New Haven: College and University Press, 1958), p. 81.

15. Note that at the very moment at which Being is crossed out, Heidegger stresses that man is far from being excluded from Being.

16. Heidegger makes a great deal out of the word "as," and has a discussion of the "as" in understanding, interpretation, and so forth, *Being and Time*, p. H150f.

17. See Heidegger's *Unterwegs zur Sprache* (Pfullingen: Neske, 1959) (*On the Way to Language*, trans. Peter D. Hertz [New York: Harper & Row, 1971]), and see Robert Bernasconi's *The Question of Language in Heidegger's History of Being* (Atlantic Highlands: Humanities Press, 1985).

18. Jean-François Lyotard, *La Condition postmoderne* (Paris: Minuit, 1979) (*The Postmodern Condition*, trans. Geoff Bennington and Brian Massumi [Manchester: Manchester University Press, 1984]).

19. For an excellent discussion of this opening, see John Sallis, "Where Does *Being and Time* Begin?" in his *Delimitations: Phenomenology and the End of Metaphysics* (Bloomington: Indiana University Press, 1986): pp. 98–118.

20. This position is given a more positive form in the idea of a hermeneutic circle. See Heidegger, *Being and Time*, pp. H314–15.

21. See the very difficult first chapter of Husserl's *Ideas* (1913): "Fact and Essence."

22. "This is a productive logic in the sense that it leaps ahead, as it were, into some area of Being, discloses it for the first time in the constitution of its Being, and after thus arriving at the structures within it, makes these available to the positive sciences as transparent assignments for their enquiry." (Heidegger, *Being and Time*, p. H10)

23. Heidegger distinguishes the terms *existenzial* ("existential") and *existenziell* ("existentiell"). I follow him in using this terminology, and yet will question his ability to keep these two separate. His most explicit account of the difference can be found on H12–13. The "existentiell" relates to an ongoing understanding of oneself in terms of one's possibilities of existence. The "existential" refers to an analytical or theoretical study of the structure of existence. The "existential" presupposes the "existentiell."

24. Sallis, "Where Does *Being and Time* Begin?" p. 98f.

25. "Language," in Heidegger, *On the Way to Language*, p. 190.

26. "Provisionally," because for Heidegger the historical is already existential and vice versa. See esp. Heidegger, *Being and Time*, pp. H372–403.

27. See Friedrich Nietzsche, "The Uses and Disadvantages of History for Life" (1874), in *Untimely Meditations*, trans. R. J. Hollingdale (Cambridge: Cambridge University Press, 1983).

28. See Heidegger, *Being and Time*, pp. 424–55.

29. This is taken up in Martin Heidegger, *Kant und das Problem der Metaphysik* (1929) (Frankfurt: Klostermann, 1973) (*Kant and the Problem of Metaphysics*, trans. James S. Churchill [Bloomington: Indiana University Press, 1962]).

30. See Heidegger, *Basic Problems of Phenomenology*.

31. Heidegger, *Being and Time*, pp. H432–33. See also my n. 6.

32. See n. 29.

33. Friedrich Nietzsche, *Jenseits von Gut und Böse: Vorspiel einer Philosophie der Zukunft* (1886) (*Beyond Good and Evil*, trans. Walter Kaufmann [New York: Vintage, 1967]), section 20.

34. See, for example, "Von Wesen der Wahrheit" (1943), in Martin Heidegger,

Wegmarken, Gesamtausgabe, vol. 9 (Frankfurt: Klostermann, 1976), pp. 177–202 ("On the Essence of Truth," trans. John Sallis in *Martin Heidegger: Basic Writings,* ed. David F. Krell [New York: Harper & Row, 1977], pp. 117–41).

35. See Jacques Derrida, "Le Retrait de la Métaphor," *PO&SIE,* 1979: pp. 103–26 ("The *Retrait* of Metaphor," *Enclitic* 2, no. 2 (Fall 1978): pp. 6–33.

PART 3, CHAPTER 2

1. Wittgenstein begins his *Tractatus Logico-Philosophicus,* trans. G. K. Ogden and F. P. Ramsey (bilingual edition) (London: Routledge & Kegan Paul, 1922), with the now famous lines: "Die Welt ist alles, was der Fall ist. Die Welt ist die Gesamtheit der Tatsachen, nicht die Dinge." ("The world is everything that is the case. The world is the totality of facts, not of things.")

2. The parallel with Hegel here must not be missed. For Hegel, Spirit has to make the detour of historical articulation. And to the extent that the parallel holds, one can ask of Heidegger what has often been put to Hegel: Does not the need for the detour presuppose the validity of the destination? It is quite true that neither self-understanding, self-knowledge, nor self-appropriation yield to immediacy. But is there not a danger that the articulation of modes of mediation will only reinforce the value of the ideal in each case, rather than put it in question? And the scheme of Heidegger's thought is surely such as to warrant questioning. It is because everydayness is only the necessary *mediation* of authenticity that the danger of losing ourselves in it arises. But what if the very value of selfhood, and hence authenticity itself, were ultimately bound up with a philosophy of identity that, after Heidegger, we have come to call metaphysical?

3. I refer of course to the English translators of *Being and Time* (Oxford: Blackwell, 1962). This geometrical analogy is offered on p. 185, n. 1.

4. "To any willing there belongs something willed, which has already made itself definite in terms of a 'for-the-sake-of-which.' If willing is to be possible ontologically, the following items are constitutive for it: (1) the prior disclosedness of the 'for-the-sake-of-which' in general (Being-ahead-of-itself); (2) the disclosedness of something with which one can concern oneself (the world as the 'wherein' of Being-already); (3) Dasein's projection of itself understandingly upon a potentiality-for-Being towards a potentiality-for-Being towards a possibility of the entity willed." (H194)

PART 3, CHAPTER 3

1. No general account of Derrida's reading of Heidegger could dispense with the idea of a double reading. He acknowledges Heidegger's enormous achievements at the boundary of philosophy and nonphilosophy, and yet will maintain that Heidegger's is perhaps the most powerful rearguard action in defense of *presence.* See Jacques Derrida, *Positions* (Paris: Minuit, 1972) (*Positions,* trans. Alan Bass [Chicago: University of Chicago Press]). Derrida's treatment of Heidegger in *De la grammatologie* is perhaps exemplary. There a critical reading is progressively deepened by a series of defenses of Heidegger against prematurely negative diagnoses.

2. See David Farrell Krell, "Death and Interpretation," in *Heidegger's Existential*

Analytic, ed. F. Elliston (The Hague: Mouton, 1978).

3. This is of course the title of a 1959 essay by Heidegger, translated by J. M. Anderson and E. Hans Freund as "Memorial Address," in *Discourse on Thinking* (New York: Harper & Row, 1966).

4. He seems to be arguing for what Gilbert Ryle in his *The Concept of Mind* (London: Hutchinson, 1949) (not altogether uninfluenced by Heidegger and by phenomenology) called a "category mistake," although Heidegger would say that what is at stake is not a distinction between one category and another, but rather between a category and an existentiale. An interesting general account of the Ryle/Heidegger relationship can be found in Michael Murray's "Heidegger and Ryle: Two Versions of Phenomenology," in *Heidegger and Modern Philosophy*, ed. Michael Murray (New Haven and London: Yale University Press, 1978).

5. See his *L'être et le néant* (Paris: Gallimard, 1943) (Jean-Paul Sartre, *Being and Nothingness*, trans. Hazel Barnes [London: Methuen, 1957]).

6. Many distinctions need to be made here, most notably between two senses of possibility. We can distinguish (1) inductive mortality: it *seems*, on the basis of others' longevity and their successful conquest of fatal disease and replacement of organs, that no one has to die. We may be immortal, but we cannot be sure. Living "forever" never comes; (2) immortality bar accident: we come to believe that death is no longer necessary, even if a fatal accident might in some cases be hard to avoid.

7. The *Shorter Oxford English Dictionary* gives us much food for thought here: "Exuberant a. 1503. [ad. L. *exuberantum, exuberare, f. ex-* + *uberare* to be fertile, f. *uber* adj., conn. w. *uber* udder.] 1. Luxuriantly fertile or prolific; abundantly productive. Also *fig.* 1645. 2. Growing or produced in superabundance 1513. 3. Overflowing, as a fountain, etc. 1678. Also *fig.* 1503."

8. Further notes from underground can be found in John Sallis, "Tunnellings," and David Farrell Krell, "The Mole: Philosophical Burrowing in Kant, Heidegger, and Nietzsche," *Boundary 2* 9, no. 3, and 10, no. 1 (*Why Nietzsche Now?*) [Spring/Fall 1981]).

9. See Emmanuel Levinas, *Totality and Infinity* (Pittsburgh: Duquesne University Press: 1969), for example, "The I breaks free from itself in paternity without thereby ceasing to be an I, for the I *is* still its son." (p. 278)

10. See Husserl's *Cartesian Meditations* (1929/1933), trans. Dorion Cairns (The Hague: Nijhoff, 1960), section 44f.

11. See, for example, "Brief über den Humanismus" (1947), in Heidegger, *Wegmarken, Gesamtausgabe* 9, Frankfurt: Klostermann, 1976) ("Letter on Humanism," 1962), trans. Frank Capuzzi and J. Glenn Gray, in *Martin Heidegger: Basic Writings*, ed. David F. Krell [New York: Harper & Row, 1977]); and "Zeit und Sein," (1968) In *Zur Sache des Denkens* (Tübingen: Niemeyer, 1969) ("Time and Being").

12. Here specifically, and elsewhere more generally, we are clearly indebted to the pathbreaking work of Jacques Derrida, and particularly to his *La voix et le phénomène*.

13. A critical allusion might be made here to Kierkegaard's discussion of a *Power*, in his *Sickness Unto Death* (Garden City, New York: Anchor, 1954), pp. 146–47.

14. Ibid., p. 147. If man were to be his own ground, the second kind of despair he distinguishes—"despairingly willing to be oneself"—would not be possible. Kierkegaard seems to ignore the possibility that this despair might simply be unfounded, or founded on an error.

15. This may have other implications. It may, for example, be thought to provide some basic answers to the questions with which other philosophers have wrestled: Why be good? Why ought we to do our duty?
16. Of course many others now make this point.
17. Death has been understood as "the possibility of the impossibility of existence," and this transformation of the future into a "toward-which" might seem to betray a certain denial of the reality of death rather than taking it seriously. Does he not say:

> When, in anticipation, resoluteness has caught up [Eingeholt] the possibility of death into its potentiality-for-Being, Dasein's authentic existence can no longer be outstripped [überholt] by anything. (Being and Time, H307)

18. Kierkegaard offers accounts of such moments of projected intensity in his *Concluding Unscientific Postscript*, trans. D. Swenson (Princeton: Princeton University Press, 1941).
19. Sartre, *Being and Nothingness*.
20. A common cause of complaint by authors against reviewers. Derrida in particular laments such moves in his *Positions*, p. 52.
21. No doubt this reference could be made at an even more telling point, but consider this remark of Heidegger's: "The inner relationship of my own work to the Black Forest and its people comes from a centuries-long and irreplaceable rootedness in the Alemannian-Swabian soil." ("Why Do I Stay in the Provinces?" [1934], trans. Thomas Sheehan, in *Heidegger: the Man and the Thinker*, ed. Thomas Sheehan [Chicago: Precedent, 1981], pp. 27–30). I would also like to acknowledge a debt to the excellent bibliography with which Professor Sheehan has endowed this book.
22. Ibid.
23. Heidegger has no time for speed. He laments the fact that people today value only what goes fast and can be grasped with one's hands. Clearly, he would have little in common with those who called themselves Futurists.
24. I have suggested that "ambiguity" might not be as resolvable as Heidegger suggests. The brief essay being considered here is a case in point. Heidegger castigates the sentimentalizing of peasant life, and yet one is more than once tempted to accuse him of it in this very passage. Indeed, it can seem hard to avoid. Is not much of the silence of the peasant the result of having nothing to say, rather then being silent about deep truth? Is there not a silence of dullness? Heidegger's extraordinary preference for the old woman who calls him Herr Professor, over those who read his books, suggests a curious status for his chosen life activity—writing—of which the woman has no comprehension.
25. Though Heidegger's commitment to the primacy of the future hardly survives the year *Being and Time* was published.
26. In *Being and Time*, section 32, "Understanding and Interpretation."
27. See "Was ist Metaphysik?" (1929) in Martin Heidegger, *Wegmarken (Gesamtausgabe* 9, 1976) ("What Is Metaphysics?" trans. David F. Krell [*Basic Writings*, 1977]), pp. 95–112.
28. See Jacques Derrida, pp. 141–64.
29. He rejects other specific points too: (1) the temporality of spatiality, and (2) the possibility of a fundamental ontology.

PART 3, CHAPTER 4

1. "At the very start . . . our analytic was oriented rather by the average way of existing, which has nothing conspicuous about it." (Heidegger, *Being and Time*, H370)
2. Indeed, Heidegger at one point even goes so far as to equate them: " . . . at bottom we mean by the term 'everydayness' nothing else than temporality, while temporality is made possible by Dasein's Being . . ." (Ibid., H372)
3. As we have already suggested, its value may not lie in its possible accomplishment, but in the way it disturbs the seeds of complacency.
4. See, for example, Derrida's essay "Différance" (1968) in his *Marges de la philosophie*, pp. 1–28.
5. " . . . in laying hold of an item of equipment, we come back to it from whatever work-world has already been disclosed." (Heidegger, *Being and Time*, H352)
6. "Gelassenheit" (1959) in *Gelassenheit* (Pfullingen: Neske, 1960), pp. 11–28 ("Memorial Address," trans. J. M. Anderson and E. H. Freund, in *Discourse on Thinking* [New York: Harper & Row, 1966]), pp. 43–57.
7. Is not this whole level of discourse (projection, disclosedness) an expansion of Kant's reference to schemata?
8. How would these doubts about authenticity affect such a laminated temporality? I am predisposed to encourage the awareness of multiple levels and recursivity in "temporal" "structures" (see the discussion of "nesting" on pp. 338–9). Words such as "nesting," "levels," and "structures" are all *representations* of multiple temporality. The question is whether "forgetting" would not itself have to be forgotten after a systematically skeptical treatment of authenticity. Arguably not. For even if risk, danger, and openness to fragmentation are counterposed to the ultimately self-fulfilling sense of "coming-towards-oneself," a questioning "relation to self" remains, which a totally unselfconscious absorption in the world of things could be said to be "forgetting."
9. For example: " . . . when one is making present something ready-to-hand by *awaiting*, the possibility of one's getting surprised by something is based on one's not awaiting something else which stands in a possible context of involvement which is lost." (Heidegger, *Being and Time*, H355) We might ask, however, how he would handle *ontological* surprise. This is surely not covered by his remarks at H264.
10. Such expressions as "a priori" open up the most difficult question of the relationship between the logical and the chronological. On the one hand, logical priority clearly cannot be reduced to temporal priority, and yet seems to draw on the sense of irreversible linearity it provides; on the other, the necessity of one-way temporal order seems to approach the logical, and for someone like Freud could be thought to be the result of the imposition of an unconscious demand for order.
11. Such a distinction could be drawn from Roland Barthes, *Le Plaisir du texte* (Paris: Seuil, 1973).
12. Consider Barthes on Fourier: "What I get from Fourier's life is his liking for *mirlitons* [little Parisian spice cakes]. . .," preface to his *Sade/ Fourier/ Loyola* (Paris: Seuill, 1971).
13. See Derrida's "*Ousia* and *Gramme*."
14. See, for example, his discussion of the "Grammar and Etymology of *Being*," in

Martin Heidegger, *An Introduction to Metaphysics*, trans. Ralph Manheim (Garden City, N.Y.: Doubleday, 1961), p. 48.

15. See p. 300f. And see the discussion in Erasmus Schöfer, "Heidegger's Language: Metalogical Forms of Thought and Grammatical Specialties," in *On Heidegger and Language*, ed. Joseph J. Kockelmans (Evanston: Northwestern University Press, 1972); and Schöfer's book *Die Sprache Heideggers* (Pfullingen: Neske, 1962).

16. See, for example, his *Positiors*, p. 55: "I sometimes have the feeling that the Heideggerean problematic is the most 'profound' and 'powerful' defense of what I try to put in question under the rubric of the *thought of presence*."

17. Derrida makes the same claim about the ideal repeatability of the present. See p. 111ff.

18. Derrida, "*Ousia* and *Gramme*," throughout.

19. Heidegger, *Die Grundprobleme der Phänomenologie* (*The Basic Problems of Phenomenology*).

20. L. E. J. Brouwer, for example, thought the relation was precisely the reverse: that arithmetical succession was based on temporal succession. See his "Intuitionism and Formalism," in *Bulletin of the American Mathematical Society* 20 (1913), p. 85.

21. Close at hand here are the sorts of problem attached to Kant's idea of a *schematism*, by which he meant something like a temporalized concept.

22. It echoes in this respect and at one level, part of J. M. E. McTaggart's famous argument for the derivativeness of the "B-series" (earlier than/later than) from the "A-series" (past/present/future). See his *The Nature of Existence*, vol. 2 (Cambridge: Cambridge University Press, 1927), pp. 10ff.

PART 3, CHAPTER 5

1. Heidegger, *Kant und das Problem der Metaphysik* (*Kant and the Problem of Metaphysics*).

2. See the much fuller discussion in David Farrell Krell's seminal paper, "Rapture: the Finitude of Time in Heidegger's Thought," in *Time and Metaphysics*, eds. David Wood and Robert Bernasconi (Warwick: Parousia Press, 1982).

3. Heidegger, *Einführung in die Metaphysik* (1953) (Tübingen: Niemeyer, 1966) (*An Introduction to Metaphysics*), p. 172.

4. Ibid., p. 169.

5. Ibid., pp. 171–72.

6. Ibid., p. 172.

7. See Derrida, *Of Grammatology*, p. 17.

8. Werner Marx, *Heidegger und die Tradition* (Stuttgart: Kohlhammer, 1961) (*Heidegger and the Tradition*, trans. Theodore Kisiel and Murray Greene [Evanston: Northwestern University Press, 1971]).

9. Heidegger, *Being and Time*, H69ff.

10. Heidegger, "Zeit und Sein" (1968) in *Zur Sache des Denkens* ("Time and Being," in *On Time and Being*).

11. Ibid., p. 7.

12. Marx, *Heidegger*, p. 145.

13. Heidegger, "Von Wesen der Wahrheit," pp. 177–202.

14. Heidegger, "Brief über den Humanismus."

15. Heidegger, "*Logos* (Heraclitus, Fragment B 50)" (1951) in Martin Heidegger, *Vorträge und Aufsätze* (Pfullingen: Neske, 1967) ("*Logos* Heraclitus, Fragment B 50," trans. David Farrell Krell, in Martin Heidegger, *Early Greek Thinking*.

16. Marx, *Heidegger*, p. 155.
17. "Die Onto-theo-logische Verfassung der Metaphysik," in Martin Heidegger, *Identität und Differenz* (Pfullingen: Neske, 1957) ("The Onto-theo-logical Constitution of Metaphysics," in *Identity and Difference* [New York, Harper and Row, 1969]).
18. Martin Heidegger, "Zur Seinsfrage" (1955) in *Wegmarken, Gesamtausgabe*, vol. 9 (*The Question of Being*, trans. William Klubach and Jean Wilde [London: Vision Press, 1959]).
19. Heidegger, *Nietzsche*.
20. "Der Spruch des Anaximander," in *Holzwege* (1950) (Frankfurt: Klostermann, 1972) ("The Anaximander Fragment").
21. Heidegger, *Die Grundprobleme der Phänomenologie* (*The Basic Problems of Phenomenology*).
22. See p. 111ff.
23. See Saussure's *Cours de Linguistique Generale* (*Course in General Linguistics*).
24. See p. 111ff.
25. Derrida, "*Ousia* and *Gramme*," p. 88.
26. *Of Grammatology*, pp. 4ff.
27. Derrida, "Différance."
28. The first English translation is not quite accurate here (included in *Speech and Phenomena*, trans. David B. Allison [Evanston: Northwestern University Press, 1973]).
29. Derrida, *Speech and Phenomena*, pp. 142–43.
30. Ibid., p. 147.
31. Ibid.
32. Derrida's mixture of charity and critique toward Heidegger is well illustrated in "The End of the Book and the Beginning of Writing," in *Of Grammatology*. The issue is whether Heidegger's interest in the *question* of Being is sufficiently different from a pursuit of Being as such to enable him to escape the charge of reinscribing presence, in a metaphysical way, in Being.
33. Surely here there is a startling convergence with Husserl's most extreme conclusions (see pp. 107–9), in which he talks of a "one dimensional quasi-temporality," a "pre-phenomenal, pre-imminent temporality."
34. See Derrida, "*Ousia* and *Gramme*," p. 42.

PART 4, CHAPTER 1

1. See "Linguistics and Grammatology" (hereinafter LG) translated from Derrida, *De la grammatologie* (*Of Grammatology*) p. 103. "Grammatology as a Positive Science" (GPS) is also to be found in this volume.
2. Am I myself not guilty of a psychologizing interpretation of Derrida? Sometimes this is just what his words invite. But whether references to *desire* are ultimately psychological is disputable. The history of its metaphysical and ontological embroilment (from Aristotle to Hegel to Lacan) would suggest otherwise.
3. See my "Derrida and the Paradoxes of Reflection," *Journal of the British Society for Phenomenology* 11, no. 3 (October 1980), pp. 225–38, and, slightly modified, pp. 279–292 in this book.
4. Derrida, "*Ousia* and *Gramme*."
5. Heidegger, *Einführung in die Metaphysik* (*An Introduction to Metaphysics*).

6. "Time *is* not. There is, it gives time. The giving that gives time is determined by denying and withholding reserve. It grants the openness of time-space, and preserves what remains denied in what-has-been, what is withheld in approach. We call the giving which gives true time an extending, which opens and conceals. An extending is itself a giving, the giving of a giving is concealed in true time." Heidegger, "Time and Being," in *On Time and Being*, p. 16.

7. If the parallels with Kant's transcendental imagination and Heidegger's *es gibt* hold up, and if Derrida is right in thinking that logical principles break down when describing the trace structure or différance, how did Kant and Heidegger avoid this consequence? In the case of Heidegger, the answer is that he creatively distorts grammar and exhorts his reader or listener "not to listen to a series of propositions, but rather to follow the movement of showing." (ibid., p. 3) Kant, as I read him, utilizes the fiction that the operation of the fundamental faculties of the mind can be described like the workings of machinery.

8. Derrida, "Différance," in *Margins*, p. 13.

9. The usual "superficial criticism of Derrida . . . overlooks the invisible erasure," writes Gayatri Spivak in a translator's footnote to *Of Grammatology*, p. 318, n. 13.

10. It will be said that this is a notorious error. Ordinary language, Nietzsche has told us, is no less metaphysical than grand theory. It is just more complacent and naive. Does not this view inflate "metaphysics" into a mere vapor, an empty term? Surely only a *deployment* of language can be metaphysical. Consider Derrida's important remark, "I have never believed that there were metaphysical concepts in and of themselves. No concept is by itself metaphysical, outside all the textual work in which it is inscribed." *Positions*, p. 57.

11. Derrida, LG, p. 72.

12. Derrida, GPS pp. 85–87.

13. Derrida, LG p. 72. Fascinatingly, Derrida outlines not just Jakobson's preference of the model of the musical staff over the line, but suggests that in his *Anagrams*, Saussure was himself raising a question about the fundamental linearity of language.

   Jacques Lacan's discussion of Saussure and linearity is instructive: "All our experience runs counter to this linearity . . . The linearity that Saussure holds to be constitutive of the chain of discourse, in conformity with its emission by a single voice and with its horizontal position in our writing—if this linearity is necessary, in fact, it is not sufficient . . . But one has only to listen to poetry, which Saussure was no doubt in the habit of doing, for a polyphony to be heard, for it to become clear that all discourse is aligned along the several staves of a score." Lacan, *Écrits*, p. 154.

14. Speech can indeed be seen as a sequence of sounds. But on such a model, distinctions between semantic units may not be apparent, the temporality of meaningful speech production and comprehension will not be registered, and there is no scope for acknowledging the different levels of temporal order in speech other than that of the succession of sounds.

15. See my "Prolegomena to a New Theory of Time," *Research in Phenomenology* 10 (1980): pp. 177–91, and pp. 335–348 in this book.

16. See Derrida's remarks in n. 10.

PART 4, CHAPTER 2

1. While the phrase "cashed out" has obtained considerable currency in English-speaking philosophy, it is interesting to note that Husserl too—Derrida's first philosophical concern—gives a central place to redeeming meaning-claims, as one might take vouchers to a bank.
2. See "Hors livre," in Jacques Derrida, La Dissémination (Paris: Seuil, 1972) ("Outwork" in Dissemination, trans. Barbara Johnson [Chicago: University of Chicago Press, 1981]).
3. See "The End of the Book and the Beginning of Writing," in Derrida, Of Grammatology.
4. Derrida, "Différance."
5. Derrida, Speech and Phenomena.
6. For a more systematically introductory account of Derrida's philosophy, in which some of these issues are developed, see my "Introduction to Derrida," Radical Philosophy 21 (Spring 1979), pp. 17–28; reprinted in Radical Philosophy Reader, ed. Roy Edgeley and Richard Osborne (London: Verso, 1985) pp. 18–42.
7. The question of the ontological commitments of our language has of course been an active concern of such Anglo-American philosophers as Whorf, Whitehead, Quine, Strawson, and Davidson.
8. From Friedrich Nietzsche, "On Truth and Falsity in Their Ultramoral Sense," trans. M. A. Mugge, in The Existentialist Tradition, ed. Nino Langiulli (Garden City, N.Y.: Doubleday, 1971).
9. See Jacques Derrida, "Structure, Sign, and Play in the Discourse of the Human Sciences," in Writing and Difference, p. 280.
10. This procedure is explained at greater length in his Positions, p. 41. Note 8 on this page helpfully points to some of Derrida's other general formulations of his strategy.
11. Derrida, "Différance," in Margins of Philosophy, pp. 11f.
12. See the paper by Christopher Macann, "Jacques Derrida's Theory of Writing and the Concept of Trace," in Journal of the British Society for Phenomenology 3, no. 2 (May 1972): pp. 197–200. But more particularly, see pp. 311–18.
13. "Différance," pp. 26–27.
14. Compare Newton Garver's "Derrida and Rousseau on Writing," Journal of Philosophy (November 1977), p. 671, n. 10.
15. See Derrida, Of Grammatology, p. 12.
16. See the title interview in Derrida, Positions, pp. 37–96.
17. Derrida, "Structure, Sign, and Play," pp. 280–81.
18. We reserve for another occasion a discussion of Derrida's Éperons: les styles de Nietzsche (Paris: Flammarion, 1978) (Spurs: Nietzsche's Styles, trans. Barbara Harlow [Chicago: University of Chicago Press, 1979]).
19. See pp. 293–310.
20. Without endorsing the rest of his paper, one must admit that Foucault does have a point when he accuses Derrida ("Mon corps, ce papier, ce feu" [1972], translated in the Oxford Literary Review 4, no. 1 [1979]) of failing to analyze "the modes of implication of the subject in discourse." I take this from Gayatri Spivak's excellent introduction to her translation of Of Grammatology, p.lxi. It lies behind Paul de Man's essay on Rousseau, "The Rhetoric of Blindness: Jacques Derrida's Reading of Rousseau," in his Blindness and Insight (London: Methuen, 1971). Derrida is perhaps blind to Rousseau's own self-deconstruction.

21. Some will dispute the description of deconstruction as a "critique." It is, however, a word Derrida himself uses. For some of *his* reservations about its implications, see *Positions*, pp. 46–47.
22. See Jean-Marie Benoist, "The End of Structuralism," *Twentieth Century Studies* 1970, p. 52.
23. Derrida, *Speech and Phenomena*, p. 31.
24. Derrida, *Of Grammatology*, pp. 34ff.
25. Derrida, *Speech and Phenomena*, p. 103.
26. The question of course is whether there would be such a thing as philosophy in the wake of these moves.
27. See Derrida, "Structure, Sign, and Play," pp. 278–79.
28. See n. 18.
29. I have already made this point. See n. 20.

PART 4, CHAPTER 3

1. For a fuller account, see Newton Garver's preface to the translation of Derrida's *La voix et le phénomène* (*Speech and Phenomena*), p. xi.
2. Nietzsche, "On Truth and Falsity in Their Ultramoral Sense."
3. I refer to both textuality and intentionality because each offers an account of how the relation between word and thing (when it occurs) is mediated. While I contrast them here, I would also hope to be able to relate these often opposed perspectives on some other occasion.
4. See Derrida's *Éperons: les styles de Nietzsche* (*Spurs: Nietzsche's Styles*).
5. Heidegger understands language in the broadest possible way so as to include dreams, actions, and gestures. Man is always speaking. See, for example, "Die Sprache" (1959) in Martin Heidegger, *Unterwegs zur Sprache* (Pfullingen: Neske, 1971) ("Language" in *Poetry, Language, Thought*), pp. 189–210.
6. From Heidegger, "Language," pp. 191–92.
7. Heidegger, "What Are Poets For?" in *Poetry, Language, Thought*, pp. 91–142.
8. Martin Heidegger, *Unterwegs zur Sprache* (Pfullingen: Neske, 1971) (*On the Way to Language* [New York: Harper & Row, 1971]).
9. See Martin Heidegger, *The Question of Being* (with the original *Zur Seinsfrage* [1956], trans. William Kluback and Jean T. Wilde [New Haven: College and University Press, 1958]).
10. Eramus Schöfer's valuable work is represented by his "Heidegger's Language," in *On Heidegger and Language*.
11. I draw here on Derrida's remarks at the end of "The Ends of Man" (1968), in *Margins of Philosophy*, pp. 111–36.
12. Consider this: "Heidegger's text is extremely important to me, and constitutes a novel, irreversible advance all of whose critical resources we are far from having exploited." Derrida, *Positions*, p. 54. I have pursued the general question of Derrida's relation to Heidegger in "Heidegger after Derrida," *Research in Phenomenology* 18 (1988).
13. In "The Ends of Man" and in "Structure, Sign, and Play in the Discourse of the Human Sciences," in Derrida, *Writing and Difference*.
14. In part, the thematics of strategy is only a recognition of the limited *economy* of such texts.
15. In Derrida, *Marges de la philosophie* (*Margins of Philosophy*).

16. For an expanded treatment, see my "Introduction to Derrida" (1979) in *Radical Philosophy Reader*, pp. 18–42.
17. See Derrida's list of varieties of presence from *Of Grammatology*, p. 12, already quoted p. 285. Another list can be found in Heidegger, "Time and Being," p. 7.
18. For Derrida's account of this procedure, see *Positions*, p. 41.
19. A more detailed account would plot the development in Derrida's general strategy of deconstruction from (1) the display of what has been called the infrastructure of a text, "leaving everything as it is" (see, for example, *Speech and Phenomena*), through (2) disruptive intervention (parts of *Of Grammatology*, for example), to (3) parasitical production (for example, *Marges de la philosophie* [*Margins of Philosophy*]) and *Glas* (Paris: Galilee, 1974) (*Glas*, trans. John P. Leavy and Richard Rand [Lincoln: University of Nebraska, 1987]).
20. My second thoughts on this question of textual idealism can be found in "Beyond Deconstruction?" in *Contemporary French Philosophy*, ed. A. Phillips Griffiths (Cambridge University Press, 1988).
21. See Gayatri Spivak, Introduction, *Of Grammatology*.
22. Derrida talks of it as a strategy without a telos.
23. In *Twilight of the Idols*, pp. 40–41.

POSTSCRIPT TO THE QUESTION OF STRATEGY

1. See Rorty, *Philosophy and the Mirror of Nature*, pp. 372,377, and elsewhere.
2. This section, for example, gives center stage to the term *"différance"* in part because Derrida's paper by that name was the focus of the colloquium in which it was originally presented. Also see Michel Foucault, "Mon corps, ce papier, ce feu" (1972), translated in the *Oxford Literary Review* 4, no. 1 (1979); and Pierre Macherey, *A Theory of Literary Production* (1966), trans. G. Wall (London: Routledge and Kegan Paul, 1978).
3. See my "Following Derrida," in *Deconstruction and Philosophy: The Texts of Jacques Derrida*, ed. John Sallis (Chicago: University of Chicago Press, 1987).
4. Derrida, "Différance."
5. Derrida writes, "There is no such thing as a 'metaphysical concept' ... The 'metaphysical' is a certain determination taken by a sequence or 'chain.' It cannot as much be opposed by a concept, but rather by a process of textual labour and a different sort of articulation." Derrida, "Hors Livre," p. 12 ("Outwork," p. 6).
6. Derrida, *Positions*, pp. 26ff.
7. To be fair, Derrida, elsewhere in *Positions*, distinguishes playing the role of a concept from producing conceptual effects. It is the latter that, more guardedly, *"différance"* can produce, pp. 40ff.
8. See Derrida, *Of Grammatology*.
9. See Derrida, "Différance," in *Margins*, p. 157. Derrida uses the word "différence" here because he is working at this point with Heidegger's ontico-ontological difference.
10. Derrida, "Structure, Sign, and Play in the Discourse of the Human Sciences," p. 292.
11. See David Farrell Krell's translation in *Martin Heidegger: Basic Writings*, pp. 100f.
12. Ibid., pp. 101f.
13. A longer discussion of this question of strategy can be found in my "Différance

and the Problem of Strategy" in Wood and Bernasconi, *Derrida and Différance*.

14. Jacques Derrida, "De l'économie restreinte à l' économie générale: une Hegelianisme sans réserve," in *L'ècriture et la différence* (Paris: Seuil, 1967) ("From Restricted to General Economy: A Hegelianism Without Reserve," in *Writing and Difference*.

15. Derrida, *Positions*, pp. 39–96.

16. Jacques Derrida, "From Restricted to General Economy," pp. 255–56.

17. Derrida, "*Ousia* and *Gramme*," p. 62.

18. Derrida, "Différance," in *Margins*, p. 20.

19. Derrida, "Structure, Sign, and Play," p. 280.

20. In his paper "Joining the Text: From Heidegger to Derrida" (in *The Yale Critics: Deconstruction in America*, ed. Jonathan Arac et al. [Minneapolis: Minnesota University Press, 1983]), Rodolphe Gasché reminds us of Derrida's claim that the thought of the trace can no more "break with a transcendental phenomenology than be reduced to it," and argues, in a way intended as an explication of Derrida's position, that the concept of "text" allows something like an appropriative displacement of the value of transcendentality. Thus, "the transcendental gesture in Derrida simultaneously seems to escape the danger of naive objectivism and the value of transcendentality itself." Or again, "The notion of text, as already in the Heideggerean notion of Being, literally 'occupies' the locus of the transcendental concept, which is to say that the former is not identical with the latter. Thus the notion of the text corresponds to a transformation of the transcendental concept and of the very locus it represents . . . the notion of the text in Derrida can be understood only if one is aware of its function and its effects with regard to the transcendental." (pp. 160–61) Nothing Gasché writes can be ignored, and this in particular seems like a definitive reply to our attempt to circumscribe Derrida within a new transcendentalism. What it would require of us is that we abandon any attempt to attribute transcendental causality to particular operations, functions, or activities (such as *différance*) and concentrate on the "text," the field in which such "operations" would "take place." The question then is how successfully one can explain a concept of text that is neither an empirical object, nor an ideal object, nor a dialectical concept, and so on. Gasché's solution proceeds via the idea of a displacement. The text "'occupies' the locus of the transcendental concept which is to say that the former is not identical with the latter." We would make three responses to this approach: (1) that the question remains, what kind of acquiesence or acceptance is required by the reader for the concept of text to have the *force* that derives from its occupying such a position without satisfying the condition (of being "transcendental") that the "position" requires; (2) Gasché says that the text "literally" "occupies" the "locus." What sort of schema is being deployed here? Is it not *transcendental space?* (3) Gasché denies that the text supplies "a priori conditions of possibility . . . for meaning." But how does he deal with those remarks of Derrida in which "presence" is said to be the "product" of "*différance*"? And surely Gasché *is* committed to textuality as the condition (in *some* sense) of meaning? Is the argument over a "priori"?

In my view, Gasché correctly relocates the difficulty (from "*différance*" to "textuality"), but the problems travel too.

I must reserve for another occasion a considered response to Rodolphe Gasché's quite excellent *The Tain of the Mirror* (Cambridge: Harvard, 1986), which has fallen into my hands too late in the day.

21. Derrida, "*Ousia* and *Gramme*," p. 65.
22. Ibid., pp. 46ff.

PART 4, CHAPTER 4

1. See Part 4, chapter 5.
2. I say imperfect because calendars always have a nonarbitrary zero point, which introduces a new asymmetry in the series, and because the linearity of calendars is enriched by the addition of various annual cycles (year, month) and is itself based on a unit that is a cycle—the day.
3. Martin Heidegger, "The Concept of Time in the Science of History" (1916) *Journal of the British Society for Phenomenology* 9, no. 1, (January 1978): pp. 3–10.
4. See, for example, Bertrand Russell, *Introduction to Mathematical Philosophy* (London: Allen & Unwin, 1919), p. 33. He describes the features of an ordered series as 1. aliorelative (or asymmetrical), 2. transitive, 3. connected.
5. "It may further be an interesting study to establish how time which is posited in a time-consciousness as Objective is related to real Objective time . . ." Husserl, *PITC*, p. 23.
6. See the preface to Barthes, *Sade, Fourier, Loyola*, pp. 8–9.
7. The mouthpiece for this discussion being Roquentin in Sartre's novel *Nausea* (Jean-Paul Sartre, *La nausée* [Paris: Gallimard, 1938] *Nausea*, trans. L. Alexander [Harmondsworth: Penguin, 1965]).
8. See Heidegger, *Being and Time*, H233, on the need for an account that treats "Dasein . . . as a whole."
9. This remark appears as part of the answer to a question in the introduction to the collection of translations, Martin Heidegger, *The End of Philosophy*, trans. Joan Stambaugh (New York: Harper & Row, 1973), p. xii.
10. For example: "what characterizes metaphysical thinking which grounds the ground for beings is the fact that metaphysical thinking departs from what is present in its presence, and thus represents it in terms of its ground as something grounded," in "The End of Philosophy and the Task of Thinking" (1966), in Heidegger, *On Time and Being*.
11. Derrida, "Linguistics and Grammatology," in *Of Grammatology*, and "Structure, Sign, and Play in the Discourse of the Human Sciences."
12. For example, Derrida remarks that "the concept of time belongs entirely to metaphysics and designates the domination of presence." "*Ousia* and *Gramme*" (1968).
13. See our discussion, pp. 111–134.
14. See Derrida, "Grammatology as a Positive Science," p. 87.
15. This point is made at length in the next chapter.
16. "There will not be books in the running brooks until the advent of hydro-semantics." J.L. Austin, "Truth," in his *Philosophical Papers* (Oxford: Clarendon, 1961), p. 94, n. 2.
17. Nietzsche wrote: "When it is trodden on a worm will curl up. This is prudent. It thereby reduces the chance of being trodden on again. In the language of morals: *humility*." "Maxims and Arrows," in *Twilight of the Idols*, p. 26.
18. It is worth noting that Heidegger's increasing distance from the existential and the ontic is accompanied by the growth of a range of expressions (opening, clearing, dwelling, house, giving, granting) in which an existential content seems

to be preserved metaphorically. But in a number of places Heidegger raises certain obstacles to a straightforward metaphorical reading. The best discussion of this is to be found in Derrida's "The Retrait of Metaphor," pp. 5–34.

19. It is true that Heidegger himself discusses these remarks of Nietzsche. In brief, he interprets them as accurate reports of the nihilistic state of Western culture. See Heidegger, *Einführung in die Metaphysik* (*An Introduction to Metaphysics*), pp. 29f. Nietzsche's remarks are to be found in *Twilight of the Idols*, p. 37.

PART 4. CHAPTER 5

1. See p. 319f.
2. Peter Geach, "Some Problems About Time," *Logic Matters*, pp. 302f.
3. Introduction, Michel Foucault, *The Archaeology of Knowledge* (London:Tavistock, 1972), p. 3.
4. "Textual Reflexivity and Totalization in Hegel," in my *Philosophy and Style* (London: Hutchinson, 1988 [forthcoming]).
5. Maurice Merleau-Ponty, *Le visible et l'invisible* (Paris: Gallimard, 1964) (*The Visible and the Invisible*, trans. Alphonso Lingis [Evanston: Northwestern University Press, 1968]).
6. See pp. 293–318.
7. See Alfred Schutz, "Making Music Together: A Study in Social Relationship," in his *Collected Works*, vol. 2, ed. Maurice Natanson (The Hague: Nijhoff, 1966).
8. The following two papers are extremely similar: Paul Ricoeur, "The Human Experience of Time and Narrative," *Research in Phenomenology* 9 (1979), pp. 17–34; and Ricoeur, "Narrative Time," in *On Narrative*, ed. W.J.T. Mitchell (Chicago: University of Chicago Press, 1981). I have not, I regret, been able to take into account Ricoeur's more recent volumes, *Temps et Récit I* (Paris: Seuil, 1983); *Temps et Récit II* (Paris: Seuil, 1984); *Temps et Récit III* (Paris: Seuil, 1985); nor David Carr's stimulating *Time, Narrative and History* (Bloomington: Indiana University Press, 1986).
9. See W. B. Gallie, *Philosophy and Historical Understanding* (London: Chatto and Windus, 1964).
10. Frederic Jameson, *The Political Unconscious: Narrative as a Socially Symbolic Act* (London: Methuen, 1981), p. 93.
11. Lyotard, *La Condition Postmoderne* (*The Postmodern Condition*).
12. Roland Barthes, "Introduction à l'analyse structurale des récits, " in *Communications 8* (Paris: Seuil, 1966), p. 1 ("Introduction to the Structural Analysis of Narrative," in *Image-Music-Text*, trans. Stephen Heath [London: Fontana, 1977]).
13. W.J.T. Mitchell, "Foreword" to *On Narrative*, pp. vii–x.
14. Barthes, "Structural Analysis," pp. 98–99.
15. Barthes, "Structural Analysis," pp. 99–100.
16. Claude Lévi-Strauss, *Le cru et le cuit: Mythologiques 1* (Paris: Plon, 1964) (*The Raw and the Cooked*, trans. John Weightman and Doreen Weightman [New York: Harper & Row 1970]); *Du miel aux cendres: Mythologiques 2* (Paris: Plon, 1967) (*From Honey to Ashes*, trans. John Weightman and Doreen Weightman [New York: Harper & Row, 1973]); *L'Origine des manières de table: Mythologiques 3* (Paris: Plon, 1968) (*The Origin of Table Manners*, trans. John Weightman and Doreen Weight-

man [New York: Harper & Row, 1978]); *L'homme nu: Mythologiques 4* (Paris: Plon, 1971).

17. Roland Barthes, *Essais Critiques* (Paris: Seuil, 1964) (*Critical Essays*, trans. Richard Howard [Evanston: Northwestern University Press, 1972]).

18. Barthes, *Critical Essays*, p. 158.

19. Ibid.

20. Lévi-Strauss, *The Raw and the Cooked*, p. 24.

21. Ibid.

22. Ricoeur, "Human Experience," p. 18.

23. Ibid, p. 22.

24. Both Roland Barthes (in "Structural Analysis") and Gerard Genette (in his "Time and Narrative in *À la recherche du temps perdue*" in *Aspects of Narrative*, ed. J. Hillis Miller [New York: Columbia University Press, 1971]) suggest that narratives are expansions of simple sentences (such as "Marcel becomes a writer," "I am walking . . ."). Barthes writes that "a narrative is a long sentence."

25. Sartre, *La nausée*.

26. Jean-Paul Sartre, *Les Mots* (Paris: Gallimard, 1964) (*Words*, trans. Irene Clephane [London: Hamish Hamilton, 1964]); *Saint Genet* (Paris: Gallimard, 1954) (*Saint Genet*, trans. Bernard Frechtman [New York: Brazillier, 1963]); and *L'idiot de la famille: Gustav Flaubert de 1821 à 1857*, 3 vols. (Paris: Gallimard, 1971; 1971; 1972).

27. Hugh J. Silverman, "The Time of Autobiography," in *Time and Metaphyhsics*, ed. David Wood and Robert Bernasconi (Warwick: Parousia Press, 1982), pp. 39–65.

28. Claude Lévi-Strauss, *Tristes Tropiques* (Paris: Plon, 1955) (*Tristes Tropiques*, trans. J. Russell [New York: Atheneum, 1961]).

29. See William Golding's *Pincher Martin* (Harmondsworth: Penguin, 1959) and the film *Incident at Owl Creek*.

30. Barthes, "Structural Analysis," p. 113.

31. Barthes, "Structural Analysis."

32. Genette, "Time and Narrative."

33. See the film *Rashomon*, directed by Akira Kurosawa.

34. A good example would be Shevek in Ursula Le Guin's novel *The Dispossessed* (London: Gollancz, 1974).

35. See Roman Jakobson, "Linguistics and Poetics," in *Style and Language*, ed. T. A. Sebeok (New York: 1960).

36. See Jacques Derrida's discussion of Blanchot's *Le folie du jour*, in his "The Law of Genre," trans. Avital Ronell, in *On Narrative*, pp. 51–78.

37. Ibid.

38. An analogous claim is harbored both in Barthes's claim (turning Saussure inside out) that semiology is enclosed within a more general linguistics (*Elements of Semiology* [1964], trans. Annette Lavers and Colin Smith [London: Cape, 1967]), and Lacan's denial of the possibility of a metalanguage. (See Lacan, *Écrits*, p. 150.) To the reappropriating power of narrativity there corresponds a theoretical and practical necessity to explain, introduce, and locate one's models, structures, and matrixes via "ordinary language." This is no temporary requirement.

39. Derrida writes: "There is no madness without the law: madness cannot be conceived before its relation to the law. Madness is law: the law is madness."

"The Law of Genre," p. 77. Compare Pascal's "Men are so necessarily mad, that not to be mad would be another form of madness," cited by Foucault in his preface to *Madness and Civilisation*(1961), trans. Richard Howard (London: Tavistock, 1967), p. xi.

PART 4, CHAPTER 6

1. Ludwig Feuerbach, *Grundsätze der Philosophie der Zukunft* (1943) (*Principles of the Philosophy of the Future*, trans. M. H. Vogel [New York: Bobbs-Merrill, 1966]) section 65.
2. Ibid.
3. I "adopt" this terms for public expediency; I do not attempt to give it an analytical treatment.
4. Cornel West, "Nietzsche's Prefiguration of Postmodern American Philosophy," *Boundary 2* 9, no. 3; and 10, no. 1 [Why Nietzsche Now?], ed. Daniel O'Hara (Spring/Fall 1981): p. 265.
5. Ibid.
6. Nietzsche, *The Will to Power*, p. 3
7. Ibid.
8. Ibid.
9. Ibid.
10. Ibid., p. 48.
11. See the discussion following Derrida's presentation of "Structure, Sign, and Play in the Discourse of the Human Sciences," in *The Structuralist Controversy*, ed. Richard Macksey and Eugenio Donato (Baltimore: Johns Hopkins University Press, 1972), p. 271.
12. Does Levinas have the obvious response to this position? His use of "finite" and "infinite" is distinctly idiosyncratic; at some critical point my actual finitude is surely presupposed by the importance of fecundity.
13. See Thomas McCarthy's introduction to: Habermas, *Legitimation Crisis* (London: Heinemann, 1976). He is in fact referring to Horkheimer here.
14. Husserl, "Philosophy as a Rigorous Science," in *Phenomenology and the Crisis of Philosophy*, p. 71.
15. Ibid.
16. Ibid.
17. An English translation of this 1795 essay can be found in *Kant's Political Writings*, ed. Hans Reiss (Cambridge: Cambridge University Press, 1971).
18. See the end of the "Exergue" to Derrida's *Of Grammatology*, p. 5.
19. If biological death is not the issue, the structure of such a paradigm shift, or epistemological break, would *mirror* Heidegger's reading of the existential bearing of death.
20. This is the guiding thematic of Vincent Descombes's *Modern French Philosophy*. The French title—*Le Même et l'Autre* (Paris: Minuit, 1979)—is more revealing.
21. See Lyotard, *La condition postmoderne* (The Posmodern Condition).
22. Derrida, "Exergue," p. 4.
23. Ibid.
24. Derrida, "Structure, Sign, and Play" p. 280.
25. Derrida, *Positions*, pp. 6–7.
26. Ibid.

27. See Derrida's *Éperons: les styles de Nietzsche (Spurs: Nietzsche's Styles)*.
28. Ibid.
29. Derrida, "Structure, Sign, and Play," p. 293.
30. See "Die Sprache" (1959) in Heidegger, *Unterwegs zur Sprache* ("Language," in *Poetry, Language, Thought*, p. 190).
31. Derrida, "Structure, Sign, and Play" (see n. 11 above), p. 267.
32. See pp. 9–36.
33. See Derrida, "Différance," p. 27.
34. See Derrida, "*Ousia* and *Gramme*," p. 63.
35. In Derrida, "Of Grammatology," pp. 85ff.
36. Derrida, *Positions*, p. 58.
37. Ibid., pp. 57, 59.
38. Ernst Bloch, *Tübinger Einleitung in die Philosophie*, vol. 1 (Frankfurt: Suhrkamp, 1963) (*A Philosophy of the Future*, trans. J. Cumming [New York: Herder & Herder, 1970], p. 143).
39. See, for example, Lyotard, *The Postmodern Condition*, and Rorty, *Philosophy and the Mirror of Nature*, esp. Part 3.
40. Immanuel Kant, *Kritik der reinen Vernunft* (Hamburg: Felix Meiner, 1956), p. 70 Bxix. (*Critique of Pure Reason*, trans. Norman Kemp Smith [London: Macmillan,1964], p. Bxix).
41. Ibid., pp. Axii, Axiii.
42. G. W. F. Hegel, *Phänomenologie des Geistes* (*Phenomenology of Spirit*), p.
43. Husserl, "Philosophy as a Rigorous Science," p. 70.
44. Ibid, p. 72.
45. Immanuel Kant, "Von Einem neuerdings erhobenen Vornehmen Ton in der Philosophie" (1796) ("On a Genteel Tone Recently Adopted in Philosophy" [untranslated]).
46. I have Kierkegaard to thank for this reference. See his *Concluding Unscientific Postscript*, p. 97f.
47. Maurice Merleau-Ponty, *Phénoménologie de la Perception* (Paris: Gallimard, 1945) (*Phenomenology of Perception*, trans. Colin Smith [London: Routledge and Kegan Paul, 1962, preface]).
48. Ibid.
49. The parallel with the remark from Derrida quoted in n. 31 is worth noticing.
50. Merleau-Ponty, "Phenomenology of Perception," p. xxi.
51. Although Hegel's "circular" understanding of his thought and method already puts this into question. Tom Rockmore's *Hegel's Circular Epistemology* (Bloomington: Indiana University Press, 1986) pursues this question thematically.
52. Nietzsche cannot always be interpreted along these lines. Consider: "Formula of my happiness: a Yes, a No, a straight line, a *goal*" (*Twilight of the Idols*, p. 27).
53. For example, John Sallis, "End(s)," in his *Delimitations* (Bloomington: Indiana University Press, 1986), pp. 128–38.
54. Jacques Derrida, "D'un ton apocalyptique adopté naguère en philosophie," in *Les Fins de l'homme* (proceedings of the 1980 Cerisy Colloquium) (Paris: Galilee, 1981). It is translated "An Apocalyptic Tone Recently Adopted in Philosophy," in the *Oxford Literary Review*, and by John P. Leavey in *Semeia* 23 (1982), pp. 63–97 from which I am quoting here.
55. Ibid., p. 80.
56. Ibid., p. 81.
57. Ibid., p. 90.

58. This should be unpacked as a "deconstruction" of the textual deployment of the term "future" and its cognates.
59. See the last page of Derrida, "Différance."
60. Derrida, "D'un ton," p. 87.
61. Derrida's focus on the word "come" could be said to embody the very power he attributes to it. The question of reflexive exemplification is developed in "Philosophy as Performance" in my *Philosophy and Style* (London: Hutchinson, 1988).
62. See p. 293ff.
63. See chapter 1.
64. Derrida makes a great deal, as I have not, of what Kant disapprovingly calls the "tone" of apocalyptic discourse. And the apocalyptic "come" is inseparable from a kind of tone. *Tone*, for Derrida, could be said to relate to the performative, ethical aspects of "come," in the way that "style" relates to "content." In isolating the tone of apocalyptic utterance, Kant unsuspectingly made quite a revolutionary move; tone is unrepresentable.
65. The *es gibt* is "only" metaphorically interpersonal, and Heidegger does not, to my knowledge, exploit this.
66. See Emmanuel Levinas, *Totalité et infini* (The Hague: Nijhoff, 1961) (*Totality and Infinity*, trans. Alphonso Lingis [The Hague: Nijhoff, 1969]).
67. Again, David Farrell Krell's retrieval of Heidegger's Marburg lectures makes it abundantly clear how easily these remarks could be grafted onto Heidegger's lifelong preoccupation with openness in its manifold forms. See his *Intimations of Mortality* (University Park: Pennsylvania University Press, 1986).

Bibliography

Alderman, Harold. *Nietzsche's Gift*. Athens: Ohio University Press, 1977.

Allison, David B., ed. *The New Nietzsche*. New York: Dell, 1977.

Arac, Jonathan, et al., eds. *The Yale Critics: Deconstruction in America*. Minneapolis: Minnesota University Press, 1983.

Aristotle. *Physics*. Edited by W. D. Ross. Oxford: Clarendon, 1936.

Austin, J. L. " A Plea for Excuses." In his *Philosophical Papers*. Oxford: Clarendon, 1961.

Austin, J. L. "Truth." In his *Philosophical Papers*. Oxford: Clarendon, 1961.

Barthes, Roland. *Essais critiques*. Paris: Seuil. 1964. (*Critical Essays*. Translated by Richard Howard. Evanston: Northwestern University Press, 1972).

———. "Introduction à l'analyse structurale des récits." *Communication* 8, 1966. ("Introduction to the Structural Analysis of Narrative." Translated by Stephen Heath. In *Image-Music-Text*. London: Fontana, 1977).

———. *Sade/ Fourier/ Loyola*. Paris: Seuil, 1971. (*Sade/ Fourier/ Loyola*. Translated by Richard Miller. New York: Hill & Wang, 1976).

———. *Le Plaisir du texte*. Paris: Seuil, 1973. (*The Pleasure of the Text*. Translated by Richard Miller. New York: Hill and Wang, 1975).

Bernasconi, Robert. *The Question of Language in Heidegger's History of Being*. Atlantic Highlands: Humanities Press, 1985.

Bloch, Ernst. *Tübinger Einleitung in die Philosophie*. Vol. 1. Frankfurt Suhrkampf, 1963. (*A Philosophy of the Future*. Translated by John Cumming. New York: Herder & Herder, 1970).

Brough, John B. "The Emergence of an Absolute Consciousness in Husserl's Early Writings on Time Consciousness." In *Husserl: Expositions and Appraisals*, edited by Frederic Elliston and Peter MacCormick. Notre Dame/London: Notre Dame University Press, 1977.

Carr, David. *Time, Narrative, and History*. Bloomington: Indiana University Press, 1986.

Casey, Edward. *Remembering: A Phenomenological Study*. Bloomington: Indiana University Press, 1987.

Danto, Arthur. "The Eternal Recurrence." In *Nietzsche: A Collection of Critical Essays*, edited by Robert Solomon. Garden City, N.Y.: Anchor, 1973.

Deleuze, Gilles. *Nietzsche et la philosophie*. Paris: Presses Universitaires de France, 1962. (*Nietzsche and Philosophy*. Translated by Hugh Tomlinson. Athlone: London, 1983).

De Man, Paul. "The Epistemology of Metaphor." In *On Metaphor*, edited by Sheldon Sacks, pp. 11–28. Chicago: University of Chicago Press, 1979.

Derrida, Jacques. *Translation and Introduction to Edmund Husserl's L'Origine de la Géometrie*. Paris Presses Universitaires de France, 1962. (*Edmund Husserl's Origin of Geometry: An Introduction*. Translated by John P. Leavy and edited by David B.

Allison. New York: Hays, 1977).

————. *L'écriture et la différence*. Paris: Seuil, 1967. (*Writing and Difference*. Translated by Alan Bass. Chicago: University of Chicago Press, 1978).

————. *De la Grammatologie*. Paris: Minuit, 1967. (*Of Grammatology*, translated by Gayatri C. Spivak. Baltimore: Johns Hopkins University Press, 1976).

————. *La voix et le phénomène*. Paris: Presses Universitaires de France, 1967. (*Speech and Phenomena, and Other Essays on Husserl's Theory of Signs*. Translated by David B. Allison. Evanston: Northwestern University Press, 1973).

————. *La dissémination*. Paris: Seuil, 1972. (*Dissemination*. Translated by Barbara Johnson. Chicago: University of Chicago Press, 1981).

————. *Éperons: les styles de Nietzsche*. Paris: Flammarion, 1978. (*Spurs: Nietzsche's Styles*. Translated by Barbara Harlow. Chicago: University of Chicago Press, 1979).

————. *Marges de la philosophie*. Paris: Minuit, 1972. (*Margins of Philosophy*. Translated by Alan Bass. Chicago: University of Chicago Press, 1978).

————. *Glas*. Paris: Galilee, 1974. (*Glas*. Translated by John P. Leavy and Richard Rand. Lincoln: University of Nebraska, 1987).

————. "Le retrait de la metaphor." PO&SIE (1979): 103–26. ("The *Retrait* of Metaphor." *Enclitic* 2, no. 2, Fall 1978: 6–33).

————. "D'un ton apocalyptique adopté naguère en philosophie." In *Les Fins de l'homme: à partir du travail de Jacques Derrida*, edited by Philippe Lacoue-Labarthe and Jean-Luc Nancy, pp. 445–86. Paris: Galilee, 1981. (Translated by John P. Learey "On an Apocalyptic Tone Recently Adopted in Philosophy." *Oxford Literary Review* and *Semeia* 23 (1982): pp. 63–97).

Descartes, René. *The Philosophical Works of Descartes*. Vol. 1, *Rules for the Direction of the Mind*. Translated by Elizabeth S. Haldane and G. R. T. Ross. Cambridge: Cambridge University Press, 1911.

Descombes, Vincent. *Le Même et l'Autre*. Paris: Minuit, 1979. (*Modern French Philosophy*. Translated by L. Scott-Fox and J. M. Harding. Cambridge: Cambridge University Press, 1980).

Elliston, Frederick. *Heidegger's Existential Analytic*. The Hague: Mouton, 1973.

Feuerbach, Ludwig. *Grundsätze der Philosophie der Zukunft* (1943). (*Principles of the Philosophy of the Future*. Translated by M. H. Vogel. New York: Bobbs-Merrill, 1966).

Foucault, Michel. *L'archéologie du savoir*. Paris: Gallimard, 1969. (*The Archaeology of Knowledge*. Translated by Alan Sheridan. London: Tavistock, 1972).

————. "Mon corps, ce papier, ce feu." In preface to *Folie et déraison, Histoire de la folie à l'âge classique*. 2d ed. Paris: Gallimard, 1972. Translated in *Oxford Literary Review* 4, no. 1 (1979).

Gadamer, Hans-Georg. "Concerning Empty and Ful-filled Time." *Southern Journal of Philosophy* 8 (1970): 341–53.

Gallie, William. *Philosophy and the Historical Understanding*. London: Chatto & Windus, 1964.

Garver, Newton. "Derrida and Rousseau on Writing." *Journal of Philosophy*, 1977.

Gasché, Rodolphe. "Joining the Text: From Heidegger to Derrida," In *The Yale Critics: Deconstruction in America*, edited by Jonathan Arac et al. Minneapolis: Minnesota University Press, 1983.

————. *The Tain of the Mirror: Derrida and the Philosophy of Reflection*. Cambridge: Harvard, 1987.

Geach, Peter. "Some Problems about Time." In his *Logic Matters*. Oxford: Blackwell, 1972.

Genette, Gerard. "Time and Narrative in *À la recherche du temps perdue*." In *Aspects of Narrative*, edited by J. Hillis Miller. New York: Columbia University Press, 1971.

Golding, William. *Pincher Martin*. Harmondsworth: Penguin, 1959.

Habermas, Jurgen. *Legitimationsprobleme in Spätkapitalismus*. Frankfurt: Suhrkampf, 1973. (*Legitimation Crisis*. Translated by Thomas McCarthy. London: Heinemann, 1976).

Harrison, Ross. "The Concept of Pre-predicative Experience." In *Philosophy and Phenomenological Understanding*, edited by Edo Pivcevic. Cambridge: Cambridge University Press, 1975.

Hayman, Ronald. *Nietzsche: A Critical Life*. London: Quartet, 1980.

Hegel, G. W. F. *Phänomenologie des Geistes*. Hamburg: Felix Meiner, 1952. (*Phenomenology of Spirit*. Translated by A. V. Miller. Oxford: Clarendon, 1977).

Heidegger, Martin. "Der Zeitbegriffe in der Geschichtswissenschaft." *Zeitschrift für Philosophie und philosophische Kritik* 161 (1916): 173–88. ("The Concept of Time in the Science of History." *Journal of the British Society for Phenomenology* 9, no. 1 [January 1978]: 3–10).

———. *Sein und Zeit*. In *Gesamtausgabe*, Vol. 2. Frankfurt: Klostermann, 1977. (*Being and Time*, translated by John Macquarrie and Edward Robinson. Oxford: Blackwell, 1967).

———. *Die Grundprobleme der Phänomenologie* (1927). Frankfurt: Klostermann, 1975. (*The Basic Problems of Phenomenology*. Translated by Albert Hofstadter. Bloomington, Indiana University Press, 1982).

———. *Kant und das Problem der Metaphysik* (1929). Frankfurt: Klostermann, 1973. (*Kant and the Problem of Metaphysics*. Translated by James S. Churchill. Bloomington: Indiana University Press, 1962).

———. *Was ist Metaphysik?* (1929). Bonn: Cohen, 1930. ("What Is Metaphysics?" Translated by David Farrell Krell. In *Martin Heidegger: Basic Writings*, edited by David Farrell Krell. London: Routledge, Kegan and Paul, 1977).
Note: the important "Postscript" (1943) and "Preface" (1949), translated by Walter Kaufmann, can be found in Kaufmann's *Existentialism from Dostoevsky to Sartre*, New York: New American Library, 1975.

———. *Nietzsche*. 2 vols. Pfullingen: Neske, 1961. (*Nietzsche*. Vol. 1: *The Will to Power as Art*. Translated by David Farrell Krell. New York: Harper & Row, 1979; *Nietzsche*. Vol. 2: *The Eternal Recurrence of the Same*. Translated by David Farrell Krell. New York: Harper & Row, 1984; *Nietzsche*. Vol. 4: *Nihilism*. Translated by Frank Capuzzi. New York: Harper & Row, 1982).

———. "Schöpferishe Landschaft: Warum bleiben wir in der Provinz?" (1934). In *Nachlese zu Heidegger*, edited by Guido Schneeberger. Berne: Francke, 1962. ("Why Do I Stay in the Provinces?" Translated by Thomas Sheehan. In *Heidegger, the Man, and the Thinker*. Chicago: Precedent, 1981).

———. "Brief über den Humanismus." (1947) In Martin Heidegger, *Wegmarken*. Frankfurt: Klostermann, 1976. ("Letter on Humanism" [1962]. Translated by Frank Capuzzi and J. Glenn Gray. In *Martin Heidegger: Basic Writings*. London: Routledge and Kegan Paul, 1978).

———. "Der Ruckgang in der Grund der Metaphysik" (1949). In Martin Heidegger, *Wegmarken*. Frankfurt: Klostermann, 1976. ("The Way Back into the Fundamental Ground of Metaphysics." Translated by Walter Kaufmann. In

Existentialism from Dostoevsky to Sartre, edited by Walter Kaufmann. New York: New American Library, 1975).

———. "Der Ursprung des Kunstwerkes" (1950). In Martin Heidegger, *Holzwege*. Frankfurt: Klostermann, 1977. ("The Origin of the Work of Art" [1965]. Translated by Albert Hofstadter. In *Martin Heidegger: Basic Writings*, edited by David Farrell Krell. New York: Harper & Row, 1977).

———. "Überwindung der Metaphysik" (largely published as "Seinsverlassenheit und Irrnis" [1951]). In Martin Heidegger, *Vorträge und Aufsätze*. Pfullingen: Neske, 1967. ("Overcoming Metaphysics." Translated by Joan Stambaugh. In Martin Heidegger, *The End of Philosophy*. New York: Harper & Row, 1973).

———. *Was heisst Denken?* (1954). Tübingen: Niemeyer, 1971. (*What Is Called Thinking?* [1968]. Translated by Fred. D. Wieck and J. Glenn Gray. New York: Harper & Row, 1972).

———. *Vorträge und Aufsätze* (1954). Pfullingen: Neske, 1967.

———. "Zur Seinsfrage" (1955). In Martin Heidegger, *Wegmarken*. Frankfurt: Klostermann, 1976. (*The Question of Being* [1958]. Translated by William Klubach and Jean Wilde. London: Vision, 1959).

———. *Identität und Differenz*. Pfullingen: Neske, 1957. (*Identity and Difference*. Translated by Joan Stambaugh. New York: Harper & Row, 1969).

———. "Gelassenheit" (1959). In Martin Heidegger, *Gelassenheit*. Pfullingen: Neske, 1960. ("Memorial Address." In *Discourse on Thinking*. Translated by John M. Anderson and E. Hans Freund. New York: Harper & Row, 1966).

———. "Zeit und Sein" (1962). In *Zur Sache des Denkens*. Tübingen: Niemeyer, 1969. ("Time and Being" [1972]. Translated by Joan Stambaugh. In Martin Heidegger, *On Time and Being*. New York: Harper & Row, 1972).

———. "Das Ende der Philosophie und die Aufgabe des Denkens" (1966). In *Zur Sache des Denkens*. Tübingen: Neimeyer, 1969. ("The End of Philosophy and the Task of Thinking." Translated by Joan Stambaugh. In *Martin Heidegger: Basic Writings*. London: Routledge & Kegan Paul, 1978).

———. "Die Sprache." In Martin Heidegger, *Unterwegs zur Sprache*. Pfullingen: Neske, 1971. ("Language." Translated by Albert Hofstadter. In Martin Heidegger, *Poetry, Language, Thought*. New York: Harper & Row, 1971).

———. *Unterwegs zur Sprache*. Pfullingen: Neske, 1971. (*On the Way to Language*. Translated by Peter Hertz. New York: Harper & Row, 1971).

———. *Early Greek Thinking*. Translated by David Farrell Krell and Frank A. Capuzzi. New York: Harper & Row, 1975.

———. *Metaphysische Anfangsgrunde der Logik im Ausgang von Leibniz* (1928). Frankfurt: Klostermann, 1978. (*The Metaphysical Foundations of Logic*. Translated by Michael Heim. Bloomington: Indiana University Press, 1984).

Husserl, Edmund. *Logische Untersuchungen*. 2 vols. Edited by U. Panzer. The Hague, Nijhoff, 1982. (*Logical Investigations*. 2 vols. Translated by J. N. Findlay. London: Routledge & Kegan Paul, 1970).

———. *Vorlesungen zur Phänomenologie des inneren Zeitbewusstseins* (1905–10). Edited by Martin Heidegger. In *Jahrbuch für Phänomenologie* 9 (1928). (*The Phenomenology of Internal Time-Consciousness*. Translated by James S. Churchill. Bloomington: Indiana University Press, 1964).

———. "Philosophie als Strenge Wissenschaft." *Logos* 1 (1910–11). ("Philosophy as a Rigorous Science." Translated by Quentin Lauer. In Edmund Husserl, *Phenomenology and the Crisis of Philosophy*, pp. 71–147. New York: Harper & Row, 1965).

————. *Ideen zu einer reinen Phänomenologie und phänomenologischen Philosophie.* In *Jahrbuch für Philosophie und Phänomenologischen Forschung,* vol. 1. Halle: Niemeyer, 1913. (*Ideas: General Introduction to Pure Phenomenology.* Translated by W. R. Boyce Gibson. London: Routledge & Kegan Paul, 1931).

————. *Cartesianische Meditationen und Pariser Vorträge.* The Hague: Nijhoff, 1950. (Material dates from 1929/1933; there was already a French translation by Gabrielle Peiffer and Emmanuel Levinas in 1931.) (*Cartesian Meditations.* Translated by Dorion Cairns. The Hague: Nijhoff, 1960).

————. "Die Frage nach dem Ursprung der Geometrie als intentional-historische Problem." Edited by Eugen Fink. In *Revue internationale de philosophie* 1, no. 2 (1939). ("The Origin of Geometry." In Edmund Husserl, *The Crisis of European Sciences and Transcendental Phenomenology.* Translated by David Carr, pp. 353–78. Evanston: Northwestern University Press, 1970). See also Jacques Derrida, *Translation and Introduction to Edmund Husserl's L'Origine de la Géométrie.* Paris: Presses Universitaires de France, 1962. (*Edmund Husserl's Origin of Geometry: An Introduction.* Translated by John P. Leavy and edited by David B. Allison. New York: Hays, 1977).

————. *Die Krisis der europäischen Wissenschaften und die tranzendentale Phänomenologie* (1936). The Hague: Nijhoff, 1954. (*The Crisis of European Sciences and Transcendental Phenomenology.* Translated by David Carr, pp. 353–78. Evanston: Northwestern University Press, 1970).

Jakobson, Roman (with Morris Halle). *The Fundamentals of Language.* The Hague: Mouton, 1956.

Jakobson, Roman. "Linguistics and Poetics." In *Style and Language.* edited by Thomas A. Sebeok. Cambridge: MIT Press, 1960.

Jameson, Frederic. *The Political Unconscious: Narrative as a Socially Symbolic Act.* London: Methuen, 1981.

Kant, Immanuel. *Critik der reinen Vernunft* (1787). Hamburg: Felix Meiner, 1956. *Critique of Pure Reason,* translated by Norman Kemp Smith. London: Macmillan, 1964).

————. *Political Writings.* Edited and translated by Hans Reiss. Cambridge: Cambridge University Press, 1971.

Kierkegaard, Sören. *Concluding Unscientific Postscript* (1846). Translated by David F. Swenson and Walter Lowrie. Princeton: Princeton University Press, 1941.

Klossowski, Pierre. *Nietzsche et le cercle vicieux.* Paris: Mercure de France, 1969.

Kockelmans, Joseph J. "Heidegger on Time and Being." In *Martin Heidegger in Europe and America,* edited by Edward Ballard and Charles Scott. The Hague: Nijhoff, 1973.

Kraus, Oskar. "Towards a Phenomenognosy of Time-Consciousness." In *The Philosophy of Brentano,* edited by Linda L. McAlister. London: Duckworth, 1976.

Krell, David Farrell. "From Fundamental – to Frontalontologie." *Research in Phenomenology* 10 (1980): 208–34.

————. "Rapture: The Finitude of Time in Heidegger's Thought." In *Time and Metaphysics,* edited by David Wood and Robert Bernasconi, pp. 121–60. Warwick: Parousia Press, 1982.

————. *Intimations of Mortality.* University Park and London: Pennsylvania University Press, 1986.

Lacan, Jacques. *Écrits.* Paris: Seuil, 1966. (*Écrits.* Translated by Alan Sheridan. London: Tavistock, 1977).

LeGuin, Ursula. *The Dispossessed.* London: Panther, 1974.

Levinas, Emmanuel. *Totalité et Infini*. The Hague: Nijhoff, 1961. (*Totality and Infinity*. Translated by Alphonso Lingis. Pittsburgh: Duquesne University Press, 1969).

Lévi-Strauss, Claude. *Tristes Tropiques*. Paris: Plon, 1955. (*Tristes Tropiques*, translated by J. Russell. New York: Atheneum, 1961).

————. *Le Cru et le cuit: Mythologiques 1*. Paris: Plon, 1964. (*The Raw and the Cooked*. Translated by John Weightman and Doreen Weightman. New York: Harper & Row, 1970).

————. *Du miel aux cendres: Mythologiques 2*. Paris: Plon, 1967. (*From Honey to Ashes*. Translated by John Weightman and Doreen Weightman [New York: Harper & Row, 1973]).

————. *L'Origine des manières de table: Mythologiques 3*. Paris: Plon, 1968. (*The Origin of Table Manners*. Translated by John Weightman and Doreen Weightman. New York: Harper & Row, 1978).

————. *L'Homme nu: Mythologiques 4*. Paris: Plon 1971.

Llewelyn, John. *Beyond Metaphysics? The Hermeneutical Circle in Contemporary Continental Philosophy*. Atlantic Highlands: Humanities Press, 1985.

————. *Derrida on the Threshold of Sense*. London: Macmillan, 1986.

Löwith, Karl. *From Hegel to Nietzsche*. Translated by David Green. Garden City, N.Y.: Anchor, 1967.

Lyotard, Jean-François. *La condition postmoderne*. Paris: Minuit, 1979. (*The Postmodern Condition*. Translated by Geoff Bennington and Brian Massumi. Manchester: Manchester University Press, 1984).

Macherey, Pierre. *Pour une théorie de la production littéraire*. Paris: Maspero, 1966. (*A Theory of Literary Production*. Translated by G. Wall. London: Routledge and Kegan Paul, 1978).

Macksey, Richard, and Donato, Eugenio, eds. *The Structuralist Controversy*. Baltimore: Johns Hopkins University Press, 1970.

Marx, Werner. *Heidegger und die Tradition*. Stuttgart: Kohlhammer, 1961. (*Heidegger and the Tradition*. Translated by Theodore Kisiel and Murray Greene. Evanston: Northwestern University Press, 1971).

Merleau-Ponty, Maurice. *Phénoménologie de la perception*. Paris: Gallimard, 1945. (*Phenomenology of Perception*. Translated by Colin Smith. London: Routledge and Kegan Paul, 1962).

————. *Le visible et l'invisible*. Paris: Gallimard, 1964. (*The Visible and the Invisible*. Translated by Alphonso Lingis. Evanston: Northwestern University Press, 1968).

Mitchell, W. J. T., ed. *On Narrative* (Chicago: University of Chicago Press, 1981).

Mohanty, J. N. *Husserl and Frege* (Bloomington: Indiana University Press, 1982).

Nietzsche, Friedrich. *Werke: Kritische Gesamtausgabe*. Edited by Giorgio Colli and Mazzino Montinari. 30 vols. Berlin: de Gruyter, 1967–78. (Hereinafter cited as *KG*).

————. "Über Wahrheit und Luge im aussermoralischen Sinn." *KG*, Bk. 3, vol. 3 (1873). ("On Truth and Lies in a Nonmoral Sense." Translated by Donald Breazeale, in *Philosophy and Truth*. New Jersey: Humanities Press, 1979).

————. "Vom Nutzen und Nachteil der Historie für das Leben" (1874). *KG*, Bk. 3, vol. 1. ("On the Uses and Disadvantages of Philosophy for Life." Translated by R. J. Hollingdale, in *Untimely Meditations*. Cambridge: Cambridge University Press, 1983)

————. *Der Wanderer und sein Schatten* (1880). *KG*, Bk. 4, vol. 3. (*The Wanderer and His Shadow*. Translated by P. V. Cohn. London: Russell and Russell, 1964).

——. "Die fröhliche Wissenschaft" (1882, 1887). *KG*, Bk. 5, vol. 2. (*The Gay Science*, translated by Walter Kaufmann. New York: Vintage, 1974).
——. *Zur Genealogie der Moral* (1887). *KG*, Bk. 6, vol. 2. (*On the Genealogy of Morals*, translated by Walter Kaufmann and R. J. Hollingdale. New York: Vintage, 1974).
——. *Die Götzendämmerung* (1889), *KG*, Bk. 6, vol. 3. (*Twilight of the Idols*. Translated by R. J. Hollingdale. Harmondsworth: Penguin, 1961).
——. *The Will to Power*. Translated by Walter Kaufmann and R. J. Hollingdale. New York: Vintage, 1968. (A translation of the notes from 1883–88, published in the *Grossoktavausgabe*, vols. 15 and 16 (1901 and 1906), and reprinted as vol. 78 of the *Taschenausgabe*, edited by Alfred Baumler, Stuttgart: Kroner, 1930).
Popper, Karl. *Conjectures and Refutations*. London: Routledge and Kegan Paul, 1972.
Poulet, Georges. *Études sur le temps humaines*. Paris: Plon, 1952–68.
——. *La distance intérieure*. Paris: Plon, 1952.
Richardson, William J. *Heidegger: Through Phenomenology to Thought*. The Hague: Nijhoff, 1963.
Ricoeur, Paul. "The Human Experience of Time and Narrative." *Research in Phenomenology* 9 (1979).
——. "Narrative Time." In *On Narrative*, edited by W. J. T. Mitchell. Chicago: University of Chicago Press, 1981.
——. *Temps et Récit*. 2 vols. Paris: Seuil, 1984. (*Time and Narrative*, vols. 1, 2, and 3. Translated by Kathleen McLaughlin and David Pellauer. Chicago: University of Chicago Press, 1984, 1985, 1986).
Rorty, Richard. *Philosophy and the Mirror of Nature*. Oxford: Blackwell, 1980.
Russell, Bertrand. *Introduction to Mathematical Philosophy*. London: Allen & Unwin, 1919.
Sallis, John. *Delimitations: Phenomenology and the End of Metaphysics*. Bloomington: Indiana, 1986.
——. *Spacings—of Reason and Imagination in Texts of Kant, Fichte and Hegel*. Chicago: University of Chicago Press, 1987.
——, ed. *Deconstruction and Philosophy: The Texts of Jacques Derrida*. Chicago: University of Chicago Press, 1987.
——. "Language and Reversal." In *Martin Heidegger in Europe and America*, edited by Edward G. Ballard and Charles E. Scott. The Hague: Nijhoff, 1973.
——. "Where Does *Being and Time* Begin?" In his *Delimitations* (1986).
Sartre, Jean-Paul. *La nausée*. Paris: Gallimard, 1938. (*Nausea*. Translated by Robert Baldick. Harmondsworth: Penguin, 1965).
——. *L'être et le néant*. Paris: Gallimard, 1943. (*Being and Nothingness*. Translated by Hazel Barnes. London: Methuen, 1957).
——. *Saint Genet*. Paris: Gallimard, 1952. (*Saint Genet*. Translated by Bernard Frechtman. New York: Braziller, 1963).
——. *Les Mots*. Paris: Gallimard, 1964. (*Words*. Translated by Irene Clephane. London: Hamish Hamilton, 1964).
——. *L'Idiot de la famille: Gustav Flaubert de 1821 à 1851*. 3 vols. Paris: Gallimard, 1971, 1971, 1972).
Saussure, Ferdinand de. *Cours de Linguistique Générale*. Edited by Tullio de Mauro. Paris: Payot, 1973. (*Course in General Linguistics*. Translated by Wade Baskin. London: Fontana, 1974).
Schmitt, Richard. *Martin Heidegger on Being Human*. New York: Random House, 1969.
Schöfer, Erasmus. "Heidegger's Language." In *On Heidegger and Language*, edited by

Joseph J. Kockelmans. Evanston: Northwestern University Press.

Schutz, Alfred. "Making Music Together." In *Collected Works*, vol. 2. The Hague: Nijhoff, 1966.

Sheehan, Thomas. "The Original Form of *Sein und Zeit*: Heidegger's *Der Begriff der Zeit* (1924)." *British Journal of Phenomenology* 10, no. 2 (May 1979): 78–83.

Sherover, Charles. *Heidegger and Kant on Time*. Bloomington: Indiana University Press, 1971.

———. *The Human Experience of Time*. New York: New York University Press, 1975.

Silverman, Hugh J. "The Time of Autobiography." In *Time and Metaphysics*, edited by David Wood and Robert Bernasconi. Warwick: Parousia Press, 1982.

———. *Inscriptions: Between Phenomenology and Structuralism*. New York: SUNY Press, 1987.

Small, Robin. "Three Interpretations of Eternal Return." *Dialogue* 22 (1983): 91–112.

Sokolowski, Robert. *The Formation of Husserl's Concept of Constitution*. The Hague: Nijhoff, 1964.

Stambaugh, Joan. *Nietzsche's Thought of Eternal Return*. Baltimore: Johns Hopkins University Press, 1972.

West, Cornel. "Nietzsche's Prefiguration of Postmodern American Philosophy." *Boundary 2* (Why Nietzsche Now?) (Spring/Fall 1981): 241–70.

Willard, Dallas. "Logical Psychologism: Husserl's Way Out." *American Philosophical Quarterly* 9, no. 1 (January 1972): 94–100.

Wood, David "Introduction to Derrida." *Radical Philosophy* 21 (Spring 1979): 17–28. Reprinted in *Radical Philosophy Reader*, edited by Roy Edgeley and Richard Osborne, pp. 18–42. London: Verso, 1985.

———. "Prolegomena to a New Theory of Time." *Research in Phenomenology* 10 (1980): 177–91.

———. "Derrida and the Paradoxes of Reflection." *Journal of the British Society for Phenomenology* 11, no. 3 (October 1980): 225–38.

———. "Style and Strategy at the Limits of Philosophy: Heidegger and Derrida." *The Monist* 63, no. 3 (September 1980): 493–511.

———, ed. *Heidegger and Language*. Warwick: Parousia, 1981.

———. "Time and the Sign." *Journal of the British Society for Phenomenology* 13, no. 2 (May 1982): 143–53.

———, ed. (with Robert Bernasconi). *Time and Metaphysics*. Warwick: Parousia, 1982.

———, ed. (with Robert Bernasconi). *Derrida and Différance*. Warwick: Parousia, 1985. Republished Evanston: Northwestern University Press, 1988.

———, ed. (with David Farrell Krell). *Exceedingly Nietzsche: Essays in Contemporary Nietzsche Interpretation*. London: Routledge, 1988.

———. "Textual Reflexivity and Totalization in Hegel." In David Wood, *Philosophy and Style*. London: Hutchinson, 1988 (forthcoming).

———. *Philosophy and Style*. London: Hutchinson, 1989 (forthcoming).

Index

423